Stochastic Modelling and Applied Probability

Formerly:
Applications of Mathematics

54

Stochastic Mechanics
Random Media
Signal Processing and Image Synthesis
Mathematical Economics and Finance
Stochastic Optimization
Stochastic Control
Stochastic Models in Life Sciences

Stochastic Modelling and Applied Probability

1 Fleming/Rishel, **Deterministic and Stochastic Optimal Control** (1975)
2 Marchuk, **Methods of Numerical Mathematics,** Second Ed. (1982)
3 Balakrishnan, **Applied Functional Analysis,** Second Ed. (1981)
4 Borovkov, **Stochastic Processes in Queueing Theory** (1976)
5 Liptser/Shiryayev, **Statistics of Random Processes I: General Theory,** Second Ed. (1977)
6 Liptser/Shiryayev, **Statistics of Random Processes II: Applications,** Second Ed. (1978)
7 Vorob'ev, **Game Theory: Lectures for Economists and Systems Scientists** (1977)
8 Shiryayev, **Optimal Stopping Rules** (1978)
9 Ibragimov/Rozanov, **Gaussian Random Processes** (1978)
10 Wonham, **Linear Multivariable Control: A Geometric Approach,** Third Ed. (1985)
11 Hida, **Brownian Motion** (1980)
12 Hestenes, **Conjugate Direction Methods in Optimization** (1980)
13 Kallianpur, **Stochastic Filtering Theory** (1980)
14 Krylov, **Controlled Diffusion Processes** (1980)
15 Prabhu, **Stochastic Storage Processes: Queues, Insurance Risk, Dams, and Data Communication,** Second Ed. (1998)
16 Ibragimov/Has'minskii, **Statistical Estimation: Asymptotic Theory** (1981)
17 Cesari, **Optimization: Theory and Applications** (1982)
18 Elliott, **Stochastic Calculus and Applications** (1982)
19 Marchuk/Shaidourov, **Difference Methods and Their Extrapolations** (1983)
20 Hijab, **Stabilization of Control Systems** (1986)
21 Protter, **Stochastic Integration and Differential Equations** (1990)
22 Benveniste/Métivier/Priouret, **Adaptive Algorithms and Stochastic Approximations** (1990)
23 Kloeden/Platen, **Numerical Solution of Stochastic Differential Equations** (1992)
24 Kushner/Dupuis, **Numerical Methods for Stochastic Control Problems in Continuous Time,** Second Ed. (2001)
25 Fleming/Soner, **Controlled Markov Processes and Viscosity Solutions** (1993)
26 Baccelli/Brémaud, **Elements of Queueing Theory** (1994)
27 Winkler, **Image Analysis, Random Fields, and Dynamic Monte Carlo Methods: An Introduction to Mathematical Aspects** (1994)
28 Kalpazidou, **Cycle Representations of Markov Processes** (1995)
29 Elliott/Aggoun/Moore, **Hidden Markov Models: Estimation and Control** (1995)
30 Hernández-Lerma/Lasserre, **Discrete-Time Markov Control Processes: Basic Optimality Criteria** (1996)
31 Devroye/Györfi/Lugosi, **A Probabilistic Theory of Pattern Recognition** (1996)
32 Maitra/Sudderth, **Discrete Gambling and Stochastic Games** (1996)
33 Embrechts/Klüppelberg/Mikosch, **Modelling Extremal Events** (1997)
34 Duflo, **Random Iterative Models** (1997)

(continued after index)

Suresh Sethi
Hanqin Zhang
Qing Zhang

Average-Cost Control of Stochastic Manufacturing Systems

With 10 Illustrations

Suresh Sethi
School of Management
University of Texas at Dallas
Richardson, TX 75083
USA

Qing Zhang
Department of Mathematics
University of Georgia
Athens, GA 30602
USA

Hanqin Zhang
Academy of Mathematics and System Sciences
Chinese Academy of Sciences
Beijing 100080
China

Cover illustration: Cover pattern courtesy of Rick Durrett, Cornell University, Ithaca, New York

ISBN 978-1-4419-1954-0 eISBN 978-0-387-27615-1

Mathematics Subject Classification (2000): MSCM13070

Average-cost control of stochastic manufacturing systems / Suresh P. Sethi, Qing Zhang, Han-Qin Zhang.
p. cm.
Includes bibliographical references and index.

1. Production control—Statistical methods. 2. Production management—Statistical methods. 3. Cost accounting. I. Zhang, Qing, 1959– II. Zhang, Hanqin. III. Title.
TS155.8.S48 2004
658.5—dc22 2004049170

9 8 7 6 5 4 3 2 1

springeronline.com

To our wives, Andrea, Ruoping, and Qian,

and our children, Chantal, Anjuli, Rui, Sheena, and Sean

Preface

This book is concerned with production planning and control in stochastic manufacturing systems. These systems consist of machines that are subject to breakdown and repair. The objective is to control the rate of production over time in order to meet the demand at a minimum long-run average cost. The exact optimal solution of such a problem is very difficult and in many cases impossible to obtain. To reduce the complexity, we consider the case in which the rates at which machine failure and repair events occur are much larger than the rate of fluctuation in the product demand. The idea behind our near-optimal decision-making approach is to derive a limiting control problem which is simpler to solve than the given original problem. The limiting problem is obtained by replacing the stochastic machine capacity process by the average total capacity of machines and by appropriately modifying the objective function. We use the optimal control of the limiting problem to construct the piecewise deterministic controls for the original problem, and show that these constructed controls are asymptotically optimal under certain assumptions on the cost functions involved.

Increasingly complex and realistic models of manufacturing systems with failure-prone machines facing uncertain demands are formulated as stochastic optimal control problems. Partial characterization of their solutions are provided when possible along with their decomposition based on event frequencies. In the latter case, two-level decisions are constructed in the manner described above and these decisions are shown to be asymptotically optimal as the average time between successive short-term events becomes much smaller than that between successive long-term events. The striking novelty of this approach is that this is done without solving for the optimal solution, which as stated earlier is an insurmountable task.

This book is a sequel to Sethi and Zhang [125]. It focuses on the long-run average-cost criteria in contrast to Sethi and Zhang who deal with discounted cost objectives. A discounted cost criterion emphasizes near-term system behavior, whereas a long-run average cost measures system performance in terms of the corresponding stationary distributions. Such criteria are often more desirable in practice for long-term production planning. In addition, from a mathematical point of view, analysis of control policies of long-run average cost problems are typically more involved than those with discounted cost. This book explores the relationship between control problems with a discounted cost and those with a long-run average cost in connection with near-optimal control.

The material covered in the book cuts across the disciplines of Applied Mathematics, Operations Management, Operations Research, and System and Control Theory. It is anticipated that the book would encourage development of new models and techniques in these disciplines. The book is written for operations researchers, system and control theorists, applied mathematicians, operations management specialists, and industrial engineers. Although some of the proofs require advanced mathematics, as a rule the final results are accessible to most of them.

We wish to thank Wendell Fleming, John Lehoczky, Ruihua Liu, Ernst Presman, Mete Soner, Wulin Suo, Michael Taksar, Houmin Yan, George Yin, and Xun-Yu Zhou, who have worked with us in the area of optimal and near-optimal controls of manufacturing systems. We are indebted to W. Fleming for invaluable discussions and advice over the years. We want to thank many of our students and associates including Yongjiang Guo, Jiankui Yang, and Yuyun Yang for their careful reading of the manuscript and assistance at various stages. We appreciate the assistance provided by Barbara Gordon in the preparation of the manuscript. Finally, we are grateful to the Natural Sciences and Engineering Research Council of Canada, the National Natural Sciences Foundation of China, the Hundred Talents Program of the Chinese Academy of Sciences, the Office of Naval Research, and the University of Texas at Dallas for their support of the research on which a large part of this book is based.

Richardson, Texas, USA — Suresh P. Sethi
Beijing, China — Hanqin Zhang
Athens, Georgia, USA — Qing Zhang

Contents

Notation

This book is divided into eleven chapters and a set of six appendices. Each of the eleven chapters is divided into sections. In any given chapter, say Chapter 4, sections are numbered consecutively as 4.1, 4.2, 4.3, and so on. Similarly, mathematical expressions in Chapter 4, such as equations, inequalities, and conditions, will be numbered consecutively as (4.1), (4.2), (4.3), ..., throughout each chapter. Also, figures and tables in that chapter are numbered consecutively as Figure 4.1, Figure 4.2, ... On the other hand, theorems are numbered consecutively in each of the sections. Thus, in any given chapter, say Chapter 3, the third theorem in Section 4 would be stated as Theorem 4.3. In Chapter 3, this theorem would be cited also as Theorem 4.3, while in all other chapters, it would be referred to as Theorem 3.4.3. The same numbering scheme is used for lemmas, corollaries, definitions, remarks, algorithms, and examples.

Each appendix, say Appendix B, has no numbered sections. Mathematical expression in Appendix B will be numbered consecutively as (B.1), (B.2), (B.3), ... Theorems are numbered consecutively as Theorem B.1, Theorem B.2, ... The same numbering scheme is used for lemmas, corollaries, definitions, and remarks. Items in Appendix B will be cited throughout the book, just as labeled in that appendix.

All deterministic and stochastic processes considered in this book are assumed to be measurable processes.

We provide clarification of some frequently used terms in this book. By ε sufficiently small (or ε small enough), we mean an $\varepsilon \in (0, \varepsilon_0]$ for some $\varepsilon_0 > 0$. The term "Hamilton-Jacobi-Bellman equation" is abbreviated as the "HJB equation." The term Hamilton-Jacobi-Bellman equation in terms of directional derivatives is abbreviated as "HJBDD." These terms, without any qualification, shall mean an average cost version of the corresponding equations. Their discounted cost version, when needed, will be so qualified, or simply referred to as dynamic programming equations. The term "open-loop controls" refers to "nonfeedback controls." The terms "surplus," "inventory/shortage," and "inventory/backlog" are used interchangeably. The terms "control," "policy," and "decision" are used interchangeably.

We make use of the following notation in this book:

$\square$	indicates the end of a proof, example, definition, or remark
#	denotes "the number of"
$\mapsto$	a mapping from one set to another
$\Rightarrow$	denotes "implies"
$\Re^n$	n-dimensional Euclidean space
$\langle \boldsymbol{x}, \boldsymbol{y} \rangle$	the scalar product of any two vectors $\boldsymbol{x}$ and $\boldsymbol{y}$ in $\Re^n$
$\lvert A \rvert$	$= \sum_{i,j} \lvert a_{ij} \rvert$ for a matrix $A = (a_{ij})$
$\lvert \boldsymbol{x} \rvert$	$= \lvert x_1 \rvert + \cdots + \lvert x_n \rvert$ for a vector $\boldsymbol{x} = (x_1, \ldots, x_n)$
A'	the transpose of a vector or matrix A
$\mathbf{0}$	$= (0, 0, \ldots, 0)$ or $(0, 0, \ldots, 0)'$
$\mathbf{1}$	$= (1, 1, \ldots, 1)$ or $(1, 1, \ldots, 1)'$
1_F	the indicator function of a set F
$\mathcal{A}, \mathcal{A}^0, \mathcal{A}^\varepsilon, \ldots$	sets of admissible controls
$\{\mathcal{F}_t\}$	filtration $\{\mathcal{F}_t, t \geq 0\}$
$\mathcal{F}_t, \mathcal{F}_t^\varepsilon, \ldots$	σ-algebras
$\mathcal{N}(x)$	the neighborhood of x
$\mathcal{S}$	set of stable controls
$C^1(\mathcal{O})$	set of all continuously differentiable functions on $\mathcal{O}$
$C, C_g, C_h, \ldots$	positive multiplicative constants
$C_1, C_2, \ldots$	positive multiplicative constants
$f_{\boldsymbol{x}}$	the gradient of a scalar function f at $\boldsymbol{x}$ if it exists
$D^+ f(\boldsymbol{x})$	the superdifferential of f at $\boldsymbol{x}$
$D^- f(\boldsymbol{x})$	the subdifferential of f at $\boldsymbol{x}$
$E\xi$	the expectation of a random variable ξ
F^c	the complement of a set F
$F_1 \cap F_2$	the intersection of sets F_1 and F_2
$F_1 \cup F_2$	the union of sets F_1 and F_2
$L^2([s, T])$	the space of all square-integrable functions on $[s, T]$
$P(\xi \in \cdot)$	the probability distribution of a random variable ξ
V	the value function
W	the potential function

$(a_1,\ldots,a_l) > \mathbf{0}$	means $a_1 > 0, \ldots, a_l > 0$
$(a_1,\ldots,a_l) \geq \mathbf{0}$	means $a_1 \geq 0, \ldots, a_l \geq 0$
$\boldsymbol{a} \geq \boldsymbol{b}$	means $\boldsymbol{a} - \boldsymbol{b} \geq \mathbf{0}$ for any vectors $\boldsymbol{a}$ and $\boldsymbol{b}$
a^+	$= \max\{a, 0\}$ for a real number a
a^-	$= \max\{-a, 0\}$ for a real number a
$a_1 \wedge \cdots \wedge a_\ell$	$= \min\{a_1,\ldots,a_\ell\}$ for any real numbers a_i, $i = 1,\ldots,\ell$
$a_1 \vee \cdots \vee a_\ell$	$= \max\{a_1,\ldots,a_\ell\}$ for any real numbers a_i, $i = 1,\ldots,\ell$
a.e.	almost everywhere
a.s.	almost surely
$\mathrm{diag}(\boldsymbol{a})$	the diagonal matrix of $\boldsymbol{a} = (a_1,\ldots,a_n)$
$\mathrm{co}(F)$	the convex hull of a set F
$\overline{\mathrm{co}}(F)$	the convex closure of a set F
$\dfrac{df(x_0)}{dx}$	$= \left.\dfrac{df(x)}{dx}\right\vert_{x=x_0}$
$\exp(Q)$	$= e^Q$ for any argument Q
$f^{-1}(\cdot)$	the inverse of a scalar function $f(\cdot)$
$\mathrm{ri}(F)$	the relative interior of a set F
$\log x$	the natural logarithm of x
$\ln x$	the logarithm of x with base e
$\boldsymbol{k}, \boldsymbol{u}, \boldsymbol{x}, \boldsymbol{z}, \ldots$	all Latin boldface italic letters stand for vectors
$\boldsymbol{u}(\cdot)$	the process $\{\boldsymbol{u}(t) : t \geq 0\}$ or simply $\boldsymbol{u}(t)$, $t \geq 0$
$\boldsymbol{x} \mapsto A\boldsymbol{x}$	linear transformation from $\Re^n$ to $\Re^m$; here $\boldsymbol{x} \in \Re^n$ and A is an $(m \times n)$ matrix

$\beta_g, \beta_h, \ldots$	positive exponential constants
λ	the minimum average cost
$\boldsymbol{\nu}$	equilibrium distribution vector of a Markov process
$\dfrac{\partial g(\boldsymbol{x},t)}{\partial \boldsymbol{x}}$	the partial derivative of $g(\boldsymbol{x},t)$ with respect to $\boldsymbol{x}$
$\dfrac{\partial g(\boldsymbol{x}_0,t)}{\partial \boldsymbol{x}}$	$= \left.\dfrac{\partial g(\boldsymbol{x},t)}{\partial \boldsymbol{x}}\right\vert_{\boldsymbol{x}=\boldsymbol{x}_0}$
$\partial_{\boldsymbol{p}} g(\boldsymbol{x},t)$	the directional derivative of $g(\boldsymbol{x},t)$ at $(\boldsymbol{x},t)$ in direction $\boldsymbol{p}$
$\nabla f(\boldsymbol{x})$	gradient of function f at $\boldsymbol{x}$
$(\Omega, \mathcal{F}, P)$	the probability space
ε	a capacity process parameter (assumed to be small)
$\rho > 0$	the discount rate
$\sigma\{k(s) : s \leq t\}$	the σ-algebra generated by the process $k(\cdot)$ up to time t

$O(y)$	a function of y such that $\sup_y \vert O(y)\vert/\vert y\vert < \infty$
$o(y)$	a function of y such that $\lim_{y\to 0} o(y)/y = 0$

Part I:

Introduction and Models of Manufacturing Systems

1 Concept of Near-Optimal Control

1.1 Introduction

This book is concerned with manufacturing systems involving machines that are subject to breakdown and repair. The systems under consideration range from single or parallel-machine systems to flowshops and jobshops. These systems exhibit an increasing complex structure of processing of products being manufactured. The objective is to control the rate of production over time in order to meet the demand at the minimum long-run average cost that includes the cost of production and the cost of inventory/shortage.

The exact optimal solution of such a problem is quite complex and difficult, perhaps impossible, to obtain. To reduce the complexity, we consider the case in which the rates, at which machine failure and repair events occur are much larger than the rate of fluctuation in the product demand. The idea behind our near-optimal decision-making approach is to derive a limiting control problem which is simpler to solve than the given original problem. The limiting problem is obtained by replacing the stochastic machine capacity process by the average total capacity of machines and by appropriately modifying the objective function. We use the optimal control of the limiting problem to construct the piecewise deterministic controls for the original problem, and we show these constructed controls are asymptotically optimal under certain assumptions on the cost functions involved.

The specific points to be addressed in this book are results on the asymptotic optimality of the constructed solution and the extent of the deviation of its average cost from the optimal average cost for the original problem. While this approach could be extended for applications in other areas, the purpose of this book is to model a variety of representative manufacturing systems in which some of the exogenous processes, deterministic or stochastic, are changing much faster than the remaining ones and to apply our methodology of near-optimal decision making to them.

In Sethi and Zhang [125] we considered systems with discounted cost criteria. Here we analyze these systems in the average-cost context. A discounted cost criterion emphasizes near-term system behavior, whereas a long-run average cost measures system performance in terms of the corresponding stationary distributions. In some situations, long-run average-cost criteria may be more appropriate. In particular, when the discount rates are small, the average-cost optimal policies, which often have a simpler form, may provide a good approximation of the discounted-cost optimal policies. In addition, from a mathematical point of view, analysis of average-cost optimal control problems is typically more involved than those with discounted cost. This can been seen from the construction of the near-optimal control from the limiting problem. This book will explore the relationship between control problems with a discounted cost and those with a long-run average cost in the context of optimal and near-optimal decisions of stochastic manufacturing systems.

In Sethi and Zhang [125] we also considered multilevel systems and a hierarchical decision-making approach for their near-optimal solutions. Doing so in the case of the average-cost criteria remains a topic for further research. Nevertheless, the models and analysis presented in this book form stepping stones for analyzing such multilevel problems.

1.2 A Brief Review of the Related Literature

There are several related approaches to near-optimal decision making in an uncertain environment. Each approach is suited to certain types of models and assumptions. We shall review these approaches briefly.

Singular Perturbations in Markov Decision Problems

Consider a Markov decision problem (MDP) such that the states of the underlying Markov chain are either subject to rather frequent changes or naturally divisible to a number of groups such that the chain fluctuates very rapidly from one state to another within a single group, but jumps less rapidly from one group to another. The structure of the given process corresponds to Markov processes admitting "strong and weak interactions" that arise in applications such as control of a queuing network, models of computer systems, and management of hydroelectric power generation. Strong interactions correspond to frequent transitions, while weak interactions correspond to infrequent transitions. For more discussions and results, see Yin and Zhang [149].

The singular perturbation approach consists of deriving, from the given problem, a simple problem called the limit MDP, which forms an appropriate asymptotic approximation to a whole family of perturbed problems containing the given MDP. Moreover, an optimal solution to the given MDP can be approximated by an optimal solution of the limit problem provided the given perturbation is small.

It should be noted that the perturbation is termed singular because it alters the ergodic structure of the Markov process. Thus, the stationary distribution of the perturbed process has a discontinuity at the zero value of the perturbation parameter. The cases that avoid the discontinuity, on the other hand, are referred to as regular perturbations.

Research dealing with singularly perturbed MDPs include works of Delebecque and Quadrat [43] and Bielecki and Filar [19] in the discrete-time case, and of Phillips and Kokotovic [98] in both the discrete- and continuous-time cases. For detailed accounts of this approach and for recent developments, see Bensoussan [14] and Yin and Zhang [149], among others.

Diffusion Approximations

The next approach we shall discuss is that of diffusion approximations; see Glynn [65] for a tutorial and a survey of the literature. The most important application of this approach concerns the scheduling of networks of queues. If a network of queues is operating under heavy traffic, that is, when the rate of customers entering some of the stations in the network is very close to the rate of service at those stations, the problem of scheduling the network can be approximated by a dynamic control problem involving diffusion processes. The controlled diffusion problem can often be solved, and the optimal policies that are obtained are interpreted in terms of the original system. A justification of this procedure is provided in Bramson and Dai [27], Harrison [66], Harrison and Van Mieghem [67], and Wein [142], for example. See also the surveys on fluid models and strong approximations by Chen and Mandelbaum [31, 32] for related research.

Kushner and Ramachandran [83] begin with a sequence of systems whose limit is a controlled diffusion process. It should be noted that the traffic intensities of the systems in sequence converge to the critical intensity of one. They show that the sequence of value functions associated with the given sequence converges to the value function of the limiting problem. This enabled them to construct a sequence of asymptotic optimal policies defined to be those for which the difference between the associated cost and the value function converges to zero as the traffic intensity approaches its critical value. For recent developments, see Kushner [82] and Whitt [143].

Krichagina, Lou, Sethi, and Taksar [80], and Krichagina, Lou, and Taksar [81] apply the diffusion approximation approach to the problem of controlling the production rate of a single product using a single unreliable machine in order to minimize the total discounted inventory/backlog costs (or to minimize long-run average total inventory/backlog costs). They embed the given system into a sequence of systems in heavy traffic. Their purpose is to obtain asymptotic optimal policies for the sequence of systems that can be expressed only in terms of the parameters of the original system.

Before concluding our discussion of the diffusion approximation approach, we should emphasize that, so far, the approach does not provide us with an estimate of how much the policies constructed for the given original system deviate from the optimal solution, especially when the optimal solution is not known, which is most often the case. As we shall see later, our near-optimal approach developed in this book enables us to provide just such an estimate in many cases.

Time-Scale Separation in Stochastic Manufacturing Systems and Hierarchical Decision Making

Gershwin [60] considers scheduling problems in a dynamic manufacturing system with machine failures, setups, demand changes, etc., and he proposes a hierarchical structure based on the frequency of occurrence of different types of events; see also Chapters 9–12 in Gershwin [61], Xie [145], and Lasserre [85]. This framework is inspired by the singular perturbation literature reviewed briefly at the beginning of this section, and it is based on the assumption that events tend to occur in a discrete spectrum which defines the hierarchical levels; see also Caromicoli, Willsky, and Gershwin [29]. In modeling the decisions at each level of the hierarchy, quantities that vary slowly (variables that correspond to higher levels of the hierarchy) are treated as static, or constant. Quantities that vary much faster (variables at lower levels of the hierarchy) are modeled in a way that ignores the variations, for example, by replacing fast changing variables with their averages. The objective of this approach is to determine an optimal control strategy for the scheduling problem under consideration. Gershwin [60] proposes the solution of one or two problems at each level to derive the control strategy. These are identified as the problems of finding the hedging point and the staircase strategies. In the hedging point strategy problem at level i, the objective is to determine level-i controls such as production rates. Constraints are imposed by the total capacity available and by the decisions made at the higher levels. The staircase strategy problem can be interpreted as the allocation of resources among activities at level i, consistent with controls or production rates determined at the previous level.

Sethi and Zhang [125] considered hierarchical control of manufacturing systems with discounted cost criteria. They show that hierarchical decision making in the context of a goal-seeking manufacturing system can lead to a near optimization of its objective. In particular, they consider manufacturing systems in which events occur at different time scales. For example, changes in demand may occur far more slowly than breakdowns and repairs of production machines. This suggests that capital expansion decisions that respond to demand are relatively longer-term decisions than decisions regarding production. Thus, longer-term decisions such as those dealing with capital expansion can be based on the average existing production capacity, and can be expected to be nearly optimal even though the short-term capacity fluctuations are ignored. Having the longer-term decisions in hand, one can then solve the simpler problem of obtaining production rates.

This completes our brief review of the existing near-optimal decision-making approaches designed for uncertain environments. This book presents a further development of the method given by Sethi and Zhang [125]. It is related to the singular perturbations approach with the difference that our framework is that of stochastic optimal control rather than the traditional one of MDPs. Our approach is based on different event frequencies. Our emphasis is in proving that the near-optimal control provided by our approach is asymptotically optimal as the frequencies of various events diverge from one another.

1.3 Plan of the Book

This book is divided into five parts: Part I – Introduction to near-optimal decision making and formulation of models of manufacturing systems; Part II – Existence and characterization of optimal control of manufacturing systems; Part III – Near-optimal decision making in manufacturing systems; Part IV – Conclusions and open research problems, and Part V – A set of appendices providing background material.

Part I consists of this chapter and Chapter 2. In this chapter, we have introduced our concept of near-optimal decision making in stochastic manufacturing systems along with a brief review of the related literature. In Chapter 2, we sketch the models of manufacturing systems that we study in this book. These include single or parallel-machine systems, dynamic flowshops, and dynamic jobshops. Machines are failure-prone and the systems face deterministic demands for their products. The purpose of these systems is to produce in a way so as to satisfy the demands at the minimum long-run average costs of production, inventories, and backlogs.

Part II consists of Chapters 3, 4, 5, and 6. In these four chapters we discuss the dynamics and the optimal controls of the manufacturing systems sketched in Sections 2.2–2.5.

In Chapter 3, we consider the optimal production control policy in the dynamic stochastic manufacturing systems consisting of single/parallel machines that are failure prone. The object is to choose the production rate over time in order to minimize the long-run average cost of production and surplus. The analysis proceeds with a study of the corresponding problem with a discounted cost. In order to use the vanishing discount method, the stable control policy that makes the Markov processes, describing the system dynamic behavior to be ergodic, is constructed. It is shown that the Hamilton-Jacobi-Bellman (HJB) equation for the average-cost problem has a solution giving rise to the minimal average cost and a potential function. The result helps in establishing a verification theorem. Finally, the optimal production control policy is specified in terms of the potential function. In this chapter, at the same time, the concept of turnpike sets is introduced to characterize the optimal inventory levels. The structure of the turnpike sets exhibits a monotone property with respect to the production capacity and demand. The property helps in solving the optimal production problem numerically and, in some cases, analytically.

In Chapter 4, we consider the stochastic manufacturing systems consisting of m machines in a flowshop configuration that must meet the demand for its product at a minimum average cost. The decision variables are input rates to the machines. We take the number of parts in the buffers of the first $(m-1)$ machines and the difference of the real and planned cumulative productions at the last machine, known as the surplus, as the states of the system. Since the number of parts in the internal buffers between any two adjacent machines must remain nonnegative, the problem is inherently a state-constrained problem. Our objective is to choose admissible input rates to minimize a long-run average of expected production and surplus costs. The stable control policy that makes the system to be ergodic is constructed. Using the vanishing discount approach, we obtain a solution of the HJB equation and, in consequence, a verification theorem is also established. At the same time, the two-machines flowshop with a limited internal buffer is also investigated.

In Chapter 5, we consider a production control problem for a general jobshop producing a number of products and subject to breakdown and repair of machines. The objective is to choose the production control policy of the final products and intermediate parts on the various machines over time, in order to meet the demand for the system's production at the minimum long-run average cost of production and surplus. We prove a verification theorem and derive the optimal feedback control policy in terms of the directional derivative of the potential function.

In Chapter 6, we consider the system with parallel machines described in Chapter 3, but with a long-run average risk-sensitive criterion. By using a logarithmic transformation, it is shown that the associated HJB equation has a viscosity solution. This leads to a dynamic stochastic game interpretation of the underlying risk-sensitive control problem.

It will be seen in Chapters 3, 4, 5, and 6 that explicit and exact optimal controls for manufacturing systems are not usually available. As a result, one must resort to obtaining near-optimal decisions. In Part III, which contains Chapters 7, 8, 9, and 10, we turn to an alternative approach that employs the idea of near-optimal decision making as described in Section

1.2 to obtain asymptotic optimal controls.

In Chapter 7, a near-optimal decision approach is used to present an asymptotic analysis of the stochastic manufacturing system consisting of single/parallel machines subject to breakdown and repair and facing a constant demand, as the rates of change of the machine states approach infinity. The analysis gives rise to a limiting problem in which the stochastic machine availability is replaced by its equilibrium mean availability. It is shown that the long-run average cost for the original problem converges to the long-run average cost of the limiting problem. Open-loop and feedback controls for the original problem are constructed from the optimal controls of the limiting problem in such a way that guarantees their asymptotic optimality. The convergence rate of the long-run average cost for the original problem to that of the limiting problem is established. This helps in providing an error estimate for the constructed open-loop asymptotic optimal control.

In Chapter 8, we present an asymptotic analysis for the stochastic manufacturing system consisting of machines in tandem subject to breakdown and repair and facing a constant demand, as the rates of change of the machine state approach infinity. We prove that the long-run average cost for the original problem converges to the long-run average cost of the limiting problem. A method of "shrinking" and "entire lifting" is introduced in order to construct the near-optimal controls for the original problem by using near-optimal controls of the limiting problem. The convergence rate of the long-run average cost for the original problem to that of the limiting problem is established. Finally, an error estimation for the constructed open-loop asymptotic optimal controls is established.

In Chapter 9, we consider a production control problem for a dynamic jobshop producing a number of products and subject to breakdown and repair of machines. As the rates of change of the machine states approach infinity, an asymptotic analysis of this stochastic manufacturing system is given. The analysis results in a limiting problem. The long-run average cost for the original problem is shown to converge to the long-run average cost of the limiting problem. The convergence rate of the long-run average cost for the original problem to that of the limiting problem, together with an error estimate for the constructed asymptotic optimal control, is established.

In Chapter 10, we derive a limiting problem when the rates of machine breakdown and repair go to infinity. It is shown that minimum cost with the long-run average risk-sensitive criterion for the original problem converges to the minimum risk-sensitive cost of the corresponding limiting problem.

Part IV consists of Chapter 11, where we describe various extensions of the existing models and open problems that remain.

Part V consists of Appendices A, B, C, D, E, and F. These provide some background material as well as technical results that are used in the book. Specifically, we state and sometimes prove, if appropriate, the required results on convergence of Markov chains, and viscosity solutions of HJB equations.

An extensive bibliography, author index, subject index, and copyright permissions follow the appendices.

We conclude this chapter by drawing a flowchart (Figure 1.1) depicting relationships between various chapters.

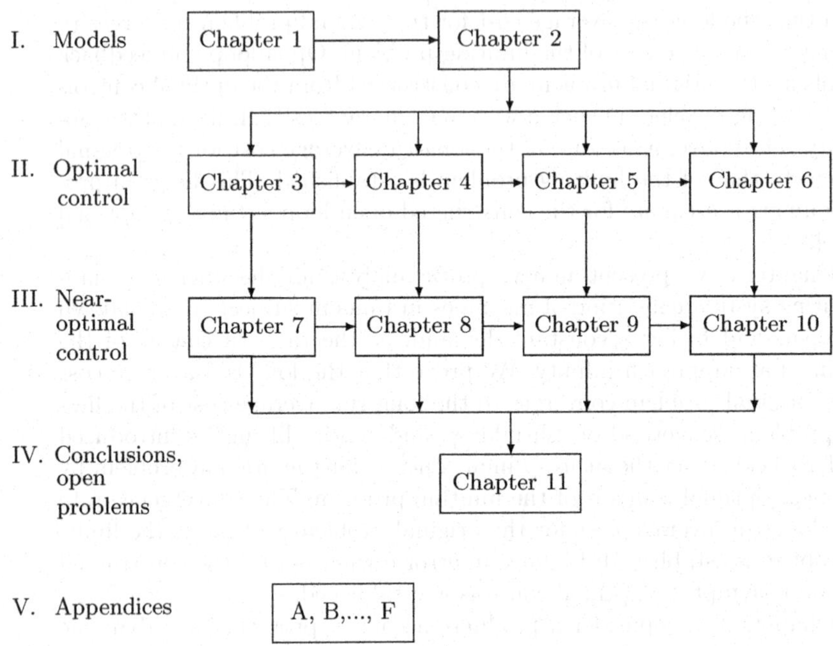

Figure 1.1. Relationships Between Chapters.

2

Models of Manufacturing Systems

2.1 Introduction

The class of convex production planning models is an important paradigm in the operations management/operations research literature. The earliest formulation of a convex production planning problem in a discrete-time framework dates back to Modigliani and Hohn [95] in 1955. They were interested in obtaining a production plan over a finite horizon in order to satisfy a deterministic demand and to minimize the total discounted convex costs of production and inventory holding. Since then the model has been further studied and extended in both continuous- and discrete-time frameworks by a number of researchers, including Johnson [76], Arrow, Karlin, and Scarf [6], Veinott [141], Adiri and Ben-Israel [1], Sprzeuzkouski [130], Lieber [91], and Hartl and Sethi [68]. A rigorous formulation of the problem along with a comprehensive discussion of the relevant literature appears in Bensoussan, Crouhy, and Proth [15].

Extensions of the convex production planning problem to handle stochastic demand have been analyzed mostly in the discrete-time framework. A rigorous analysis of the stochastic problem has been carried out in Bensoussan, Crouhy, and Proth [15]. Continuous-time versions of the model that incorporate additive white noise terms in the dynamics of the inventory process were analyzed by Sethi and Thompson [114], Bensoussan, Sethi, Vickson, and Derzko [17], and Beyer [18].

In this book, we consider production planning problems with the aim of minimizing a long-run average cost. Preceding works that relate most closely to problems under consideration here include Bai [8], Srivatsan, Bai, and Gershwin [132], Basak, Bisi, and Ghosh [12], Srivatsan and Dallery [133], Bielecki and Kumar [20], and Sethi and Zhang [125]. These works incorporate piecewise deterministic processes (PDP) either in the dynamics or in the constraints of the model. In the models considered by them, the production capacity rather than the demand for production is modeled as a stochastic process. In particular, the process of machine breakdown and repair is modeled as a birth–death process, thus making the production capacity over time a finite state Markov chain.

In what follows, we shall sketch the models of manufacturing systems that are considered in this book. These models will be formulated as continuous-time stochastic optimal control problems with a long-run average-cost criterion. Their precise formulations will appear in subsequent chapters where needed.

2.2 A Parallel-Machine, Single Product Model

Let $u(t) \geq 0$ denote the rate of production, z the rate of demand, and $x(t)$ the difference between cumulative production and cumulative demand, called the surplus, at time t. They satisfy the one-dimensional ordinary differential equation

$$\frac{d}{dt}x(t) = u(t) - z, \quad x(0) = x, \tag{2.1}$$

where x denotes the given initial surplus. Note that a positive value of surplus denotes inventory and a negative value denotes backlog or shortage.

Assume that the production capacity consists of a single or a number of parallel machines that are subject to breakdown and repair. Let $k(\cdot) = \{k(t) : t \geq 0\}$ denote the stochastic total production capacity process assumed to be a finite state Markov chain. Then, the production rate $u(t)$ must satisfy the constraint

$$0 \leq u(t) \leq k(t), \quad t \geq 0. \tag{2.2}$$

Here we are assuming without any loss in generality that a unit of capacity is required to process one unit of the product at rate 1. Let $\mathcal{A}(k)$ be the class of all admissible controls with the initial condition $k(0) = k$.

Given the initial surplus $x(0) = x$ and the initial capacity $k(0) = k$, the objective is to choose an admissible control $u(\cdot) = \{u(t),\ t \geq 0\}$ in $\mathcal{A}(k)$ so as to minimize the long-run average expected costs of surplus and production, namely,

$$J(x, k, u(\cdot)) = \limsup_{T\to\infty} \frac{1}{T} E \int_0^T g(x(t), u(t))\, dt, \tag{2.3}$$

where $g(x, u)$ is a convex function representing costs of surplus x and production u.

Remark 2.1. Before proceeding further, we should clarify the use of notation x. As $x(0) = x$, it represents the initial condition for the state equation (2.1). In defining the surplus cost function $g(x, u)$, it plays the role of a dummy variable. Finally, in the cost functional $J(x, k, u(\cdot))$, it acts as a variable representing the initial value of the state. Later, when the state is a vector, the same remarks apply to notation $\boldsymbol{x}$. □

Remark 2.2. In the system dynamics (2.1) and those appearing later in the book, we treat the material that is processed in the system as though it is continuous fluid. In addition to describing real systems with continuous materials, such a treatment is also standard in the literature directed at manufacturing systems with discrete parts; see Gershwin [61] and Dallery and Gershwin [39], for example. Lou and Van Ryzin [93] consider the flow rate approximation to be appropriate at the higher planning level as well as necessary for tractability of the problem. Once the production rate is determined, the detailed tracking of individual parts is considered at the lower scheduling level. Gershwin [61] introduces a staircase strategy that converts the production rate determined at the higher level into loading of discrete parts at discrete times. Clearly, the operation processing times on each machine must be much smaller than the mean time between machine failures and the mean repair time, for the fluid approximation to be reasonable. Indeed, Alvarez, Dallery, and David [5] have shown experimentally that the approximation is good in flowshops producing discrete parts provided the machine mean uptimes and downtimes are at least one order of magnitude larger than the processing times; see David, Xie, and Dallery [41] for some theoretical justification for the approximation. In most realistic cases, this time scale condition is satisfied; see Van Ryzin, Lou, and Gershwin [138] for a method of flow rate approximation when the condition is not satisfied. Moreover, the results based on the flow rate control models have been applied and have been shown to significantly outperform the ones used in practice; see, e.g., Akella, Choong, and Gershwin [2], Gershwin, Akella, and Choong [62], Lou and Kager [92], and Yan, Lou, Sethi, Gardel, and Deosthali [146]. □

Remark 2.3. In order to illustrate the formulation of the system dynamic (2.1), we first consider a manufacturing system consisting of a reliable machine of a unit capacity producing a single product, say, a gear. Let us assume that it takes r minutes to perform all the required operations on the unit capacity machine to produce one gear. Let d denote the demand in number of gears per hour. Let $y(t)$ denote the surplus expressed in number of gears with $y(0) = y$ and let $w(t)$ denote the production rate in gears per minute. Then it is clear that

$$\frac{d}{dt}y(t) = w(t) - \frac{d}{60}, \quad y(0) = y, \quad 0 \le w(t) \le \frac{1}{r}.$$

Define $x(t) = ry(t)$, $z = rd/60$, and $u(t) = rw(t)$, $t \ge 0$. With this change of variables, we obtain (2.1) with $0 \le u(t) \le 1$. The change of variables is equivalent to defining a "product" so that a unit of the product means r gears. Then in (2.1), $x(t)$ denotes the surplus expressed in units of the product, $u(t)$ denotes the production rate expressed in product units per minute, and z denotes the demand in product units per minute.

Finally, as described at the beginning of this section, if the system consists of a single or of a number of parallel unreliable machines and if its total capacity is $k(t)$ at time t, then the production rate constraint is modified to $0 \le u(t) \le k(t)$, $t \ge 0$, i.e., (2.2). □

Sometimes, it is desirable to obtain controls that discourage large deviation in the states, which, while occurring with low probabilities, are extremely costly. Such controls are termed robust or risk-sensitive controls. This task is accomplished by considering a risk-sensitive criterion such as

$$H(x, k, u(\cdot)) = \limsup_{T\to\infty} \frac{1}{T} \log E \exp\left(\int_0^T g(x(t), u(t))\, dt \right), \tag{2.4}$$

where $u(\cdot)$ is production control and $x(\cdot)$ is the surplus process given by

$$\frac{d}{dt}x(t) = -ax(t) + u(t) - z, \quad x(0) = x. \tag{2.5}$$

Here $a \ge 0$ is a constant, representing the deterioration rate (or spoilage rate) of the finished product. The objective is to choose an admissible control $u(\cdot) = \{u(t),\ t \ge 0\}$ in $\mathcal{A}(k)$ so as to minimize $H(x, k, u(\cdot))$.

Remark 2.4. In model (2.5), we assume a positive deterioration rate a for items in storage. This corresponds to a stability condition typically imposed for disturbance attenuation problems on an infinite time horizon (see Fleming and McEneaney [52]), and this assumption is essential in the analysis of the optimality for risk-sensitive control problems and constructing near-optimal control policies. □

Many single product models found in the literature are special cases of the model formulated here. We give one example. Bielecki and Kumar [20] considered a stochastic model with $g(x, u) = g^+x^+ + g^-x^-$, and $k(t)$, a (two-state) birth–death process. Here g^+ denotes the unit inventory carrying or holding cost per unit time, g^- denotes the unit backlog or shortage cost per unit time, $x^+ = \max\{x, 0\}$, and $x^- = \max\{-x, 0\}$.

The model given by (2.1) with the long-run average-cost criterion (2.3) has been rigorously analyzed by Sethi, Suo, Taksar, and Zhang [113], Sethi,

Yan, Zhang, and Zhang [116], and Sethi, Zhang, and Zhang [119], and will be studied in Chapters 3 and 7. The analyses of the model (2.5) with the risk-sensitive criterion (2.4) in Chapters 6 and 10 are based on Fleming and Zhang [58].

2.3 A Parallel-Machine, Multiproduct Model

Let $\Re^n$ denote the n-dimensional Euclidean space and let $\Re^n_+$ denote the subspace of $\Re^n$ with nonnegative components. Let $\boldsymbol{u}(t) = (u_1(t), \ldots, u_n(t))' \in \Re^n_+$, $\boldsymbol{x}(t) = (x_1(t), \ldots, x_n(t))' \in \Re^n$, $\boldsymbol{z} = (z_1, \ldots, z_n)' \in \Re^n_+$ denote the rates of production of n different products, the vector of their surpluses, and the rates of their demand, respectively. We assume that the production rate $\boldsymbol{u}(t)$ at time t satisfies

$$r_1 u_1(t) + \cdots + r_n u_n(t) \le k(t), \quad t \ge 0, \tag{2.6}$$

where $r_i > 0$ $(i = 1, 2, \ldots, n)$ is the amount of capacity required to process one unit of the ith product at rate 1. Then

$$\frac{d}{dt}\boldsymbol{x}(t) = \boldsymbol{u}(t) - \boldsymbol{z}, \quad \boldsymbol{x}(0) = \boldsymbol{x}, \tag{2.7}$$

with $\boldsymbol{x} \in \Re^n$ denoting the vector of initial surplus levels.

Remark 3.1. In the model formulated here, simultaneous continuous production of different products is allowed. For this model to reasonably approximate a manufacturing system producing various discrete part-types, we need setup costs and times for switching production from one product to another to be negligible, in addition to the condition described in Remark 2.2. □

Remark 3.2. Without any loss of generality, we may set $r_1 = 1$. With $r_1 = 1$, the model reduces to the single product model of Section 2.2 when we set $n = 1$. □

The capacity process $k(\cdot)$ is assumed to be a finite state Markov chain. Let $\mathcal{A}(k)$ be the class of all admissible controls. The problem is to find a control $\boldsymbol{u}(\cdot) \in \mathcal{A}(k)$ that minimizes the cost functional

$$J(\boldsymbol{x}, k, \boldsymbol{u}(\cdot)) = \limsup_{T\to\infty} \frac{1}{T} E \int_0^T g(\boldsymbol{x}(t), \boldsymbol{u}(t))\, dt, \tag{2.8}$$

where $\boldsymbol{x}(0) = \boldsymbol{x}$ and $k(0) = k$ are initial values of surplus and capacity, respectively, and $g(\boldsymbol{x}, \boldsymbol{u})$ represent convex costs of surplus $\boldsymbol{x}$ and production $\boldsymbol{u}$.

Srivatsan and Dallery [133] consider this model with $n = 2$,

$$g(\boldsymbol{x}, \boldsymbol{u}) = g_1^+ x_1^+ + g_1^- x_1^- + g_2^+ x_2^+ + g_2^- x_2^-,$$

and $k(\cdot)$, a birth–death process. The parallel machines, multiproduct model has been investigated by Sethi, Suo, Taksar, and Yan [112], and Sethi and Zhang [117]. These works will form the basis of our analysis of the model in Sections 3.7 and 7.6.

2.4 A Single Product Dynamic Flowshop

Assume that we have m machines in tandem, as shown in Figure 2.1, devoted to producing a single final product. We use $\boldsymbol{k}(t) = (k_1(t), \ldots, k_m(t))$, $t \geq 0$, to denote the vector of random capacity processes of the m machines. Let $\boldsymbol{u}(t) = (u_1(t), \ldots, u_m(t))' \in \Re_+^m$ denote the vector of production rates for the m machines. Then $0 \leq u_j(t) \leq k_j(t)$, $j = 1, 2, \ldots, m$. We use $x_1(t), \ldots, x_{m-1}(t)$ to denote the inventories in the output buffers of the first $m - 1$ machines and $x_m(t)$, the surplus of the final product. We write $\boldsymbol{x}(t) = (x_1(t), \ldots, x_m(t))'$. Then the inventories in the internal buffers should be nonnegative, i.e.,

$$x_1(t) \geq 0, \ldots, x_{m-1}(t) \geq 0. \tag{2.9}$$

Let z represent the demand rate facing the system. Then

$$\begin{aligned} \frac{d}{dt} x_1(t) &= u_1(t) - u_2(t), & x_1(0) &= x_1, \\ \frac{d}{dt} x_2(t) &= u_2(t) - u_3(t), & x_2(0) &= x_2, \\ \cdots & & \cdots & \\ \frac{d}{dt} x_m(t) &= u_m(t) - z, & x_m(0) &= x_m, \end{aligned} \tag{2.10}$$

with $x_1, x_2, \ldots, x_{m-1}$ denoting the initial inventory levels in the internal buffers and x_m denoting the initial surplus of the final product. Let $\boldsymbol{x}(t) = (x_1(t), x_2(t), \ldots, x_m(t))'$ and $\boldsymbol{x}(0) = \boldsymbol{x} = (x_1, x_2, \ldots, x_m)'$. Further, let $g(\boldsymbol{x}, \boldsymbol{u})$ denote the convex cost of inventories x_1, x_2, $\ldots$, x_{m-1} and surplus x_m when producing at rate $\boldsymbol{u}$.

The manufacturing system modeled here will be termed a *dynamic flowshop*. In order to formulate the optimization problem facing the system, let $\mathcal{A}(\boldsymbol{x}, \boldsymbol{k})$ denote the class of admissible controls. The additional dependence of the class on $\boldsymbol{x}$ is required because the admissible controls must satisfy the state constraints (2.9) for all $t \geq 0$, given the initial surplus $\boldsymbol{x}$. The class will be precisely defined in Chapter 4.

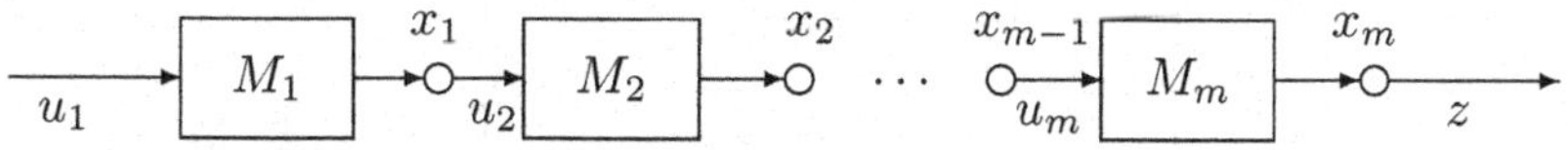

Figure 2.1. A Single Product Dynamic Flowshop with m-Machine.

The problem is to find a control $\boldsymbol{u}(\cdot) \in \mathcal{A}(\boldsymbol{x}, \boldsymbol{k})$ that minimizes the cost functional

$$J(\boldsymbol{x}, \boldsymbol{k}, \boldsymbol{u}(\cdot)) = \limsup_{T \to \infty} \frac{1}{T} E \int_0^T g(\boldsymbol{x}(t), \boldsymbol{u}(t))\, dt \tag{2.11}$$

subject to (2.10) with $\boldsymbol{x}(0) = \boldsymbol{x}$ and $\boldsymbol{k}(0) = \boldsymbol{k}$.

Van Ryzin, Lou, and Gershwin [139] use dynamic programming to approximately solve the problem with $m = 2$. Bai [8] develops a hierarchical approach to solve the problem, and Srivatsan, Bai, and Gershwin [132] apply the approach to the scheduling of a semiconductor fabrication facility.

A rigorous analysis of the model was given in Presman, Sethi, Zhang, and Bisi [102], and Presman, Sethi, Zhang, and Zhang [105, 106]. These works will be taken up in Chapter 4. An asymptotic analysis of the model carried out by Sethi, Zhang, and Zhang [122] will be reported in Chapter 8.

2.5 A Dynamic Jobshop

In this section, we generalize the models developed in Sections 2.2–2.4 and describe a model of a dynamic jobshop. Because a general description of dynamic jobshops is somewhat involved, this will be postponed to Section 5.2. Here we only formulate an illustrative example.

In Figure 2.2, we have four machines $M_1, \ldots, M_4$, two (final) product types, and five buffers. Machine M_i, $i = 1, 2, 3, 4$, has capacity $k_i(t)$ at time t, and the jth product type $j = 1, 2$, has a constant rate of demand z_j. Let $\boldsymbol{z} = (z_1, z_2)'$. As indicated in the figure, we use u_{01}, u_{02}, u_{14}, u_{15}, u_{26}, and u_{33} to denote the production rate and x_j, $j = 1, 2, \ldots, 5$, to denote the surplus.

Then the system can be described by

$$\begin{aligned}
&\frac{d}{dt}x_1(t) = u_{01}(t) - u_{14}(t) - u_{15}(t), && \frac{d}{dt}x_4(t) = u_{33}(t) + u_{15}(t) - z_1,\\
&\frac{d}{dt}x_2(t) = u_{14}(t) - u_{26}(t), && \frac{d}{dt}x_5(t) = u_{02}(t) - z_2,\\
&\frac{d}{dt}x_3(t) = u_{26}(t) - u_{33}(t), &&
\end{aligned} \tag{2.12}$$

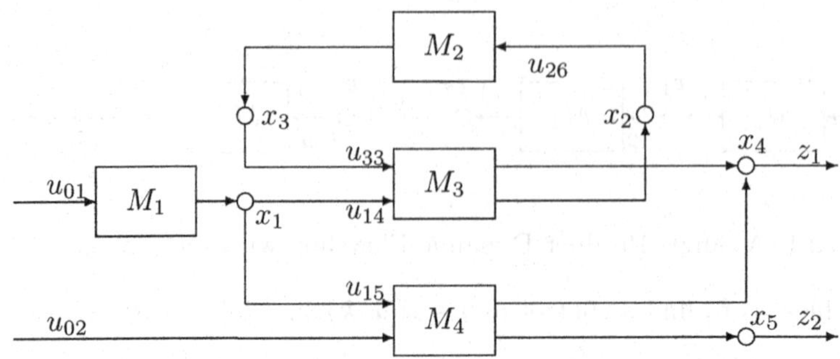

Figure 2.2. A Typical Dynamic Jobshop.

with the state constraint

$$\boldsymbol{x} \in \{(x_1, \ldots, x_5)' : x_1 \geq 0,\ x_2 \geq 0,\ x_3 \geq 0\}, \tag{2.13}$$

and the control constraints

$$\begin{aligned} r_1 u_{01}(t) \leq k_1(t), \quad r_2 u_{02}(t) + r_5 u_{15}(t) \leq k_4(t), \\ r_3 u_{33}(t) + r_4 u_{14}(t) \leq k_3(t), \quad r_6 u_{26}(t) \leq k_2(t), \end{aligned} \tag{2.14}$$

where r_j is a given constant denoting the amount of the required machine capacity to sustain the production rate of a unit of (intermediate) part-type j per unit time, $j = 1, 2, \ldots, 6$.

If each machine has only two states, say 0 or 1, then the possible states of $\boldsymbol{k}$, which is a vector of the states of M_1, M_2, M_3, and M_4, is $2^4 = 16$.

Note that the state constraint is required since the inventory in each of the internal buffers, i.e., x_1, x_2, and x_3, must be nonnegative.

Let

$$\boldsymbol{x}(t) = (x_1(t), \ldots, x_5(t))', \quad \boldsymbol{u}(t) = (u_{01}(t), \ldots, u_{26}(t))'. \tag{2.15}$$

As in the previous section, we use $\mathcal{A}(\boldsymbol{x}, \boldsymbol{k})$ to denote the class of admissible controls. Then our control problem of a general dynamic jobshop with stochastic demand can be formulated as follows:

$$\begin{cases} \min J(\boldsymbol{x}, \boldsymbol{k}, \boldsymbol{u}(\cdot)) = \limsup\limits_{T\to\infty} \dfrac{1}{T} E \displaystyle\int_0^T g(\boldsymbol{x}(t), \boldsymbol{u}(t))\, dt, \\ \text{s.t. } \dfrac{d}{dt}\boldsymbol{x}(t) = A\boldsymbol{u}(t) + B\boldsymbol{z}, \\ \qquad \boldsymbol{x}(0) = \boldsymbol{x}, \quad \boldsymbol{k}(0) = \boldsymbol{k}, \quad \text{and} \quad \boldsymbol{u}(\cdot) \in \mathcal{A}(\boldsymbol{x}, \boldsymbol{k}), \\ \text{minimum average cost } \lambda^*(\boldsymbol{x}, \boldsymbol{k}) = \inf\limits_{\boldsymbol{u}(\cdot)\in\mathcal{A}(\boldsymbol{x},\boldsymbol{k})} J(\boldsymbol{x}, \boldsymbol{k}, \boldsymbol{u}(\cdot)), \end{cases} \tag{2.16}$$

where matrices A and B can be defined appropriately; see Section 5.2. Note that for the system dynamics (2.12), we have

$$A = \begin{pmatrix} 1 & 0 & 0 & -1 & -1 & 0 \\ 0 & 0 & 0 & 1 & 0 & -1 \\ 0 & 0 & -1 & 0 & 0 & 1 \\ 0 & 0 & 1 & 0 & 1 & 0 \\ 0 & 1 & 0 & 0 & 0 & 0 \end{pmatrix}, \quad B = \begin{pmatrix} 0 & 0 \\ 0 & 0 \\ 0 & 0 \\ -1 & 0 \\ 0 & -1 \end{pmatrix}.$$

Bai and Gershwin [9, 10] have applied a heuristic approach to jobshops and flowshops with multiple part-types. A rigorous analysis of the jobshop system is carried out by Presman, Sethi, and Zhang [101]. This will be given in Chapter 5. An asymptotic analysis of the general case is carried out by Sethi, Zhang, and Zhang [123], and will be reported in Chapter 9.

It should be noted that it is the jobshop model that is employed in the context of applications to wafer fabrication scheduling. Some work in this connection appears in Srivatsan, Bai, and Gershwin [132] and Yan, Lou, Sethi, Gardel, and Deosthali [146].

Part II:

Optimal Control of Manufacturing Systems: Existence and Characterization

3

Optimal Control of Parallel-Machine Systems

3.1 Introduction

Thompson and Sethi [135] considered a deterministic single machine, single product production planning model whose purpose is to obtain the production rate over time to minimize an integral representing a discounted quadratic loss function. The model is solved both with and without nonnegative production constraints. It is shown that there exists a turnpike level of inventory, to which the optimal inventory level approaches monotonically over time. The model was generalized by Sethi and Thompson [114, 115], Bensoussan, Sethi, Vickson, and Derzko [17], and Beyer [18], by incorporating an additive white noise term in the dynamics of the inventory process. Moreover, the concept of turnpike inventory level for the stochastic production planning problem was introduced.

Kimemia and Gershwin [79] and Fleming, Sethi, and Soner [55], on the other hand, modeled uncertainty in the production capacity (consisting of unreliable machines) and the demand rates, respectively, as finite state Markov chains. Kimemia and Gershwin [79] studied a system with machines in tandem but without internal buffers, a system known also as a no-wait flowshop. Using the method of dynamic programming as in Rishel [108], they characterized optimal policy to be defined by a number of thresholds (one for each product in production) called hedging points. They used a quadratic approximation for the "cost-to-go" and suggested a linear programming approach that can be implemented in real time.

These Markov chain formulations have inspired a good deal of work that includes Akella and Kumar [3], Bielecki and Kumar [20], Boukas, Haurie, and Van Delft [23], Haurie and Van Delft [70], Fleming, Sethi, and Soner [55], Sethi, Soner, Zhang, and Jiang [111], Hu and Xiang [71], Morimoto and Fujita [96], Basak, Bisi, and Ghosh [12], and Ghosh, Aropostathis, and Marcus [63], and Sethi and Zhang [125]. Many of these papers deal with discounted cost objectives. Exceptions are Bielecki and Kumar [20], Basak, Bisi, and Ghosh [12], and Ghosh, Aropostathis, and Marcus [63]. Bielecki and Kumar [20] dealt with a single machine (with two states: up and down), single product problem. They obtained an explicit solution for the threshold inventory level, in terms of which the optimal policy is as follows: Whenever the machine is up, produce at the maximum possible rate if the inventory level is less than the threshold, produce on demand if the inventory level is exactly equal to the threshold, and do not produce at all if the inventory level exceeds the threshold. Basak, Bisi, and Ghosh [12], and Ghosh, Aropostathis, and Marcus [63] incorporated both diffusion and jump Markov processes in their production planning model, and thus generalized Kimemia and Gershwin [79] as well as Sethi and Thompson [114] and Bensoussan, Sethi, Vickson, and Derzko [17].

In this chapter, we study single or parallel machine systems producing products using machines that are subject to breakdown and repair. The product demands are assumed to be deterministic. We generalize the Bielecki and Kumar model [20] to incorporate general cost functions and machine capacity processes. We assume no state constraints, which we deal with in Chapters 4, 5, 8, and 9.

We use the vanishing discount approach to show that the average-cost Hamilton-Jacobi-Bellman (HJB) equation has a solution that provides the minimal average cost and a potential function. The result helps in establishing a verification theorem specifying the optimal control policy in terms of the potential function. We define what are known as the turnpike sets in terms of the potential function. An important result derived in this chapter is the monotonicity of the turnpike sets in terms of the capacity level or the demand level. By and large, the problem of solving the optimal production planning requires that we locate the turnpike sets. The monotonicity of the turnpike sets facilitates the solution of the production planning problem. On the one hand, the monotonicity property can be used to solve some optimal control problems in a closed form, which are otherwise difficult to handle. On the other, it can greatly reduce computations needed for numerical approaches for solving the problem.

The plan of the chapter is as follows. In Section 3.2 we precisely state the production–inventory model under consideration. In Section 3.3 we establish a systematic approach of constructing the ergodic (stable) control policies. In addition, we develop some estimates for the value function of the corresponding discounted cost problems. In Section 3.4, with the results obtained in Section 3.3 and the use of the vanishing discount approach, a solution of the HJB equation is shown to exist. The solution consists of the minimum average cost and a potential function, both of which are related to the value function of the corresponding discounted problem. A verification theorem is also proved. In Section 3.5 the optimal control policy is specified in terms of the potential function. In Section 3.6 we show that the turnpike sets possess a monotonicity property with respect to the capacity level. In Section 3.7 we extend the results obtained in Sections 3.3 and 3.4 to allow for multiple products. The chapter is concluded with endnotes in Section 3.8.

3.2 Problem Formulation

In order to specify our model, let $x(t)$, $u(t)$, $k(t)$, and z denote, respectively, the surplus level (state variable), the production rate (control variable), the capacity level, and the constant demand rate at time $t \in [0, \infty)$. Here, surplus refers to inventory when $x(t) \geq 0$ and to backlog when $x(t) < 0$. We assume that $x(t) \in \Re$ and $u(t) \in \Re_+$. The capacity $k(\cdot)$ is a finite state Markov chain defined on a probability space $(\Omega, \mathcal{F}, P)$ such that $k(t) \in \mathcal{M} = \{0, 1, \ldots, m\}$. The representation for $\mathcal{M}$ stands, usually, but not necessarily, for the case of m identical machines, each with a unit capacity having two states: up and down. This is not an essential assumption. In general, $\mathcal{M}$ could be any finite set of nonnegative numbers representing production capacities in the various states of the system. Let $Q = (q_{ij})$ denote an $(m+1) \times (m+1)$ matrix such that $q_{ij} \geq 0$ if $i \neq j$ and $q_{ii} = -\sum_{j \neq i} q_{ij}$. For any functions $\varphi(\cdot)$ on $\mathcal{M}$ and $k \in \mathcal{M}$, write

$$Q\varphi(\cdot)(k) = \sum_{j \neq k} q_{kj}[\varphi(j) - \varphi(k)]. \tag{3.1}$$

We sometimes use notation $Q\varphi(k)$ to represent $Q\varphi(\cdot)(k)$. The Markov chain $k(\cdot)$ is generated by Q, i.e., for all bounded real-valued functions $\varphi(\cdot)$ defined on $\mathcal{M}$,

$$\varphi(k(t)) - \int_0^t Q\varphi(\cdot)(k(s))\, ds$$

is a martingale. Matrix Q is known as an *infinitesimal generator* or simply a *generator*.

Definition 2.1. A control process (production rate) $u(\cdot) = \{u(t) \in \Re_+ : t \geq 0\}$ is called *admissible* with respect to the initial capacity $k(0) = k$, if: (i) $u(\cdot)$ is adapted to the filtration $\{\mathcal{F}_t\}$ with $\mathcal{F}_t = \sigma\{k(s) : 0 \leq s \leq t\}$, the σ-field generated by $k(\cdot)$; and (ii) $0 \leq u(t)(\omega) \leq k(t)(\omega)$ for all $t \geq 0$ and $\omega \in \Omega$. □

Let $\mathcal{A}(k)$ denote the set of admissible control processes with the initial condition $k(0) = k$. For any $u(\cdot) \in \mathcal{A}(k)$, the dynamics of the system is

$$\frac{d}{dt}x(t) = u(t) - z, \quad x(0) = x, \quad t \geq 0. \tag{3.2}$$

Definition 2.2. A function $u(\cdot,\cdot)$ defined on $\Re \times \mathcal{M}$ is called an *admissible feedback control*, or simply *feedback control*, if: (i) for any given initial surplus x and production capacity k, the equation

$$\frac{d}{dt}x(t) = u(x(t), k(t)) - z$$

has a unique solution; and (ii) the process $u(\cdot) = \{u(t) = u(x(t), k(t)),\ t \geq 0\} \in \mathcal{A}(k)$. With a slight abuse of notation, we shall express the admissibility condition (ii) simply as the function $u(\cdot,\cdot) \in \mathcal{A}(k)$. □

Let $h(x)$ and $c(u)$ denote the surplus cost and the production cost functions, respectively. For every $u(\cdot) \in \mathcal{A}(k)$, $x(0) = x$, $k(0) = k$, define

$$J(x, k, u(\cdot)) = \limsup_{T\to\infty} \frac{1}{T} E \int_0^T [h(x(t)) + c(u(t))]\, dt. \tag{3.3}$$

The problem is to choose an admissible $u(\cdot)$ that minimizes the cost functional $J(x, k, u(\cdot))$. We define the average-cost function as

$$\lambda^*(x, k) = \inf_{u(\cdot)\in\mathcal{A}(k)} J(x, k, u(\cdot)). \tag{3.4}$$

We will show in the sequel that $\lambda^*(x, k)$ is independent of (x, k). So $\lambda^*(x, k)$ is simply written as λ^* hereafter. Now let us make the following assumptions on the cost functions $h(x)$ and $c(u)$, generator Q, and set $\mathcal{M}$.

(A1) $h(x)$ is a nonnegative, convex function with $h(0) = 0$. There are positive constants C_{h1}, C_{h2}, and $\beta_{h1} \geq 1$ such that

$$h(x) \geq C_{h1}|x|^{\beta_{h1}} - C_{h2}, \quad x \in \Re.$$

Moreover, there are constants C_{h3} and $\beta_{h2} \geq \beta_{h1} \geq 1$ such that

$$|h(x) - h(y)| \leq C_{h3}(1 + |x|^{\beta_{h2}-1} + |y|^{\beta_{h2}-1})|x - y|, \quad x, y \in \Re.$$

(A2) $c(u)$ is a nonnegative twice continuously differentiable function defined on $[0,\ m]$ with $c(0) = 0$. Moreover, for simplicity in exposition, $c(u)$ is either strictly convex or linear.

(A3) We assume that Q is *strongly irreducible*; see Definition A.1.

(A4) Let $(\nu_0, \nu_1, \ldots, \nu_m)$ be the equilibrium distribution of $k(\cdot)$. The average capacity level $\bar{k} = \sum_{j=0}^m j\nu_j > z$.

(A5) $z \notin \mathcal{M}$.

Remark 2.1. Assumption (A1) is the usual assumption on the growth rate of the surplus cost function to ensure the existence of a solution to the HJB equation of the optimal control problem specified in Section 3.4. From Assumption (A2) it is clear that $c(u)$ and $dc(u)/du$ are nondecreasing. This assumption is required in order for the optimal control policy to be written explicitly in a simple form. Assumption (A4) is necessary in the sense that if this were not true, then even if the system is always producing at its maximum capacity, it still would not meet the demand and the backlog would build up without bound over time, with the consequence that no optimal solution with a finite average cost would exist. From Assumptions (A4) and (A5), it is easily seen that there is a unique $k_0 \in \mathcal{M}$ such that $0 \leq k_0 < z < k_0 + 1 \leq m$. Assumption (A5) is innocuous and is used for the purpose of ensuring the differentiability of the value function in the discounted cost case. This would then translate into the differentiability of the potential function in the average-cost case. □

In (3.3), we have defined our average-cost function to be the limit superior (lim sup) of the finite horizon average costs as the horizon increases, and our optimal control problem to be one of minimizing the cost over the class of admissible controls. In general, the limit of the finite horizon costs may not exist for all controls in this admissible class. Then it may be also of interest to know the class of controls over which the optimal control also minimizes the limit inferior (lim inf) of the finite horizon costs. For this purpose, we define a smaller class of controls as follows:

Definition 2.3. A control $u(\cdot) \in \mathcal{A}(k)$ is called *stable* if it satisfies the condition

$$\lim_{T\to\infty} \frac{E|x(T)|^{\beta_{h2}+1}}{T} = 0,$$

where $x(\cdot)$ is the surplus process corresponding to the control $u(\cdot)$ with $(x(0), k(0)) = (x, k)$ and β_{h2} is as defined in Assumption (A1). □

Let $\mathcal{S}(k)$ denote the class of stable controls with the initial condition $k(0) = k$. It is clear that $\mathcal{S}(k) \subset \mathcal{A}(k)$. It can be seen in the next section that the set of stable admissible controls $\mathcal{S}(k)$ is nonempty.

We will show in Sections 3.4 and 3.5 that there exists a stable Markov control policy $u^*(\cdot) \in \mathcal{S}(k)$ such that $u^*(\cdot)$ is optimal, i.e., it minimizes the average cost defined by (3.3) over all $u(\cdot) \in \mathcal{A}(k)$ and, furthermore,

$$\lim_{T\to\infty} \frac{1}{T} E \int_0^T [h(x^*(t)) + c(u^*(t))]\, dt = \lambda^*,$$

where $x^*(\cdot)$ is the surplus process corresponding to $u^*(\cdot)$ with $(x^*(0), k(0)) = (x, k)$. Moreover, for any other (stable) control $u(\cdot) \in \mathcal{S}(k)$,

$$\liminf_{T\to\infty} \frac{1}{T} E \int_0^T [h(x(t)) + c(u(t))]\, dt \geq \lambda^*.$$

In this chapter, our main focus is on a single product manufacturing system with stochastic machine capacity and constant demand. While the results to be derived in Sections 3.3 and 3.4 can be extended to a multiproduct framework, the monotonicity properties obtained later in Sections 3.5 and 3.6 make sense only in single product cases. It is for this reason that we have chosen in Sections 3.2–3.6 to deal only with a single product model. It should also be noted that the classical literature on the convex production planning problem is concerned mainly with single product problems. The multiproduct case will be dealt with in Section 3.7.

3.3 Estimates for Discounted Cost Value Functions

We will use the vanishing discount approach to study our problem, we begin with an analysis of the discounted problem.

In order to derive the HJB equation for the average-cost control problem formulated above and study the existence of its solutions, we introduce a corresponding control problem with the cost discounted at a rate $\rho > 0$. For $u(\cdot) \in \mathcal{A}(k)$, we define the expected discounted cost as

$$J^{\rho}(x,k,u(\cdot)) = E\int_0^{\infty} e^{-\rho t}[h(x(t)) + c(u(t))]\,dt. \tag{3.5}$$

Define the value function of the discounted cost problem as

$$V^{\rho}(x,k) = \inf_{u(\cdot)\in\mathcal{A}(k)} J^{\rho}(x,k,u(\cdot)). \tag{3.6}$$

The dynamic programming equation associated with this problem (see Sethi and Zhang [125]) is

$$\rho\phi^{\rho}(x,k) = F\left(k, \frac{\partial\phi^{\rho}(x,k)}{\partial x}\right) + h(x) + Q\phi^{\rho}(x,\cdot)(k), \tag{3.7}$$

where $\partial\phi^{\rho}(x,k)/\partial x$ is the partial derivative of $\phi^{\rho}(x,k)$ with respect to its first variable x,

$$F\left(k, \frac{\partial\phi^{\rho}(x,k)}{\partial x}\right) = \inf_{0\le u\le k}\left\{(u-z)\frac{\partial\phi^{\rho}(x,k)}{\partial x} + c(u)\right\}. \tag{3.8}$$

Theorem 3.1. *Under Assumptions* (A1), (A2), (A3), *and* (A5), *the value function $V^{\rho}(x,k)$ has the following properties:*

(i) *$V^{\rho}(x,k)$ is continuously differentiable and convex for any fixed $k \in \mathcal{M}$. Moreover, there are positive constants $C_{\rho1}$, $C_{\rho2}$, and $C_{\rho3}$ such that, for any k,*

$$C_{\rho1}|x|^{\beta_{h1}} - C_{\rho2} \le V^{\rho}(x,k) \le C_{\rho3}(1 + |x|^{\beta_{h2}}),$$

where β_{h1}, β_{h2} are defined in Assumption (A1).

(ii) $V^\rho(x,k)$ *is the unique solution of the* HJB *equation* (3.7).

(iii) $V^\rho(x,k)$ *is strictly convex if the cost function* $h(x)$ *is also.*

Proof. See Lemmas 3.1 and 3.2, and Theorem 3.1 in Chapter 3 of Sethi and Zhang [125]. □

Remark 3.1. Note that the constants $C_{\rho 1}$, $C_{\rho 2}$, and $C_{\rho 3}$ may depend on the discount rate ρ. □

Let $c_u(u) = dc(u)/du$ and $V_x^\rho(x,k) = \partial V^\rho(x,k)/\partial x$. Now define

$$u^\rho(x,k) = \begin{cases} 0, & \text{if } V_x^\rho(x,k) > -c_u(0), \\ (c_u(u))^{-1}(-V_x^\rho(x,k)), & \text{if } -c_u(k) \le V_x^\rho(x,k) \le -c_u(0), \\ k, & \text{if } V_x^\rho(x,k) < -c_u(k), \end{cases} \tag{3.9}$$

when $c(u)$ is strictly convex, and

$$u^\rho(x,k) = \begin{cases} 0, & \text{if } V_x^\rho(x,k) > -c, \\ k \wedge z, & \text{if } V_x^\rho(x,k) = -c, \\ k, & \text{if } V_x^\rho(x,k) < -c, \end{cases} \tag{3.10}$$

when $c(u) = cu$ for some constant $c \ge 0$. It follows from the convexity of $V^\rho(x,k)$ that $u^\rho(x,k)$ is nonincreasing in x. From Lemma F.1, the differential equation

$$\frac{d}{dt}x(t) = u^\rho(x(t), k(t)(\omega)) - z, \quad x(0) = x,$$

has a unique solution $x^\rho(t)$, $t \ge 0$, for each sample path of $k(t)(\omega)$, $\omega \in \Omega$. Therefore, it follows from (3.8) and Theorem 3.1 that $u^\rho(x,k)$ is an optimal control policy.

In order to study the long-run average-cost control problem using the vanishing discount approach, we must first obtain some estimates for the value function $V^\rho(x,k)$ defined in (3.6) for small values of ρ.

Lemma 3.1. *Let Assumptions* (A3) *and* (A4) *hold. Let*

$$\sigma = \inf\left\{t : \int_0^t [k(s) - z]\, ds = \Delta\right\}$$

for any fixed $\Delta > 0$. *Then, for any* $r \ge 1$, *there is a constant* C_r *independent of* Δ *such that* $E\sigma^r \le C_r(\Delta^r + 1)$.

Proof. It can be shown as Lemma B.5, that there exists a constant $C_1 > 0$

such that, for any $t > 0$,

$$E\left[\exp\left(\frac{1}{\sqrt{t+1}}\left|\int_0^t [k(s)-\bar{k}]\,ds\right|\right)\right] \le C_1, \tag{3.11}$$

where $\bar{k}$ is the average capacity level defined in Assumption (A4).

Note by the definition of σ that

$$\begin{aligned} P(\sigma \ge t) &\le P\left(\int_0^t [k(s)-z]\,ds \le \Delta\right) \\ &= P\left(\int_0^t [k(s)-\bar{k}]\,ds + (\bar{k}-z)t \le \Delta\right). \end{aligned}$$

Since $\bar{k} > z$ by Assumption (A4), we have, for $t \ge \Delta/(\bar{k}-z)$,

$$\begin{aligned} P(\sigma \ge t) &\le P\left(\left|\int_0^t [k(s)-\bar{k}]\,ds\right| \ge t(\bar{k}-z)-\Delta\right) \\ &= P\left(\frac{1}{\sqrt{t+1}}\left|\int_0^t [k(s)-\bar{k}]\,ds\right| \ge \frac{t(\bar{k}-z)-\Delta}{\sqrt{t+1}}\right) \\ &\le \exp\left(-\frac{t(\bar{k}-z)-\Delta}{\sqrt{t+1}}\right)\cdot E\exp\left(\frac{1}{\sqrt{t+1}}\left|\int_0^t [k(s)-\bar{k}]\,ds\right|\right). \end{aligned}$$

In view of this and inequality (3.11), we have that, for $t \ge 2\Delta/(\bar{k}-z)$,

$$\begin{aligned} P(\sigma \ge t) &\le C_1\exp\left(-\frac{t(\bar{k}-z)}{\sqrt{t+1}} + \frac{\Delta}{\sqrt{t+1}}\right) \\ &= C_1\exp\left\{-\frac{t(\bar{k}-z)}{2\sqrt{t+1}} + \left(-\frac{t(\bar{k}-z)}{2\sqrt{t+1}} + \frac{\Delta}{\sqrt{t+1}}\right)\right\} \\ &\le C_1\exp\left(-\frac{t(\bar{k}-z)}{2\sqrt{t+1}}\right). \end{aligned}$$

Therefore,

$$\begin{aligned} E\sigma^r &= r\int_0^\infty t^{r-1}P(\sigma \ge t)\,dt \\ &\le r\int_0^{2\Delta/(\bar{k}-z)} t^{r-1}\,dt + rC_1\int_{2\Delta/(\bar{k}-z)}^\infty t^{r-1}\exp\left(-\frac{t(\bar{k}-z)}{2\sqrt{t+1}}\right)dt \\ &\le C_r(\Delta^r+1), \end{aligned}$$

for some constant $C_r > 0$, which is independent of Δ. □

Corollary 3.1. *Let Assumptions* (A3) *and* (A4) *hold. Let* $\widehat{\sigma}$ *be a stopping time with respect to* $\mathcal{F}_t$, *and let*

$$\sigma = \inf\left\{t : \int_{\widehat{\sigma}}^{\widehat{\sigma}+t} [k(s)-z]\,ds = \Delta\right\},$$

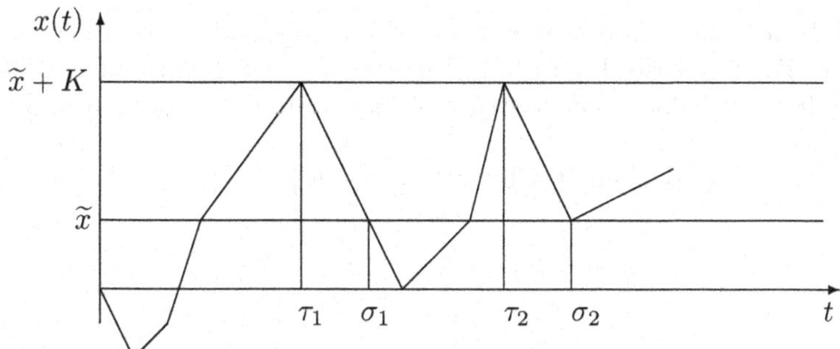

Figure 3.1. Surplus Process $x(t)$ under Control $u(\cdot)$.

for any fixed $\Delta > 0$. *Then, for any* $r \geq 1$, *there is a constant* $\widehat{C}_r$ *independent of* Δ *such that* $E\sigma^r \leq \widehat{C}_r(\Delta^r + 1)$.

Proof. The proof is similar to the proof of Lemma 3.1. Here only the inequality

$$E\left[\exp\left(\frac{1}{\sqrt{t+1}}\left|\int_{\widehat{\sigma}}^{\widehat{\sigma}+t}[k(s)-\bar{k}]\,ds\right|\right)\right] \leq C_1,$$

for some $C_1 > 0$, is used instead of (3.11), which is given by Corollary B.2. □

Lemma 3.2. *Let Assumptions* (A3) *and* (A4) *hold. For any fixed* $r \geq 1$, *there exists a constant* $\widetilde{C}_r > 0$ *such that, for any* $(\widetilde{x}, \widetilde{k}), (x, k) \in \Re \times \mathcal{M}$, *we can find an admissible control* $u(\cdot) \in \mathcal{A}(k)$ *such that*

$$E\tau^r \leq \widetilde{C}_r\left(1 + |\widetilde{x} - x|^r\right), \tag{3.12}$$

where

$$\tau = \inf\{t > 0 : \ (x(t), k(t)) = (\widetilde{x}, \widetilde{k})\}, \tag{3.13}$$

and $x(\cdot)$ *is the surplus process corresponding to the control* $u(\cdot)$ *and the initial condition* $(x(0), k(0)) = (x, k)$.

Proof. We provide a proof only in the case when $(x, k) = (0, 0)$. The proofs in all other cases are similar. Recall that $\{\nu_k : k = 0, 1, \ldots, m\}$ is the stationary distribution of the Markov chain $k(\cdot)$, i.e.,

$$\lim_{t\to\infty} P(k(t) = k | k(0) = i) = \nu_k, \quad k, i \in \mathcal{M}.$$

Notice that, by Assumption (A3), $k(\cdot)$ is a strongly irreducible Markov chain, so we have $\nu_k > 0$, $k = 0, 1, \ldots, m$. Since $\mathcal{M}$ is a finite set, we have a $t_0 > 0$ such that, for all $t \geq t_0$,

$$P(k(t) = k | k(0) = i) \geq \nu_k / 2, \quad k, i \in \mathcal{M}. \tag{3.14}$$

Take K to be a number such that $K > t_0 z$. We first consider the case when $\widetilde{x} + K > 0$. For this case, Figure 3.1 is plotted for $\widetilde{x} > 0$ without loss of generality. In what follows, we construct an appropriate control used in the proof. Define

$$\tau_1 = \inf\{t > 0 : x(t) = \widetilde{x} + K\}.$$

Let

$$u(t) = k(t), \quad \text{for } t \leq \tau_1,$$

so that

$$x(t) = \int_0^t [u(s) - z]\, ds = \int_0^t [k(s) - z]\, ds, \quad t \leq \tau_1,$$

and

$$u(t) = 0, \quad x(t) = \widetilde{x} + K - (t - \tau_1)z, \quad \text{for } \tau_1 < t \leq \tau_1 + K/z.$$

For convenience in notation, we write $\sigma_1 = \tau_1 + K/z$. Proceeding in this manner, we can define the required control $u(\cdot) \in \mathcal{A}(0)$ inductively:

$$u(t) = \begin{cases} k(t), & \text{if } \sigma_{n-1} < t \leq \tau_n, \\ 0, & \text{if } \tau_n < t \leq \sigma_n, \end{cases}$$

where

$$x(t) = \int_0^t [u(s) - z]\, ds, \quad 0 \leq t \leq \tau_n,$$
$$\tau_n = \inf\{t > \sigma_{n-1} : x(t) = \widetilde{x} + K\} \quad \text{and} \quad \sigma_n = \tau_n + K/z, \quad n > 1.$$

We set $\sigma_0 = 0$ and $\tau_0 = 0$. The control $u(\cdot)$ can be characterized as follows: Use the maximum available production rate $u(t) = k(t)$ to move the surplus process from 0 or $\widetilde{x}$ to $\widetilde{x}+K$, and then use the zero production rate until the surplus process drops to the level $\widetilde{x}$. A sample path of the surplus process for $\widetilde{x} + K > 0$ is graphed in Figure 3.1.

It is obvious by our construction that $x(\sigma_n) = \widetilde{x}$, $n = 1, 2, \ldots$. Furthermore, by the strong Markov property of $k(\cdot)$, we have

$$\begin{aligned} P(\tau > \sigma_n) &\leq P(k(\sigma_n) \neq \widetilde{k}, k(\sigma_{n-1}) \neq \widetilde{k}, \ldots, k(\sigma_1) \neq \widetilde{k}) \\ &= P(k(\sigma_1) \neq \widetilde{k}) P(k(\sigma_2) \neq \widetilde{k} | k(\sigma_1) \neq \widetilde{k}) \\ &\quad \times \cdots \times P(k(\sigma_n) \neq \widetilde{k} | k(\sigma_{n-1}) \neq \widetilde{k}). \end{aligned} \tag{3.15}$$

Note that $\sigma_i - \sigma_{i-1} \geq K/z > t_0$, $i = 1, 2 \ldots$. By (3.14) we have

$$P(k(\sigma_1) \neq \widetilde{k}) \leq 1 - \delta,$$

$$P(k(\sigma_i) \neq \widetilde{k} | k(\sigma_{i-1}) \neq \widetilde{k}) \leq 1 - \delta, \quad i = 2, 3, \ldots,$$

where $\delta \equiv \min\{\nu_k/2 : k \in \mathcal{M}\}$. Then from (3.15) we have

$$P(\tau > \sigma_n) \leq (1-\delta)^n. \tag{3.16}$$

Recall that $\sigma_n - \sigma_{n-1} = \tau_n - \sigma_{n-1} + K/z$,

$$\tau_1 - \sigma_0 = \inf\left\{t > 0 : \int_0^t [k(s) - z]\, ds \geq \widetilde{x} + K\right\},$$

and, for $n = 2, 3, ...,$

$$\tau_n - \sigma_{n-1} = \inf\left\{t > 0 : \widetilde{x} + \int_{\sigma_{n-1}}^{\sigma_{n-1}+t} [k(s) - z]\, ds \geq \widetilde{x} + K\right\}.$$

Apply Corollary 3.1 for any $r \geq 1$ to obtain that, for $n \geq 1$,

$$\begin{aligned} E|\tau_n - \sigma_{n-1}|^r &\leq C_r \left(1 + |\widetilde{x} + K|^r\right) \\ &\leq C_1 \left(1 + |\widetilde{x}|^r + |K|^r\right), \end{aligned} \tag{3.17}$$

where $C_1 > 0$ is a constant independent of $\widetilde{x}$. Consequently, there exists a positive constant C_2 such that

$$\begin{aligned} E|\tau_n - \tau_{n-1}|^r &= E|(\tau_n - \sigma_{n-1}) + (\sigma_{n-1} - \tau_{n-1})|^r \\ &\leq 2^{r-1} E\left(|\tau_n - \sigma_{n-1}|^r + |\sigma_{n-1} - \tau_{n-1}|^r\right) \\ &\leq C_2 \left(1 + |\widetilde{x}|^r\right). \end{aligned}$$

Noting that $\tau_n \geq \sigma_0 + \sigma_1 + \cdots + \sigma_{n-1} \geq (n-1)K/z$, therefore,

$$\begin{aligned} E\tau^r &= r \int_0^\infty t^{r-1} P(\tau \geq t)\, dt \\ &= r \sum_{n=1}^\infty E \int_{\tau_{n-1}}^{\tau_n} t^{r-1} P(\tau \geq t)\, dt \\ &\leq r \sum_{n=1}^\infty E \int_{\tau_{n-1}}^{\tau_n} t^{r-1} P(\tau \geq \tau_{n-1})\, dt \\ &\leq r \sum_{n=1}^\infty E \int_{\tau_{n-1}}^{\tau_n} t^{r-1} P(\tau \geq \sigma_{(n-2)\vee 0})\, dt \\ &\leq \sum_{n=1}^\infty E\left(\tau_n^r - \tau_{n-1}^r\right)(1-\delta)^{(n-2)\vee 0} \\ &\leq \sum_{n=1}^\infty n^r E\left(\max_{1\leq j\leq n}\{\tau_j - \tau_{j-1}\}\right)^r (1-\delta)^{(n-2)\vee 0} \\ &\leq C_2(1 + |\widetilde{x}|^r) \sum_{n=1}^\infty n^{r+1}(1-\delta)^{(n-2)\vee 0} \\ &\leq C_3 \left(1 + |\widetilde{x}|^r\right), \end{aligned}$$

for some positive constant C_3 independent of $\widetilde{x}$. This implies the lemma when $\widetilde{x} + K > 0$.

When $\widetilde{x} + K \leq 0$, the proof is the same except that we define $u(t) = 0$ until time τ_1, which in this case exactly equals $|\widetilde{x} + K|/z$. □

For any constant M, define the feedback policy

$$\widetilde{u}(x,k) = \begin{cases} 0, & \text{if } x > M, \\ k \wedge z, & \text{if } x = M, \\ k, & \text{if } x < M, \end{cases} \tag{3.18}$$

and $\widetilde{x}(t)$ to be the corresponding surplus process with the initial condition $(\widetilde{x}(0), k(0)) = (x, k)$.

Corollary 3.2. *Under Assumptions* (A3) *and* (A4), *for any constant* M, *the feedback policy* $\widetilde{u}(x,k)$ *defined by* (3.18) *is stable. That is,*

$$\lim_{T\to\infty} \frac{E|\widetilde{x}(T)|^{\beta_{h2}+1}}{T} = 0.$$

Proof. Let

$$u(t) = k(t), \quad \text{for } t \leq \tau_1 = \inf\{t > 0 : x(t) = M + K\}$$

such that

$$x(t) = x + \int_0^t [u(s) - z]\, ds = x + \int_0^t [k(s) - z]\, ds, \quad t \leq \tau_1,$$

and

$$u(t) = 0, \quad x(t) = M + K - z(t - \tau_1), \quad \text{for } \tau_1 < t \leq \tau_1 + K/z.$$

For convenience in notation, we write $\sigma_1 = \tau_1 + K/z$. Proceeding in this manner, we can define the required control $u(\cdot) \in \mathcal{A}(k)$ inductively:

$$u(t) = \begin{cases} k(t), & \text{if } \sigma_{n-1} < t \leq \tau_n, \\ 0, & \text{if } \tau_n < t \leq \sigma_n, \end{cases}$$

where

$$x(t) = x + \int_0^t [u(s) - z]\, ds, \quad 0 \leq t \leq \tau_n,$$

$$\tau_n = \inf\{t > \sigma_{n-1} : x(t) = M + K\} \quad \text{and} \quad \sigma_n = \tau_n + K/z, \quad n > 1.$$

Then, we have that, for any $t \geq 0$,

$$|\widetilde{x}(t)| \leq |x(t)| + K. \tag{3.19}$$

Furthermore, for $n = 1, 2, ...$,

$$\begin{aligned}
&|x(t)| \leq |M| + K, \quad \text{for } t \in [\tau_{n-1},\ \sigma_{n-1}],\\
&|x(t)| \leq |M| + (m+z)(\tau_n - \sigma_{n-1}), \quad \text{for } t \in [\sigma_{n-1},\ \tau_n],
\end{aligned}$$

where $\tau_0 = \sigma_0 = 0$. Therefore, the corollary follows from (3.17) and (3.19).
□

Theorem 3.2. *Let Assumptions* (A1)–(A4) *hold. There exists a constant* $\rho_0 > 0$ *such that* $\rho V^\rho(0,0)$ *for* $0 < \rho \leq \rho_0$ *is bounded.*

Proof. By Lemma 3.2 we know that there exists a control $u(\cdot) \in \mathcal{A}(0)$ such that, for each $r \geq 1$,

$$E\left[\tau_0^r\right] \leq C_1, \tag{3.20}$$

where $C_1 > 0$ is a constant (which depends on r) and

$$\tau_0 = \inf\left\{t > 0 :\ (x(t), k(t)) = (0,0)\right\},$$

with $x(\cdot)$ being the surplus process corresponding to the control process $u(\cdot)$ and the initial condition $(x(0), k(0)) = (0,0)$. By the optimality principle, we have

$$\begin{aligned}
V^\rho(0,0) &\leq E\bigg\{\int_0^{\tau_0} e^{-\rho t}[h(x(t)) + c(u(t))]\,dt\\
&\qquad\qquad + e^{-\rho\tau_0}V^\rho(x(\tau_0), k(\tau_0))\bigg\}\\
&= E\bigg\{\int_0^{\tau_0} e^{-\rho t}[h(x(t)) + c(u(t))]\,dt + e^{-\rho\tau_0}V^\rho(0,0)\bigg\}\\
&= E\left\{\int_0^{\tau_0} e^{-\rho t}[h(x(t)) + c(u(t))]\,dt\right\} + V^\rho(0,0)E\left[e^{-\rho\tau_0}\right].
\end{aligned}$$

Note that $|x(t)| \leq (m+z)t$ for $0 \leq t \leq \tau_0$, where we recall that m is the largest possible production capacity. Thus by Assumptions (A1) and (A2), for $0 \leq t \leq \tau_0$,

$$\begin{aligned}
&h(x(t)) \leq C_{h3}(1 + |x(t)|^{\beta_{h2}}) \leq C_2(1 + t^{\beta_{h2}}),\\
&c(u(t)) \leq c(m),
\end{aligned}$$

where C_2 is a positive constant. It follows from (3.20) that

$$\begin{aligned}
&[1 - Ee^{-\rho\tau_0}]V^\rho(0,0)\\
&\qquad \leq E\int_0^{\tau_0} [c(m) + C_2(1 + t^{\beta_{h2}})]\,dt\\
&\qquad = [c(m) + C_2]\,E\tau_0 + \frac{C_2}{\beta_{h2}+1}E\left[\tau_0^{\beta_{h2}+1}\right] \leq C_3,
\end{aligned} \tag{3.21}$$

for some positive constant C_3 (independent of ρ). Now, using the inequality $1 - e^{-\rho\tau_0} \geq \rho\tau_0 - \rho^2\tau_0^2/2$, we can get

$$[1 - Ee^{-\rho\tau_0}]V^\rho(0,0) \geq \left[E\tau_0 - \frac{\rho E[\tau_0^2]}{2}\right] \cdot \rho V^\rho(0,0). \tag{3.22}$$

From the definition of τ_0 and (3.20), we know that $\tau_0 > 0$ and both $E\tau_0$ and $E\tau_0^2$ are finite. Therefore, we have

$$0 < E\tau_0 < \infty \quad \text{and} \quad 0 < E[\tau_0^2] < \infty.$$

Take $\rho_0 = E\tau_0/E[\tau_0^2]$. By (3.21) and (3.22), we have that, for $0 < \rho \leq \rho_0$,

$$\rho V^\rho(0,0) \leq \frac{C_3}{E\tau_0 - \rho_0 E[\tau_0^2]/2} = \frac{2C_3}{E\tau_0},$$

which yields the theorem. □

Let us define the function

$$\widetilde{V}^\rho(x,k) = V^\rho(x,k) - V^\rho(0,0), \tag{3.23}$$

for which the following results can be derived.

Theorem 3.3. *Let Assumptions* (A1)–(A4) *hold. The function* $\widetilde{V}^\rho(x,k)$ *is convex in* x. *For each* k, *it is locally uniformly bounded, i.e., there exists a constant* $C > 0$ *such that*

$$|\widetilde{V}^\rho(x,k)| \leq C(1 + |x|^{\beta_{h2}+1}), \quad (x,k) \in \Re \times \mathcal{M}, \quad \rho \geq 0. \tag{3.24}$$

Proof. The convexity of $\widetilde{V}^\rho(\cdot,k)$ follows from that of $V^\rho(\cdot,k)$. Thus, we need only to show inequality (3.24). We first consider an upper bound for $\widetilde{V}^\rho(x,k)$. By Lemma 3.2, there exists a constant $C_1 > 0$ and a control $u(\cdot) \in \mathcal{A}(k)$ such that

$$E\left[\tau_0^{\beta_{h2}+1}\right] \leq C_1\left(1 + |x|^{\beta_{h2}+1}\right), \tag{3.25}$$

with

$$\tau_0 = \inf\{t > 0 : (x(t), k(t)) = (0,0)\},$$

where $x(\cdot)$ is the state corresponding to $u(\cdot)$ and the initial condition

$(x(0), k(0)) = (x, k)$. Then, from the optimality principle, we have

$$\begin{aligned} V^\rho(x,k) &\le E\Big\{\int_0^{\tau_0} e^{-\rho t}[h(x(t)) + c(u(t))]\,dt \\ &\qquad + e^{-\rho\tau_0} V^\rho(x(\tau_0), k(\tau_0))\Big\} \\ &= E\Big\{\int_0^{\tau_0} e^{-\rho t}[h(x(t)) + c(u(t))]\,dt \\ &\qquad + e^{-\rho\tau_0} V^\rho(0,0)\Big\} \\ &\le E\Big\{\int_0^{\tau_0} e^{-\rho t}[h(x(t)) + c(u(t))]\,dt + V^\rho(0,0)\Big\}. \end{aligned} \tag{3.26}$$

Note that $|x(t)| \le |x| + (m+z)t$ for $0 \le t \le \tau_0$. Thus by Assumptions (A1) and (A2), for $0 \le t \le \tau_0$,

$$\begin{aligned} h(x(t)) &\le C_2\left(1 + |x|^{\beta_{h2}} + t^{\beta_{h2}}\right), \\ c(u(t)) &\le c(m), \end{aligned}$$

for some positive constant C_2. Therefore, by (3.25) and (3.26), there exists a positive constant C_3 (independent of ρ) such that

$$\begin{aligned} \widetilde{V}^\rho(x,k) &= V^\rho(x,k) - V^\rho(0,0) \\ &\le E\int_0^{\tau_0} \left[c(m) + C_2(1 + |x|^{\beta_{h2}} + t^{\beta_{h2}})\right]\,dt \\ &\le C_3(1 + |x|^{\beta_{h2}+1}). \end{aligned} \tag{3.27}$$

Next we consider the lower bound of $\widetilde{V}^\rho(x,k)$. By Lemma 3.2, there exists an admissible control $u(\cdot) \in \mathcal{A}(0)$ such that

$$E\left[\tau^{\beta_{h2}+1}\right] \le C_4\left(1 + |x|^{\beta_{h2}+1}\right), \quad x \in \Re, \tag{3.28}$$

where

$$\tau = \inf\{t > 0 : (x(t), k(t)) = (x, k)\},$$

$C_4 > 0$ is a constant independent of x, and $x(\cdot)$ is the surplus process corresponding to the control $u(\cdot)$ and the initial condition $(x(0), k(0)) = (0, 0)$. Apply again the optimality principle to obtain

$$\begin{aligned} V^\rho(0,0) &\le E\Big\{\int_0^{\tau} e^{-\rho t}[h(x(t)) + c(u(t))]\,dt \\ &\qquad + e^{-\rho\tau} V^\rho(x(\tau), k(\tau))\Big\} \\ &= E\Big\{\int_0^{\tau} e^{-\rho t}[h(x(t)) + c(u(t))]\,dt + e^{-\rho\tau} V^\rho(x,k)\Big\} \\ &\le E\Big\{\int_0^{\tau} e^{-\rho t}[h(x(t)) + c(u(t))]\,dt + V^\rho(x,k)\Big\}. \end{aligned}$$

Therefore, in view of $x(t) \le (m+z)t$ and Assumptions (A1) and (A2), we have

$$
\begin{aligned}
\widetilde{V}^{\rho}(x,k) &= V^{\rho}(x,k) - V^{\rho}(0,0) \\
&\ge E\Big\{(1-e^{-\rho\tau})V^{\rho}(x,k) \\
&\qquad - \int_0^{\tau} e^{-\rho t}[h(x(t)) + c(u(t))]\,dt\Big\} \\
&\ge E\Big\{-\int_0^{\tau} e^{-\rho t}[h(x(t)) + c(u(t))]\,dt\Big\} \\
&\ge -E\Big\{\int_0^{\tau} \big[c(m) + C_{h3}(1 + ((m+z)t)^{\beta_{h2}})\big]\,dt\Big\} \\
&\ge -C_5(1 + |x|^{\beta_{h2}+1}),
\end{aligned}
\tag{3.29}
$$

for some positive constant C_5 (independent of ρ). The theorem follows from (3.27) and (3.29). □

The next corollary shows the Lipschitz continuity of $\widetilde{V}^{\rho}(x,k)$.

Corollary 3.3. *Let Assumptions* (A1)–(A4) *hold. The function* $\widetilde{V}^{\rho}(x,k)$, $\rho > 0$, *is locally uniformly Lipschitz continuous in* x. *That is, for any bounded interval* $\mathcal{I} \subset \Re$, *there exists a constant* $C > 0$ *(independent of* ρ*) such that*

$$|\widetilde{V}^{\rho}(x,k) - \widetilde{V}^{\rho}(\widetilde{x},k)| \le C\,|x - \widetilde{x}|,$$

for all $x,\ \widetilde{x} \in \mathcal{I}$, *and* $\rho > 0$.

Proof. The result follows from part (iii) of Lemma C.1 and the fact that $\widetilde{V}^{\rho}(x,k)$, $\rho > 0$, are locally uniformly bounded. □

Corollary 3.4. *Let Assumptions* (A1)–(A4) *hold. For* $(x,k) \in \Re\times\mathcal{M}$, *there is a subsequence of* ρ, *still denoted by* ρ, *such that the limits of* $\rho V^{\rho}(x,k)$ *and* $\widetilde{V}^{\rho}(x,k)$ *exist as* $\rho \to 0$. *Write*

$$\widehat{\lambda} = \lim_{\rho\to 0} \rho V^{\rho}(x,k) \quad \text{and} \quad V(x,k) = \lim_{\rho\to 0} \widetilde{V}^{\rho}(x,k). \tag{3.30}$$

Moreover, the convergence is locally uniform in (x,k) *and* $V(\cdot,\cdot)$ *is locally Lipschitz continuous.*

Proof. The proof follows immediately from Theorems 3.2 and 3.3, Corollary 3.3, and Theorem F.1 (Arzelà-Ascoli theorem). □

Remark 3.2. $V(x,k)$ is called *the relative cost function.* □

3.4 Verification Theorem

The HJB equation for the optimal control problem of Section 3.2 takes the form

$$\lambda = F\left(k, \frac{\partial \phi(x,k)}{\partial x}\right) + h(x) + Q\phi(x,\cdot)(k), \tag{3.31}$$

where λ is a constant, $\phi(\cdot,\cdot)$ is a real-valued function defined on $\Re \times \mathcal{M}$, and $F(k, \partial\phi(x,k)/\partial x)$ is defined by (3.8).

Before we define a solution to the HJB equation (3.31), we first introduce some notation. Let $\mathcal{G}$ denote the family of real-valued functions $G(\cdot,\cdot)$ defined on $\Re \times \mathcal{M}$ such that, for each $k \in \mathcal{M}$,

(i) $G(x,k)$ is convex in x;

(ii) $G(x,k)$ is continuously differentiable with respect to x; and

(iii) there is a constant $C > 0$ such that

$$|G(x,k)| \le C(1 + |x|^{\beta_{h2}+1}), \quad x \in \Re,$$

where β_{h2} is given by Assumption (A1).

Definition 4.1. A solution of the HJB equation (3.31) is a pair $(\lambda, W(\cdot,\cdot))$ with λ a constant and $W(\cdot,\cdot) \in \mathcal{G}$. The function $W(\cdot,\cdot)$ is called a *potential function* for the control problem if $\lambda = \lambda^*$, the minimum long-run average cost. □

In Theorem 4.1 we show that the limit $(\widehat{\lambda}, V(\cdot), \cdot)$ obtained in (3.30) is a viscosity solution as defined in Appendix D. Then in Theorem 4.2 we show that $V(\cdot,k) \in C^1$ and therefore $(\widehat{\lambda}, V(\cdot,\cdot))$ is indeed a solution.

Theorem 4.1. *Let Assumptions* (A1)–(A4) *hold. Then* $(\widehat{\lambda}, V(x,k))$ *is a viscosity solution to the* HJB *equation* (3.31). *Moreover, the constant* $\widehat{\lambda}$ *is unique in the following sense: If* $(\widetilde{\lambda}, \widetilde{V}(x,k))$ *is another viscosity solution to* (3.31), *then* $\widetilde{\lambda} = \widehat{\lambda}$.

Proof. By Corollary 3.4, we know that the convergence of $\widetilde{V}^\rho(x,k)$ to $V(x,k)$ in (3.30) is locally uniform in (x,k). From Theorem 3.1 we know that $V^\rho(x,k)$ is the solution, and thus a viscosity solution to (3.7). As a result, $\widetilde{V}^\rho(x,k)$ is a viscosity solution to

$$\rho\widetilde{V}^\rho(x,k) + \rho V^\rho(0,0) = F\left(k, \frac{\partial V^\rho(x,k)}{\partial x}\right) + h(x) + Q\widetilde{V}^\rho(x,\cdot)(k). \tag{3.32}$$

Note that $\widehat{\lambda} = \lim_{\rho\to 0} \rho V^\rho(0,0)$ which is easy to see from the facts that $\rho V^\rho(x,k) \to \widehat{\lambda}$, $\rho\widetilde{V}^\rho(x,k) \to 0$, and $V^\rho(x,k) = \widetilde{V}^\rho(x,k) + V^\rho(0,0)$. In

(3.32), $\rho\widetilde{V}^\rho(x,k) \to 0$ locally uniformly. Furthermore, $\partial V^\rho(x,k)/\partial x = \partial\widetilde{V}^\rho(x,k)/\partial x$. Using the properties of viscosity solutions (Appendix D), we can conclude that $(\widehat{\lambda}, V(\cdot,\cdot))$ is a viscosity solution to (3.31).

The uniqueness of $\widehat{\lambda}$ follows from Theorem D.1. □

In the next theorem, we derive the smoothness and boundedness of $V(x,k)$. Before getting to the theorem, we introduce some notation. In order to incorporate possible nondifferentiability of the value function, we consider the superdifferential and subdifferential of the function. For any function $\phi(\boldsymbol{x},k)$ defined on $\Re^n \times \mathcal{M}$, let $D^+\phi(\boldsymbol{x},k)$ and $D^-\phi(\boldsymbol{x},k)$ denote the superdifferential and subdifferential of $\phi(\boldsymbol{x},k)$ with respect to $\boldsymbol{x}$, respectively, i.e.,

$$D^+\phi(\boldsymbol{x},k) = \left\{\boldsymbol{r} \in \Re^n : \limsup_{|\boldsymbol{h}|\to 0} \frac{\phi(\boldsymbol{x}+\boldsymbol{h},k) - \phi(\boldsymbol{x},k) - \langle \boldsymbol{h},\boldsymbol{r}\rangle}{|\boldsymbol{h}|} \le 0\right\}, \tag{3.33}$$

and

$$D^-\phi(\boldsymbol{x},k) = \left\{\boldsymbol{r} \in \Re^n : \liminf_{|\boldsymbol{h}|\to 0} \frac{\phi(\boldsymbol{x}+\boldsymbol{h},k) - \phi(\boldsymbol{x},k) - \langle \boldsymbol{h},\boldsymbol{r}\rangle}{|\boldsymbol{h}|} \ge 0\right\}. \tag{3.34}$$

Theorem 4.2. *Let Assumptions* (A1)–(A5) *hold. The relative cost function* $V(x,k)$ *defined by* (3.30) *is continuously differentiable in* x, *and* $(\widehat{\lambda}, V(x,k))$ *is a classical solution to the* HJB *equation* (3.31). *Moreover,* $V(x,k)$ *is convex in* x *and*

$$|V(x,k)| \le C(1+|x|^{\beta_{h2}+1}).$$

The relative cost function $V(x,k)$ *is strictly convex if the cost function* $h(x)$ *is also.*

Proof. The convexity of $V(x,k)$ follows from the convexity of $\widetilde{V}^\rho(x,k)$. The local boundedness of $V(x,k)$ follows from Theorem 3.3.

Since the function $V(x,k)$ is convex in x, to prove that $(\widehat{\lambda}, V(x,k))$ is a classical solution to the HJB equation (3.31), it suffices to show that the subdifferential $D^-V(x,k)$ is a singleton in view of Lemma C.2.

Note, by Assumption (A5), that the map

$$p \to F(k,p) = \inf_{0\le u\le k}\{(u-z)p + c(u)\}$$

is not constant on any nontrivial interval. If $V(x,k)$ is differentiable at x_n, then

$$\widehat{\lambda} = F\left(k, \frac{\partial V(x_n,k)}{\partial x}\right) + h(x_n) + QV(x_n,\cdot)(k).$$

Taking $x_n \to x$ as $n \to \infty$, we obtain

$$\widehat{\lambda} = F(k,y) + h(x) + QV(x,\cdot)(k), \quad y \in \Gamma(x),$$

where

$$\Gamma(x) = \left\{ \lim_{n\to\infty} \frac{\partial V(x_n, k)}{\partial x} : V(x,k) \text{ is differentiable at } x_n \right.$$
$$\left. \text{and } \lim_{n\to\infty} x_n = x \right\}.$$

It follows from the concavity of $F(k,x)$ and Lemma C.2 that

$$\widehat{\lambda} \leq F(k,y) + h(x) + QV(x,\cdot)(k), \quad y \in D^- V(x,k).$$

However, the viscosity solution implies that

$$\widehat{\lambda} \geq F(k,y) + h(x) + QV(x,\cdot)(k), \quad y \in D^- V(x,k).$$

Hence,

$$\widehat{\lambda} = F(k,y) + h(x) + QV(x,\cdot)(k), \quad y \in D^- V(x,k).$$

Thus, for fixed (x,k), $F(k,\cdot)$ is constant on the convex set $D^- V(x,k)$. Therefore, $D^- V(x,k)$ is a singleton.

Finally, we prove that the relative cost function $V(x,k)$ is strictly convex if the cost function $h(x)$ is so. By Theorem 3.1, for any fixed $x_1, x_2 \in \Re$, and for each fixed $\delta \in (0,\ 1)$, there exists a positive constant C_0 (see Lemma 3.3.1 in Sethi and Zhang [125]) such that

$$V^\rho(\delta x_1 + (1-\delta)x_2, k) \leq \delta V^\rho(x_1,k) + (1-\delta)V^\rho(x_2,k) + C_0. \tag{3.35}$$

This implies that

$$\begin{aligned} &V^\rho(\delta x_1 + (1-\delta)x_2, k) - V^\rho(0,k) \\ &\leq \delta \left[V^\rho(x_1,k) - V^\rho(0,k)\right] + (1-\delta)\left[V^\rho(x_2,k) - V^\rho(0,k)\right] + C_0. \end{aligned} \tag{3.36}$$

Taking the limit with $\rho \to 0$, we have

$$V(\delta x_1 + (1-\delta)x_2, k) \leq \delta V(x_1,k) + (1-\delta)V(x_2,k) + C_0. \tag{3.37}$$

Consequently, $V(x,k)$ is strictly convex. □

We now give a verification theorem.

Theorem 4.3. *Let Assumptions* (A1)–(A4) *hold. Let* $(\lambda, W(x,k))$ *be a solution to the* HJB *equation* (3.31). *Then:*

(i) *If there is a control* $u^*(\cdot) \in \mathcal{A}(k)$ *such that*

$$\begin{aligned} &F\left(k(t), \frac{\partial W(x^*(t), k(t))}{\partial x}\right) \\ &\qquad = (u^*(t) - z)\frac{\partial W(x^*(t), k(t))}{\partial x} + c(u^*(t)), \end{aligned} \tag{3.38}$$

for a.e. $t \geq 0$ with probability 1, where $x^*(\cdot)$ is the surplus process corresponding to the control $u^*(\cdot)$, and

$$\lim_{T\to\infty} \frac{E[W(x^*(T), k(T))]}{T} = 0, \tag{3.39}$$

then

$$\lambda = J(x, k, u^*(\cdot)).$$

(ii) *For any $u(\cdot) \in \mathcal{A}(k)$, we have $\lambda \leq J(x, k, u(\cdot))$, i.e.,*

$$\limsup_{T\to\infty} \frac{1}{T} E \int_0^T [h(x(t)) + c(u(t))]\, dt \geq \lambda. \tag{3.40}$$

(iii) *Furthermore, for any (stable) control $u(\cdot) \in \mathcal{S}(k)$, we have*

$$\liminf_{T\to\infty} \frac{1}{T} E \int_0^T [h(x(t)) + c(u(t))]\, dt \geq \lambda. \tag{3.41}$$

Proof. We begin with the proof of part (i). Since $(\lambda, W(x, k))$ is a solution to the HJB equation (3.31) and $u^*(\cdot)$ satisfies (3.38), we have

$$\begin{aligned}(u^*(t) - z)\frac{\partial W(x^*(t), k(t))}{\partial x} &+ QW(x^*(t), \cdot)(k(t)) \\ &= \lambda - h(x^*(t)) - c(u^*(t)).\end{aligned} \tag{3.42}$$

Since $W(x, k) \in \mathcal{G}$, we can apply Dynkin's formula (see Fleming and Rishel [53]) and (3.42) to get

$$\begin{aligned}&E\,[W(x^*(T), k(T))] \\ &= W(x, k) + E \int_0^T \Bigg[(u^*(t) - z)\frac{\partial W(x^*(t), k(t))}{\partial x} \\ &\qquad + QW(x^*(t), \cdot)(k(t)) \Bigg] dt \\ &= W(x, k) + E \int_0^T [\lambda - h(x^*(t)) - c(u^*(t))]\, dt \\ &= W(x, k) + \lambda T - E \int_0^T [h(x^*(t)) + c(u^*(t))]\, dt.\end{aligned} \tag{3.43}$$

We can rewrite (3.43) as

$$\begin{aligned}\lambda = &\frac{1}{T}\{E[W(x^*(T), k(T))] - W(x, k)\} \\ &+ \frac{1}{T} E \int_0^T [h(x^*(t)) + c(u^*(t))]\, dt,\end{aligned}$$

and the first part of the theorem is proved by taking the limit as $T \to \infty$ and using condition (3.39).

For the proof of part (iii), if $u(\cdot) \in \mathcal{S}(k)$, then from $W(x,k) \in \mathcal{G}$, we know that

$$\lim_{T\to\infty} \frac{E[W(x(T),k(T))]}{T} = 0.$$

Moreover, from the HJB equation (3.31) we have

$$(u(t)-z)\frac{\partial W(x(t),k(t))}{\partial x} + QW(x(t),\cdot)(k(t)) \\ \geq \lambda - h(x(t)) - c(u(t)).$$

Now (3.41) can be proved similarly as before.

Finally, we apply Lemma F.3 to show part (ii), i.e., the optimality of the control $u^*(\cdot)$ in the (natural) class of all admissible controls. Let $u(\cdot) \in \mathcal{A}(k)$ be any control and let $x(\cdot)$ be the corresponding surplus process. Suppose that

$$J(x,k,u(\cdot)) < \lambda. \tag{3.44}$$

Set

$$f(t) = E[h(x(t)) + c(u(t))].$$

Without loss of generality we may assume that

$$\int_0^t f(s)\,ds < \infty,$$

for each $t > 0$, or else, we would have $J(x,k,u(\cdot)) = \infty$. Note that

$$J(x,k,u(\cdot)) = \limsup_{T\to\infty} \frac{1}{T}\int_0^T f(s)\,ds,$$

while

$$\rho J^\rho(x,k,u(\cdot)) = \rho\int_0^\infty e^{-\rho s} f(s)\,ds.$$

Therefore, we can apply Lemma F.3 and (3.44) to obtain

$$\limsup_{\rho\to 0} \rho J^\rho(x,k,u(\cdot)) < \lambda. \tag{3.45}$$

On the other hand, we know from Theorem 4.1 that

$$\lim_{\rho\to 0} \rho V^\rho(x,k) = \widehat{\lambda} = \lambda.$$

This equation and (3.45) imply the existence of a $\rho > 0$ such that

$$\rho J^\rho(x,k,u(\cdot)) < \rho V^\rho(x,k),$$

which contradicts the definition of $V^\rho(x,k)$. Thus (ii) is proved. □

Remark 4.1. From Theorems 4.2 and 4.3, we know that the relative cost function $V(x,k)$ defined by (3.30) is also a potential function. Furthermore, $\widehat{\lambda}$ defined by (3.30) is the minimal average cost, i.e., $\widehat{\lambda}=\lambda^*$. □

Remark 4.2. In the simple special case ($m=1, c(u)=0, h(x)=h_1x^+ + h_2x^-$) solved by Bielecki and Kumar [20], we note that optimality is shown only over the class of stable controls and not over the (natural) class of admissible controls. That is, they prove (i) and (iii) but not (ii). □

3.5 Existence and Characterization of Optimal Policy

From Theorem 4.2, we know that the relative cost function $V(x,k)\in\mathcal{G}$. Moreover, it is also a potential function in view of Theorems 4.2 and 4.3. In a way similar to (3.9) and (3.10), based on Theorem 4.3, let us now define a control policy $u^*(\cdot,\cdot)$ via the potential function $V(\cdot,\cdot)$ given by (3.30) as follows:

$$u^*(x,k)=\begin{cases}0, & \text{if } V_x(x,k)>-c_u(0),\\ (c_u(u))^{-1}(-V_x(x,k)), & \text{if } -c_u(k)\le V_x(x,k)\le -c_u(0),\\ k, & \text{if } V_x(x,k)<-c_u(k),\end{cases} \tag{3.46}$$

when $c(u)$ is strictly convex, and

$$u^*(x,k)=\begin{cases}0, & \text{if } V_x(x,k)>-c,\\ k\wedge z, & \text{if } V_x(x,k)=-c,\\ k, & \text{if } V_x(x,k)<-c,\end{cases} \tag{3.47}$$

when $c(u)=cu$ for some constant $c\ge 0$. Therefore, the control policy $u^*(\cdot,\cdot)$ satisfies condition (3.38). Furthermore, it follows from the discussion after Remark 3.1 that equation

$$\frac{d}{dt}x(t)=u^*(x(t),k(t))-z,\quad x(0)=x,$$

has a unique solution $x^*(t)$, $t\ge 0$, for each sample path $k(t)$.

Next we devote ourselves to proving that $u^*(\cdot,\cdot)$ is a stable control. For this, we first derive some intermediate results.

Lemma 5.1. *Let Assumptions* (A1)–(A5) *hold. For each* $k\in\mathcal{M}$, *we have*

$$\inf_{x\in\Re} V(x,k)>-\infty. \tag{3.48}$$

Proof. Let (x_ρ, k_ρ) be the minimum point of the value function $V^\rho(\cdot,\cdot)$. Then we can write

$$\begin{aligned}\widetilde{V}^\rho(x,k) &= V^\rho(x,k) - V^\rho(x_\rho,k_\rho) + V^\rho(x_\rho,k_\rho) - V^\rho(0,0)\\ &\geq V^\rho(x_\rho,k_\rho) - V^\rho(0,0).\end{aligned}$$

By Theorem 3.3, we have

$$\widetilde{V}^\rho(x,k) \geq -C\left(1 + |x_\rho|^{\beta_{h2}+1}\right).$$

Since $\widetilde{V}^\rho(x,k) \to V(x,k)$, we need only to show that $\{x_\rho\}$ is bounded. From the definition of (x_ρ, k_ρ) and Theorem 3.1, we can see that

$$\frac{\partial V^\rho(x_\rho,k_\rho)}{\partial x} = 0, \quad \rho > 0.$$

Recall that $V^\rho(\cdot,\cdot)$ is a solution of the HJB equation (3.7). Thus,

$$\begin{aligned}\rho V^\rho(x_\rho,k_\rho) = \inf_{0\leq u\leq k_\rho} &\left\{(u-z)\frac{\partial V^\rho(x_\rho,k_\rho)}{\partial x} + c(u)\right\}\\ &+ h(x_\rho) + QV^\rho(x_\rho,\cdot)(k_\rho).\end{aligned}$$

Using the fact that (x_ρ, k_ρ) is the minimum point of $V^\rho(\cdot,\cdot)$, we can conclude that $QV^\rho(x_\rho,\cdot)(k_\rho) \geq 0$. Therefore,

$$\rho V^\rho(x_\rho,k_\rho) \geq h(x_\rho). \tag{3.49}$$

From Theorem 3.2, we know that there exist constants $C_1 > 0$ and $\rho_0 > 0$ such that, for $0 < \rho \leq \rho_0$,

$$\rho V^\rho(x_\rho,k_\rho) \leq \rho V^\rho(0,0) \leq C_1, \quad \rho > 0,$$

where the definition of (x_ρ, k_ρ) is applied in the first inequality. Therefore, it follows from (3.49) that $h(x_\rho) \leq C_1$, and the boundedness of $\{x_\rho\}$ follows from Assumption (A1). □

In order to state and prove the next lemma, define

$$\mathcal{B}_k = \left\{x : \frac{\partial V(x,k)}{\partial x} > -\frac{dc(0)}{du}\right\}$$

and

$$\widetilde{\mathcal{B}}_k = \left\{x : \frac{\partial V(x,k)}{\partial x} < -\frac{dc(k)}{du}\right\}.$$

Lemma 5.2. *Let Assumptions* (A1)–(A5) *hold. The sets* $\mathcal{B}_k$ *and* $\widetilde{\mathcal{B}}_k$ *are nonempty for each* $k \in \mathcal{M}$.

Proof. Define

$$\mathcal{B}_k^0 = \left\{ x : \frac{\partial V(x,k)}{\partial x} > 0 \right\}.$$

By Assumption (A2) we know that $-(dc(0)/du) \leq 0$. Since $\partial V(x,k)/\partial x$ is nondecreasing, we have $\mathcal{B}_k^0 \subset \mathcal{B}_k$. Thus, in order to prove that $\mathcal{B}_k \neq \emptyset$, it suffices to show that $\mathcal{B}_k^0 \neq \emptyset$. If $\mathcal{B}_k^0 = \emptyset$, we will have

$$\frac{\partial V(x,k)}{\partial x} \leq 0, \qquad x \in \Re. \tag{3.50}$$

Using the fact that $V(\cdot,k)$ is a convex function bounded from below, we can conclude that

$$\frac{\partial V(x,k)}{\partial x} \to 0, \quad \text{as } x \to \infty.$$

Thus, we have that, as $x \to \infty$,

$$F\left(k, \frac{\partial V(x,k)}{\partial x}\right) = \inf_{0 \leq u \leq k} \left\{ (u-z)\frac{\partial V(x,k)}{\partial x} + c(u) \right\} \to 0.$$

Since $V(\cdot,\cdot)$ is a solution of the HJB equation (3.31) and $h(x) \to \infty$ as $x \to \infty$, we can see that

$$QV(x,\cdot)(k) \to -\infty, \quad \text{as } x \to \infty. \tag{3.51}$$

Note from (3.50) that $V(\cdot,k)$ is decreasing. Recall that

$$QV(x,\cdot)(k) = \sum_{i \neq k} q_{ki}(V(x,i) - V(x,k)).$$

Moreover, from Assumption (A3) specifying that the generator Q is strongly irreducible, there is an $i \neq k$ such that $q_{ki} > 0$. Then (3.51) leads to

$$V(x,i) \to -\infty, \quad \text{as } x \to \infty,$$

which is a contradiction to Lemma 5.1. Therefore, we have proved that $\mathcal{B}_k \supset \mathcal{B}_k^0 \neq \emptyset$.

Similarly, we can show that $\widetilde{\mathcal{B}}_k \neq \emptyset$. If $\widetilde{\mathcal{B}}_k = \emptyset$, then

$$\frac{\partial V(x,k)}{\partial x} \geq -\frac{dc(k)}{du}, \quad x \in \Re,$$

and thus $F(k, \partial V(x,k)/\partial x)$ is bounded from below for $x \to -\infty$. By letting $x \to -\infty$ and noting that $h(x) \to \infty$, we can get a contradiction as above.

□

From the convexity of the function $V(\cdot, k)$, there are x_k and $\hat{x}_k$ with $-\infty < \hat{x}_k < x_k < \infty$ such that

$$\mathcal{B}_k = (x_k, \infty) \quad \text{and} \quad \widetilde{\mathcal{B}}_k = (-\infty, \hat{x}_k).$$

The control policy $u^*(\cdot, \cdot)$ given by (3.46) and (3.47) can be written as

$$u^*(x,k) = \begin{cases} 0, & x > x_k, \\ \left(\dfrac{dc(u)}{du}\right)^{-1}\left(-\dfrac{\partial V(x,k)}{\partial x}\right), & \hat{x}_k \le x \le x_k, \\ k, & x < \hat{x}_k. \end{cases} \tag{3.52}$$

Theorem 5.1. *Let Assumptions* (A1)–(A5) *hold. The control policy* $u^*(\cdot,\cdot)$ *given by* (3.52) *is stable.*

Proof. Let $x^*(\cdot)$ denote the surplus process corresponding to control policy $u^*(\cdot,\cdot)$ with $x^*(0) = x$ and $k(0) = k$. Let

$$M = \max\{|x_k|, |\hat{x}_k| : k \in \mathcal{M}\}.$$

Then $u^*(\cdot,\cdot)$ has the following property:

$$u^*(x,k) = \begin{cases} 0, & \text{if } x \ge M+1, \\ k, & \text{if } x \le -(M+1). \end{cases}$$

Let $\widehat{x}(\cdot)$, with $\widehat{x}(0) = x$ and $k(0) = k$, be the surplus process corresponding to the control policy defined by

$$\widehat{u}(x,k) = \begin{cases} 0, & \text{if } x > M+1, \\ k \wedge z, & \text{if } x = M+1, \\ k, & \text{if } x < M+1. \end{cases}$$

Furthermore, let $\widetilde{x}(\cdot)$, with $\widetilde{x}(0) = x$ and $k(0) = k$, be the surplus process corresponding to the control policy defined by

$$\widetilde{u}(x,k) = \begin{cases} 0, & \text{if } x > -(M+1), \\ k \wedge z, & \text{if } x = -(M+1), \\ k, & \text{if } x < -(M+1). \end{cases}$$

It is easy to see that

$$\widetilde{x}(t) \le x^*(t) \le \widehat{x}(t),$$

and thus, for any $r > 0$, we have

$$E|x^*(t)|^r \leq C_1 \left(E|\widetilde{x}(t)|^r + E|\widehat{x}(t)|^r \right),$$

where $C_1 > 0$ is a positive constant. In view of Corollary 3.2, take $r = \beta_{h2} + 1$ to conclude that

$$\lim_{t \to \infty} \frac{E|x^*(t)|^{\beta_{h2}+1}}{t} = 0,$$

which implies the theorem. □

Now we are in the position to state and prove the following theorem.

Theorem 5.2. *Let Assumptions* (A1)–(A5) *hold. The control policy* $u^*(\cdot, \cdot)$, *defined in* (3.46) *or* (3.47) *as the case may be, is optimal.*

Proof. By Theorem 4.3, we need only to show that

$$\lim_{t \to \infty} \frac{E[V(x^*(t), k(t))]}{t} = 0.$$

But this is implied by Theorem 4.2 and the fact that $u^*(\cdot, \cdot)$ is a stable control policy. □

Remark 5.1. When $c(u) = 0$, i.e., there is no production cost in the model, the optimal control policy can be chosen to be the *hedging point policy*, which has the following form: There are real numbers x_k, $k = 1, \ldots, m$, such that

$$u^*(x, k) = \begin{cases} 0, & x > x_k, \\ k \wedge z, & x = x_k, \\ k, & x < x_k. \end{cases}$$

x_k $(k = 1, \ldots, m)$ are called turnpike levels or thresholds. We will provide a more detailed discussion in the next section. □

3.6 Turnpike Set Analysis

A major characteristic of the optimal policy in convex production planning with a sufficiently long horizon is that there exists a time-dependent threshold or turnpike level (see Thompson and Sethi [135]), such that production takes place in order to reach the turnpike level if the inventory level is below the turnpike level and no production takes place if the inventory is above that level. Once on the turnpike level, only necessary production takes place so as to remain on the turnpike. In a finite horizon case, the above policy holds at every instant sufficiently removed from the horizon, see Zhang and Yin [152]. In the infinite horizon case with the usual assumption of constant demand, the turnpike level will be constant and the policy will hold everywhere. It is the purpose of this section to characterize the turnpike sets of the system defined in Section 3.2. It is proved, as in the discounted case by Sethi and Zhang [125], that the turnpike sets exhibit a monotone property with respect to production capacity. Usually, the problem of solving the optimal production planning is equivalent to the problem of locating the turnpike sets. Therefore, the knowledge about the monotone property of the turnpike sets definitely helps to solve some optimal control problems in a closed form. On the other hand, it can greatly reduce the computation needed for numerical approaches for solving optimal control problems; see Sharifnia [128].

Now we introduce the definition of the turnpike sets.

Definition 6.1. Let $V(x,k)$ given by (3.30) be a potential function. We call

$$\mathcal{T}(k) = \left\{ y_k : V(y_k, k) + \frac{dc(z)}{du} y_k = \inf_{y \in \Re} \left\{ V(y,k) + \frac{dc(z)}{du} y \right\} \right\} \tag{3.53}$$

the turnpike set associated with the capacity level k ($k \in \mathcal{M}$). □

From Theorem 4.2, the turnpike set given by (3.53) can be simply written as

$$\mathcal{T}(k) = \left\{ y_k : \frac{\partial V(y_k, k)}{\partial x} + \frac{dc(z)}{du} = 0 \right\}. \tag{3.54}$$

Under Assumptions (A4) and (A5), we have $0 < z < m$, i.e., the demand can be met using the maximum available capacity. Define $i_0 \in \mathcal{M}$ to be such that $i_0 < z < i_0 + 1$. Observe that, for $j \le i_0$,

$$\frac{dx(t)}{dt} \le j - z \le i_0 - z < 0.$$

Therefore, $x(t) \to -\infty$ as $t \to \infty$ provided j is absorbing. Hence, only those $i \in \mathcal{M}$, for which $i \ge i_0 + 1$, are of special interest to us. First we have the following lemma.

Lemma 6.1. *Let Assumptions* (A1)–(A5) *hold and let the surplus cost function $h(\cdot)$ be differentiable. Define*

$$L(i) = \frac{dh(y_i)}{dx} + \sum_{k \ne i} q_{ik} \left(\frac{\partial V(y_i, k)}{\partial x} - \frac{\partial V(y_i, i)}{\partial x} \right), \quad y_i \in \mathcal{T}(i), \quad i \in \mathcal{M}.$$

Then, $L(i) \geq 0$ *for* $i \leq i_0$*, and* $L(i) = 0$ *for* $i \geq i_0 + 1$.

Proof. By Theorem 4.2 and Remark 4.1,

$$\begin{aligned}\lambda^* = &\inf_{0 \leq u \leq i}\left\{(u-z)\frac{\partial V(x,i)}{\partial x} + c(u)\right\} + h(x) \\ &+ \sum_{k \neq i} q_{ik}[V(x,k) - V(x,i)].\end{aligned} \tag{3.55}$$

From the definition of y_i given by (3.53) and the convexity of $c(u)$ and $V(x,i)$, for $y \leq y_i$ and $u \leq z$,

$$\frac{\partial V(y,i)}{\partial x} + \frac{dc(u)}{du} \leq 0.$$

Therefore, it follows from (3.55) that, for $y \leq y_i$ and $i \leq i_0$,

$$\begin{aligned}\lambda^* = &(i-z)\frac{\partial V(y,i)}{\partial x} + c(i) + h(y) \\ &+ \sum_{k \neq i} q_{ik}[V(y,k) - V(y,i)].\end{aligned} \tag{3.56}$$

Taking the derivative on both sides at points at which $\partial V(x,i)/\partial x$ is differentiable, we get that, for $y \leq y_i$ and $i \leq i_0$,

$$0 = (i-z)\frac{\partial^2 V(y,i)}{\partial x^2} + \frac{dh(y)}{dx} + \sum_{k \neq i} q_{ik}\left(\frac{\partial V(y,k)}{\partial x} - \frac{\partial V(y,i)}{\partial x}\right). \tag{3.57}$$

Note that

$$(i-z)\frac{\partial^2 V(y,i)}{\partial x^2} \leq 0, \tag{3.58}$$

if $\partial V(x,i)/\partial x$ is differentiable at y. Because $\partial V(x,i)/\partial x$ is a nondecreasing function, therefore, the set of differentiable points for $\partial V(y,i)/\partial x$ is dense in $(-\infty, y_i)$. Letting $y \to y_i$ in (3.57) and using (3.58), we have $L(i) \geq 0$ for $i \leq i_0$.

If $i \geq i_0+1$, from the definition of y_i and the convexity of $V(x,i)$ and $c(u)$, we know that for small enough $\delta > 0$, there exists a decreasing function $f(y)$ of y on interval $[y_i - \delta,\, y_i + \delta]$ such that

$$\inf_{0 \leq u \leq i}\left\{(u-z)\frac{\partial V(y,i)}{\partial x} + c(u)\right\} = [f(y) - z]\frac{\partial V(y,i)}{\partial x} + c(f(y)), \tag{3.59}$$

and

$$f(y) \to z, \quad \text{as } y \to y_i. \tag{3.60}$$

From the continuity of $\partial V(x,i)/\partial x$ (see Theorem 4.2), we know that

$$f(y) \geq z, \quad \text{for } y \leq y_i \quad \text{and} \quad f(y) \leq z, \quad \text{for } y > y_i. \tag{3.61}$$

By (3.55) and (3.59), we have

$$\lambda^* = [f(y) - z]\frac{\partial V(y,i)}{\partial x} + c(f(y)) + h(y) + \sum_{k\neq i} q_{ik}[V(y,k) - V(y,i)]. \tag{3.62}$$

Taking derivative on both sides of (3.62) at points at which $\partial V(x,i)/\partial x$ and $f(y)$ are differentiable, and noting that

$$\left[\frac{\partial V(y,i)}{\partial x} + \frac{dc(u)}{du}\right]_{u=f(y)} = 0,$$

we have

$$-[f(y) - z]\frac{\partial^2 V(y,i)}{\partial x^2} = \frac{dh(y)}{dx} + \sum_{k\neq i} q_{ik}\left(\frac{\partial V(y,k)}{\partial x} - \frac{\partial V(y,i)}{\partial x}\right). \tag{3.63}$$

Letting $y \uparrow y_i$ and using (3.60) and (3.61), we get

$$\frac{dh(y_i)}{dx} + \sum_{k\neq i} q_{ik}\left(\frac{\partial V(y_i,k)}{\partial x} - \frac{\partial V(y_i,i)}{\partial x}\right) \leq 0,$$

where we use the monotonicity of $dh(x)/dx$ and $\partial V(x,k)/\partial x$. In the same way, letting $y \downarrow y_i$ and using (3.60) and (3.61), we get

$$\frac{dh(y_i)}{dx} + \sum_{k\neq i} q_{ik}\left(\frac{\partial V(y_i,k)}{\partial x} - \frac{\partial V(y_i,i)}{\partial x}\right) \geq 0.$$

Therefore, $L(i) = 0$ for $i \geq i_0 + 1$. □

Theorem 6.1. *Let Assumptions* (A1)–(A5) *hold, and let the surplus cost function* $h(x)$ *be differentiable and strictly convex. If*

$$Q = \begin{pmatrix} -\mu_0 & \mu_0 & & & \\ \lambda_1 & -(\lambda_1+\mu_1) & \mu_1 & & \\ & \ddots & \ddots & \ddots & \\ & & \lambda_{m-1} & -(\lambda_{m-1}+\mu_{m-1}) & \mu_{m-1} \\ & & & \lambda_m & -\lambda_m \end{pmatrix}$$

with $\mu_j > 0$ $(0 \leq j \leq m-1)$ *and* $\lambda_j > 0$ $(1 \leq j \leq m)$, *then*

$$y_{i_0} \geq y_{i_0+1} \geq \cdots \geq y_m.$$

Proof. Suppose contrariwise that $y_{m-1} < y_m$. First we show that from this assumption,

$$y_{m-2} < y_{m-1}. \tag{3.64}$$

By Lemma 6.1, we have

$$0 = \frac{dh(y_m)}{dx} + \lambda_m \left(\frac{\partial V(y_m, m-1)}{\partial x} - \frac{\partial V(y_m, m)}{\partial x} \right). \tag{3.65}$$

Using $y_{m-1} < y_m$, we obtain

$$\frac{dh(y_{m-1})}{dx} < \frac{dh(y_m)}{dx}, \tag{3.66}$$

and

$$\frac{\partial V(y_m, m-1)}{\partial x} - \frac{\partial V(y_m, m)}{\partial x} > \frac{\partial V(y_{m-1}, m-1)}{\partial x} - \frac{\partial V(y_m, m)}{\partial x}. \tag{3.67}$$

Noting that by (3.54), we have

$$\frac{\partial V(y_{m-1}, m-1)}{\partial x} - \frac{\partial V(y_m, m)}{\partial x} = -\frac{dc(z)}{du} + \frac{dc(z)}{du} = 0. \tag{3.68}$$

Therefore, it follows from (3.67) that

$$\frac{\partial V(y_m, m-1)}{\partial x} - \frac{\partial V(y_m, m)}{\partial x} > 0. \tag{3.69}$$

Using (3.65) and (3.69), we have

$$\frac{dh(y_m)}{dx} < 0. \tag{3.70}$$

It follows from Lemma 6.1 that

$$\begin{aligned} L(m-1) = {} & \frac{dh(y_{m-1})}{dx} + \lambda_{m-1} \left(\frac{\partial V(y_{m-1}, m-2)}{\partial x} - \frac{\partial V(y_{m-1}, m-1)}{\partial x} \right) \\ & + \mu_{m-1} \left(\frac{\partial V(y_{m-1}, m)}{\partial x} - \frac{\partial V(y_{m-1}, m-1)}{\partial x} \right) \geq 0. \end{aligned} \tag{3.71}$$

Similar to (3.69), we have

$$\frac{\partial V(y_{m-1}, m)}{\partial x} - \frac{\partial V(y_{m-1}, m-1)}{\partial x} < 0. \tag{3.72}$$

Combining (3.66) and (3.70)–(3.72), we get

$$\frac{\partial V(y_{m-1}, m-2)}{\partial x} - \frac{\partial V(y_{m-1}, m-1)}{\partial x} > 0. \tag{3.73}$$

Using (3.54) again, we have

$$\frac{\partial V(y_{m-2}, m-2)}{\partial x} - \frac{\partial V(y_{m-1}, m-1)}{\partial x} = 0. \tag{3.74}$$

Similar to (3.69), in view of (3.74), it can be shown that (3.73) holds if and only if (3.64) holds. Hence we have (3.64). Going along the same line, we can prove that if (3.64) holds, then

$$y_{m-3} < y_{m-2}. \tag{3.75}$$

Repeating this procedure, by Lemma 6.1 with $i \geq 1$, we get

$$y_0 < y_1 < \cdots < y_{m-1} < y_m. \tag{3.76}$$

On the other hand, using Lemma 6.1 with $i = 0$, we have

$$0 \leq \frac{dh(y_0)}{dx} + \mu_0 \left(\frac{\partial V(y_0, 1)}{\partial x} - \frac{\partial V(y_0, 0)}{\partial x} \right). \tag{3.77}$$

It follows from (3.76) that

$$\frac{dh(y_0)}{dx} \leq 0 \quad \text{and} \quad \frac{\partial V(y_0, 1)}{\partial x} - \frac{\partial V(y_0, 0)}{\partial x} < 0,$$

which contradicts (3.77).

Consequently, we have $y_{m-1} \geq y_m$. In the same way, we can prove

$$y_{m-1} \leq y_{m-2} \leq \cdots \leq y_{i_0}.$$

Thus we get the theorem. □

3.7 Multiproduct Systems

For our treatment of the multiproduct problem, we begin with the corresponding discounted cost problem, which can be obtained by using the methodology reported in Sethi and Zhang [125]. Their methodology utilizes the existence of optimal piecewise-deterministic controls. They also provide a verification theorem and the existence of an optimal feedback control based on the HJB equation in terms of directional derivatives (HJBDD). We simply state their results in an analogous manner as they pertain to our problem. Then we use a vanishing discount approach to address the average-cost problem under consideration.

We consider an n-product manufacturing system with stochastic production capacity and constant demand for each product over time. In order to specify the model, let $\boldsymbol{x}(t) \in \Re^n$, $\boldsymbol{u}(t) \in \Re^n_+$, and $\boldsymbol{z} \in \Re^n_+$ denote the surplus level, the production rate, and the constant demand rate, respectively. The system equation is

$$\frac{d}{dt}\boldsymbol{x}(t) = \boldsymbol{u}(t) - \boldsymbol{z}, \quad \boldsymbol{x}(0) = \boldsymbol{x}. \tag{3.78}$$

Definition 7.1. A production control process $\boldsymbol{u}(\cdot) = \{\boldsymbol{u}(t),\, t \geq 0\}$ is *admissible*, if: (i) $\boldsymbol{u}(t)$ is adapted to the filtration $\{\mathcal{F}_t\}$ with $\mathcal{F}_t = \sigma(k(s), 0 \leq s \leq t)$; and (ii) $0 \leq u_1(t) + u_2(t) + \cdots + u_n(t) \leq k(t),\ u_i(t) \geq 0,\ i = 1, \ldots, n,\ t \geq 0$. □

Let

$$\mathcal{U}(k) = \left\{ \boldsymbol{u} : \ u_i \geq 0,\ i = 1, \ldots, n,\ \sum_{i=1}^{n} u_i \leq k \right\}$$

denote the set of feasible controls given the machine capacity k. Let $\mathcal{A}(k)$ denote the collection of all admissible controls with the initial condition $k(0) = k$.

Remark 7.1. Note here that we have assumed $r_i = 1$ $(1 \leq i \leq n)$ in (2.6). The extension of results to other cases is standard. □

Definition 7.2. A function $\boldsymbol{u}(\cdot, \cdot)$ defined on $\Re^n \times \mathcal{M}$ is called an *admissible feedback control*, or simply a *feedback control*, if: (i) for any given initial surplus $\boldsymbol{x}$ and production capacity k, the equation

$$\frac{d}{dt}\boldsymbol{x}(t) = \boldsymbol{u}(\boldsymbol{x}(t), k(t)) - \boldsymbol{z}$$

has a unique solution; and (ii) the control defined by $\boldsymbol{u}(\cdot) = \{\boldsymbol{u}(t) = \boldsymbol{u}(\boldsymbol{x}(t), k(t)), t \geq 0\} \in \mathcal{A}(k)$. With a slight abuse of notation, we simply call $\boldsymbol{u}(\cdot, \cdot)$ a feedback control when no ambiguity arises. □

Let $g(\boldsymbol{x}, \boldsymbol{u}) : \Re^n \times \Re^n_+ \mapsto \Re_+$ denote the surplus and production cost. For any $\boldsymbol{u}(\cdot) \in \mathcal{A}(k)$, define

$$J(\boldsymbol{x}, k, \boldsymbol{u}(\cdot)) = \limsup_{T \to \infty} \frac{1}{T} E \int_0^T g(\boldsymbol{x}(t), \boldsymbol{u}(t))\, dt, \tag{3.79}$$

where $\boldsymbol{x}(0) = \boldsymbol{x}$ is the initial surplus, and $k(0) = k$ is the the initial capacity. Our goal is to choose $\boldsymbol{u}(\cdot) \in \mathcal{A}(k)$ so as to minimize the cost functional $J(\boldsymbol{x}, k, \boldsymbol{u}(\cdot))$.

We assume that the cost function $g(\boldsymbol{x}, \boldsymbol{u})$, and the production capacity process $k(\cdot)$ satisfy the following:

(A6) $g(\boldsymbol{x}, \boldsymbol{u})$ is a nonnegative convex function with $g(0, 0) = 0$. There are positive constants C_{g1}, C_{g2}, and $\beta_{g1} \geq 1$ such that, for any fixed $\boldsymbol{u}$,

$$g(\boldsymbol{x}, \boldsymbol{u}) \geq C_{g1}|\boldsymbol{x}|^{\beta_{g1}} - C_{g2}, \quad \boldsymbol{x} \in \Re^n, \quad \boldsymbol{u} \in \Re^n_+.$$

Moreover, there are constants $C_{g3} > 0$ and $\beta_{g2} \geq \beta_{g1}$ such that, for $\boldsymbol{u}, \widetilde{\boldsymbol{u}} \in \Re^n_+$ and $\boldsymbol{x}, \widetilde{\boldsymbol{x}} \in \Re^n$,

$$\begin{aligned}&|g(\boldsymbol{x},\boldsymbol{u}) - g(\widetilde{\boldsymbol{x}},\widetilde{\boldsymbol{u}})| \\ &\qquad \leq C_{g3}\left[\left(1+|\boldsymbol{x}|^{\beta_{g2}-1}+|\widetilde{\boldsymbol{x}}|^{\beta_{g2}-1}\right)|\boldsymbol{x}-\widetilde{\boldsymbol{x}}| + |\boldsymbol{u}-\widetilde{\boldsymbol{u}}|\right].\end{aligned}$$

(A7) The average capacity $\bar{k} = \sum_{j=0}^{m} j\nu_j > \sum_{j=1}^{n} z_j$, where $\boldsymbol{z} = (z_1, \ldots, z_n)'$ is the demand vector.

A control $\boldsymbol{u}(\cdot) \in \mathcal{A}(k)$ is called *stable* if the condition

$$\lim_{T\to\infty} \frac{E|\boldsymbol{x}(T)|^{\beta_{g2}+1}}{T} = 0 \tag{3.80}$$

holds, where $\boldsymbol{x}(\cdot)$ is the surplus process corresponding to the control $\boldsymbol{u}(\cdot)$ with $(\boldsymbol{x}(0), k(0)) = (\boldsymbol{x}, k)$ and β_{g2} is defined in Assumption (A6). Let $\mathcal{S}(k) \subset \mathcal{A}(k)$ denote the class of stable controls. It will be shown in the next section that the set of stable admissible controls $\mathcal{S}(k)$ is nonempty.

We will show in what follows that there exists a constant λ^*, independent of the initial condition $(\boldsymbol{x}(0), k(0)) = (\boldsymbol{x}, k)$, and a stable Markov control $\boldsymbol{u}^*(\cdot) \in \mathcal{A}(k)$ such that $\boldsymbol{u}^*(\cdot)$ is optimal, i.e., it minimizes the cost defined by (3.79) over all $\boldsymbol{u}(\cdot) \in \mathcal{A}(k)$ and, furthermore,

$$\lim_{T\to\infty} \frac{1}{T} E \int_0^T g(\boldsymbol{x}^*(t), \boldsymbol{u}^*(t))\, dt = \lambda^*.$$

Moreover, for any other (stable) control $\boldsymbol{u}(\cdot) \in \mathcal{S}(k)$,

$$\liminf_{T\to\infty} \frac{1}{T} E \int_0^T g(\boldsymbol{x}(t), \boldsymbol{u}(t))\, dt \geq \lambda^*.$$

In order to study the average-cost control problem, we introduce the corresponding control problem with discounted cost rate $\rho > 0$. For $\boldsymbol{u}(\cdot) \in \mathcal{A}(k)$, we define the expected discounted cost as

$$J^\rho(\boldsymbol{x}, k, \boldsymbol{u}(\cdot)) = E \int_0^\infty e^{-\rho t} g(\boldsymbol{x}(t), \boldsymbol{u}(t))\, dt.$$

Define the value function of the discounted problem as

$$V^\rho(\boldsymbol{x}, k) = \inf_{\boldsymbol{u}(\cdot)\in\mathcal{A}(k)} J^\rho(\boldsymbol{x}, k, \boldsymbol{u}(\cdot)). \tag{3.81}$$

The following result is from Chapter 4 of Sethi and Zhang [125].

Theorem 7.1. *Let Assumptions* (A3) *and* (A6) *hold. Then the value function* $V^\rho(\boldsymbol{x}, k)$ *is convex, and satisfies the following Lipschitz condition:*

$$|V^\rho(\boldsymbol{x}, k) - V^\rho(\widehat{\boldsymbol{x}}, k)| \leq C\left(1 + |\boldsymbol{x}|^{\beta_{g2}-1} + |\widehat{\boldsymbol{x}}|^{\beta_{g2}-1}\right)|\boldsymbol{x} - \widehat{\boldsymbol{x}}| \tag{3.82}$$

for some positive constant C and all $\boldsymbol{x}, \widehat{\boldsymbol{x}} \in \Re^n, k \in \mathcal{M}$.

The dynamic programming equation for the discounted problem can be written as

$$\rho\phi^\rho(\boldsymbol{x}, k) = \inf_{\boldsymbol{u}\in\mathcal{U}(k)} \left\{ \left\langle \boldsymbol{u} - \boldsymbol{z}, \frac{\partial \phi^\rho(\boldsymbol{x}, k)}{\partial \boldsymbol{x}} \right\rangle + g(\boldsymbol{x}, \boldsymbol{u}) \right\} + Q\phi^\rho(\boldsymbol{x}, \cdot)(k), \tag{3.83}$$

where $\langle \cdot, \cdot \rangle$ denotes the inner product, and $\phi^\rho(\boldsymbol{x}, k)$ is defined on $\Re^n \times \mathcal{M}$.

Without requiring that the value function $V^\rho(\cdot, k)$ is differentiable, we follow a different approach. Recall that $V^\rho(\cdot, k)$ is convex. It is convenient to write an HJB equation in terms of *directional derivatives* (HJBDD). Such notion is convenient when dealing with control problems with state constraints, as in flowshops and jobshops. Next, we first give the notion of these derivatives and some related properties of convex functions.

A function $f(\boldsymbol{x}), \boldsymbol{x} \in \Re^n$, is said to have a *directional derivative* $\partial_{\boldsymbol{p}} f(\boldsymbol{x})$ along direction $\boldsymbol{p} \in \Re^n$, defined by

$$\lim_{\delta\to 0} \frac{f(\boldsymbol{x} + \delta \boldsymbol{p}) - f(\boldsymbol{x})}{\delta} = \partial_{\boldsymbol{p}} f(\boldsymbol{x}),$$

whenever the limit exists. If a function $f(\boldsymbol{x})$ is differentiable at $\boldsymbol{x}$, then $\partial_{\boldsymbol{p}} f(\boldsymbol{x})$ exists for every $\boldsymbol{p}$ and

$$\partial_{\boldsymbol{p}} f(\boldsymbol{x}) = \langle \nabla f(\boldsymbol{x}), \boldsymbol{p} \rangle, \tag{3.84}$$

where $\nabla f(\boldsymbol{x})$ is the gradient of $f(\boldsymbol{x})$ at $\boldsymbol{x}$. It is well known that a continuous convex function defined on $\Re^n$ is differentiable almost everywhere.

Formally we can write the HJBDD equation for our problem as

$$\rho\phi^\rho(\boldsymbol{x}, k) = \inf_{\boldsymbol{u}\in\mathcal{U}(k)} \{\partial_{\boldsymbol{u}-\boldsymbol{z}} \phi^\rho(\boldsymbol{x}, k) + g(\boldsymbol{x}, \boldsymbol{u})\} + Q\phi^\rho(\boldsymbol{x}, \cdot)(k). \tag{3.85}$$

Similar to Theorem 3.1, we can establish the smoothness of $V^\rho(x, k)$. We have the following result.

Theorem 7.2. *Let Assumptions* (A3) *and* (A6) *hold. Then:*

(i) *the value function $V^\rho(\boldsymbol{x}, k)$ is a viscosity solution of* (3.83); *and*

(ii) *the value function $V^\rho(\boldsymbol{x}, k)$ satisfies equation* (3.85) *for all $\boldsymbol{x} \in \Re^n$.*

In order to study the long-run average-cost control problem using the vanishing discount approach, just as in Section 3.3, we must first obtain some estimates for the value function $V^\rho(\boldsymbol{x}, k)$.

Lemma 7.1. *Let Assumptions* (A3) *and* (A7) *hold. For any $r \geq 1$ and any $(\boldsymbol{x}, k) \in \Re^n \times \mathcal{M}$, $\widehat{\boldsymbol{x}} \in \Re^n$, there exist a control $\boldsymbol{u}(\cdot), t \geq 0$, and a positive* *constant* C_r, *independent of* $(\boldsymbol{x}, k)$ *and* $\widehat{\boldsymbol{x}}$, *such that*

$$E[\tau^r(\boldsymbol{x}, \widehat{\boldsymbol{x}}, k)] \le C_r \left(1 + \sum_{j=1}^{n} |x_j - \widehat{x}_j|^r\right),$$

where

$$\tau(\boldsymbol{x}, \widehat{\boldsymbol{x}}, k) = \inf\{t \ge 0 : \boldsymbol{x}(t) = \widehat{\boldsymbol{x}}\},$$

and $\boldsymbol{x}(t), t \ge 0$, *is the surplus process corresponding to the control* $\boldsymbol{u}(t)$, $t \ge 0$, *and initial condition* $(\boldsymbol{x}(0), k(0)) = (\boldsymbol{x}, k)$.

Proof. We first prove the lemma for $\widehat{\boldsymbol{x}} \ge \boldsymbol{x}$, i.e., $\widehat{x}_i \ge x_i$, $i = 1, \ldots, n$. In this case, let us define a feedback control policy

$$\boldsymbol{u}(\boldsymbol{y}, k) = (u_1(\boldsymbol{y}, k), \ldots, u_n(\boldsymbol{y}, k))'$$

by

$$u_j(\boldsymbol{y}, k) = \begin{cases} \dfrac{(\widehat{x}_j - y_j)k}{\sum_{\ell=1}^{n}(\widehat{x}_\ell - y_\ell)}, & \text{if } \boldsymbol{y} \le \widehat{\boldsymbol{x}},\ \boldsymbol{y} \ne \widehat{\boldsymbol{x}},\ k < \displaystyle\sum_{\ell=1}^{n} z_\ell, \\ \dfrac{(k - \sum_{\ell=1}^{n} z_\ell)(\widehat{x}_j - y_j)}{\sum_{\ell=1}^{n}(\widehat{x}_\ell - y_\ell)} + z_j, & \text{if } \boldsymbol{y} \le \widehat{\boldsymbol{x}},\ \boldsymbol{y} \ne \widehat{\boldsymbol{x}},\ k \ge \displaystyle\sum_{\ell=1}^{n} z_\ell, \\ 0, & \text{otherwise.} \end{cases} \tag{3.86}$$

By Lemma F.1, the differential equation

$$\frac{d}{dt}\boldsymbol{x}(t) = \boldsymbol{u}(\boldsymbol{x}(t), k(t)) - \boldsymbol{z}, \quad \boldsymbol{x}(0) = \boldsymbol{x},$$

has a unique solution. Furthermore, it is easy to see that, when $\boldsymbol{y} \le \widehat{\boldsymbol{x}}$ and $\boldsymbol{y} \ne \widehat{\boldsymbol{x}}$,

$$\sum_{j=1}^{n} u_j(\boldsymbol{y}, k) = k, \quad \boldsymbol{u}(\boldsymbol{y}, k) \ge 0, \tag{3.87}$$

and thus $\boldsymbol{u}(\cdot, \cdot)$ defines a feedback control policy. Moreover, note that for all j such that $\widehat{x}_j \ge y_j$, the ratio

$$\frac{u_j(\boldsymbol{y}, k) - z_j}{\widehat{x}_j - y_j} = \frac{k - \sum_{i=1}^{n} z_i}{\sum_{\ell=1}^{n}(\widehat{x}_\ell - y_\ell)}$$

is a constant for all k, and therefore when the machine is in the *up* state, defined by $k(t) \ge \sum_{i=1}^{n} z_i$, the corresponding surplus process moves along the straight line joining $\widehat{\boldsymbol{x}}$ and $\boldsymbol{y}$.

Now let us consider

$$X(t) = \sum_{\ell=1}^{n} x_\ell(t),$$

and let $X = \sum_{\ell=1}^{n} x_\ell$, $\widehat{X} = \sum_{\ell=1}^{n} \widehat{x}_\ell$, and $Z = \sum_{\ell=1}^{n} z_\ell$. Define

$$\tau = \inf\{t \geq 0, X(t) = \widehat{X}\}.$$

It can be shown that $\tau = \tau(\boldsymbol{x}, \widehat{\boldsymbol{x}}, k)$. Note that under the feedback control policy $\boldsymbol{u}(\cdot,\cdot)$ defined by (3.86), we have that, for $t \leq \tau$,

$$\frac{dX(t)}{dt} = k(t) - Z, \quad X(0) = X.$$

Thus, by applying Lemma 3.2, we can conclude that there exists a positive constant $\widehat{C}_r$ independent of X, $\widehat{X}$, and k such that

$$E\tau^r \leq \widehat{C}_r(1 + |\widehat{X} - X|^r), \tag{3.88}$$

which implies the lemma.

We now show that Lemma 7.1 holds for any $\boldsymbol{x}$ and $\widehat{\boldsymbol{x}}$. Let

$$i_0 \in \arg\min_i \left\{ \frac{\widehat{x}_i - x_i}{z_i} \right\}.$$

If $\widehat{x}_{i_0} - x_{i_0} \geq 0$, then it reduces to the case we have just proved. Thus we may assume

$$\widehat{x}_{i_0} - x_{i_0} < 0,$$

and without loss of generality, we assume $i_0 = 1$.

Define

$$\boldsymbol{u}^0(t) = 0, \quad 0 \leq t \leq t_0 = \frac{|\widehat{x}_1 - x_1|}{z_1},$$

then the corresponding trajectory $\boldsymbol{x}(t)$, $0 \leq t \leq t_0$, satisfies

$$\begin{aligned} x_1(t_0) &= \widehat{x}_1, \\ x_i(t_0) &= x_i - \frac{x_1 - \widehat{x}_1}{z_1} z_i \\ &\leq x_i - \frac{x_i - \widehat{x}_i}{z_i} z_i = \widehat{x}_i, \quad 2 \leq i \leq n. \end{aligned}$$

Define $\widetilde{\boldsymbol{x}} = \boldsymbol{x}(t_0)$. Then as we have shown before, there is a feedback control $\boldsymbol{u}(\cdot,\cdot)$ such that

$$E[\widetilde{\tau}(\widetilde{\boldsymbol{x}}, \widehat{\boldsymbol{x}}, k(t_0))]^r \leq C\left(1 + \sum_{i=1}^{n} |\widetilde{x}_i - \widehat{x}_i|^r\right),$$

where

$$\widetilde{\tau}(\widetilde{\boldsymbol{x}}, \widehat{\boldsymbol{x}}, k(t_0)) = \inf\{t \geq 0 : \boldsymbol{x}(t_0 + t) = \widehat{\boldsymbol{x}}\}.$$

Define

$$\boldsymbol{u}(t) = \begin{cases} 0, & \text{if } 0 \leq t \leq t_0, \\ \boldsymbol{u}(\boldsymbol{x}(t), k(t)), & \text{if } t > t_0. \end{cases}$$

Then we have

$$\begin{aligned} E[\tau^r(\boldsymbol{x},\widehat{\boldsymbol{x}},k)] &= E\left[t_0+\widetilde{\tau}(\widetilde{\boldsymbol{x}},\widehat{\boldsymbol{x}},k(t_0))\right]^r \\ &\le 2^{r-1}(t_0^r + E[\widetilde{\tau}^r(\widetilde{\boldsymbol{x}},\widehat{\boldsymbol{x}},k(t_0))]) \\ &\le C_1\left(1+\sum_{i=1}^n |x_i-\widehat{x}_i|^r\right), \end{aligned}$$

for some constant $C_1>0$ independent of $\boldsymbol{x}$ and $\widetilde{\boldsymbol{x}}$. The lemma is thus proved. □

From the proof of the lemma we can prove the following theorem.

Theorem 7.3. *Let Assumptions* (A3) *and* (A7) *hold. For any* $(\boldsymbol{x},k),(\widehat{\boldsymbol{x}},\widehat{k})\in\Re^n\times\mathcal{M}$, *and* $r\ge 1$, *there exist a control* $\boldsymbol{u}(\cdot)\in\mathcal{A}(k)$ *and a positive constant* $\widehat{C}_r$ *independent of* $(\boldsymbol{x},k)$ *and* $(\widehat{\boldsymbol{x}},\widehat{k})$ *such that*

$$E[\tau^r(\boldsymbol{x},\widehat{\boldsymbol{x}},k,\widehat{k})]\le \widehat{C}_r\left(1+\sum_{i=1}^n|\widehat{x}_i-x_i|^r\right),$$

where

$$\tau(\boldsymbol{x},\widehat{\boldsymbol{x}},k,\widehat{k})=\inf\{t\ge 0:(\boldsymbol{x}(t),k(t))=(\widehat{\boldsymbol{x}},\widehat{k})\},$$

and $\boldsymbol{x}(\cdot)$ *is the surplus process corresponding to the control* $\boldsymbol{u}(\cdot)$ *and the initial condition* $(\boldsymbol{x}(0),k(0))=(\boldsymbol{x},k)$.

Proof. This theorem can be proved similarly as the proof of Lemma 3.2, and so the details are omitted. □

With Theorem 7.3 in hand, we can follow the corresponding proofs of Theorem 3.2 and Corollary 3.3 to prove the following results.

Theorem 7.4. *Let Assumptions* (A3) *and* (A6)–(A7) *hold. Then:*

(i) *There exists a constant* $\rho_0>0$ *such that* $\{\rho V^\rho(0,0):0<\rho\le\rho_0\}$ *is bounded.*

(ii) *The function*

$$\widetilde{V}^\rho(\boldsymbol{x},k)=V^\rho(\boldsymbol{x},k)-V^\rho(0,0) \tag{3.89}$$

is convex in $\boldsymbol{x}$. *It is locally uniformly bounded, i.e., there exists a constant* $C>0$ *such that, for* $(\boldsymbol{x},k)\in\Re^n\times\mathcal{M}$ *and* $\rho\ge 0$,

$$|V^\rho(\boldsymbol{x},k)-V^\rho(0,0)|\le C\left(1+|\boldsymbol{x}|^{\beta_{g2}+1}\right).$$

(iii) $\widetilde{V}^\rho(\boldsymbol{x},k)$ *is locally uniformly Lipschitz continuous in* $\boldsymbol{x}$ *with respect to* $\rho>0$, *i.e., for any* $r>0$, *there exists a constant* $\widehat{C}>0$, *independent of* ρ, *such that*

$$|\widetilde{V}^\rho(\boldsymbol{x},k)-\widetilde{V}^\rho(\widehat{\boldsymbol{x}},k)|\le\widehat{C}|\boldsymbol{x}-\widehat{\boldsymbol{x}}|$$

for all $k \in \mathcal{M}$ *and all* $|\boldsymbol{x}|, |\widehat{\boldsymbol{x}}| \leq r$.

Proof. Going along the lines of the proofs of Theorems 3.2 and 3.3, and Corollary 3.3, we can prove the theorem. The details are omitted. □

The HJB equation for the optimal control problem formulated at the beginning of the section takes the form

$$\lambda = \inf_{\boldsymbol{u} \in \mathcal{U}(k)} \left\{ \left\langle \boldsymbol{u} - \boldsymbol{z}, \frac{\partial \phi(\boldsymbol{x}, k)}{\partial \boldsymbol{x}} \right\rangle + g(\boldsymbol{x}, \boldsymbol{u}) \right\} + Q\phi(\boldsymbol{x}, \cdot)(k), \tag{3.90}$$

and the corresponding HJBDD equation can be written as

$$\lambda = \inf_{\boldsymbol{u} \in \mathcal{U}(k)} \{ \partial_{\boldsymbol{u} - \boldsymbol{z}} \phi(\boldsymbol{x}, k) + g(\boldsymbol{x}, \boldsymbol{u}) \} + Q\phi(\boldsymbol{x}, \cdot)(k), \tag{3.91}$$

where λ is a constant and $\phi(\boldsymbol{x}, k)$ is a real-valued function defined on $\Re^n \times \mathcal{M}$.

Before we define the solutions to equations (3.90) and (3.91), we first introduce some notation. Let $\mathcal{G}$ denote the family of real-valued functions $G(\boldsymbol{x}, k)$ defined on $\Re^n \times \mathcal{M}$ such that, for each $k \in \mathcal{M}$,

(i) $G(\boldsymbol{x}, k)$ is convex in $\boldsymbol{x}$; and

(ii) there is a constant $C > 0$ such that

$$|G(\boldsymbol{x}, k)| \leq C(1 + |\boldsymbol{x}|^{\beta_{g2}+1}), \quad \boldsymbol{x} \in \Re^n,$$

where β_{g2} is given in Assumption (A6).

Definition 7.3. A solution of the HJBDD equation (3.91) is a pair $(\lambda, W(\boldsymbol{x}, k))$ with λ a constant and $W(\boldsymbol{x}, k) \in \mathcal{G}$. The function $W(\boldsymbol{x}, k)$ is called the *potential function* for the control problem, if λ is the minimum long-run average cost. □

The next theorem is concerned with a solution $(\lambda, W(\boldsymbol{x}, k))$ of the HJB equation (3.90).

Theorem 7.5. *Let Assumptions* (A3) *and* (A6)–(A7) *hold. There exists a subsequence of* ρ, *still denoted by* ρ, *such that for* $(\boldsymbol{x}, k) \in \Re^n \times \mathcal{M}$, *the limits of* $\rho V^\rho(\boldsymbol{x}, k)$ *and* $\widetilde{V}^\rho(\boldsymbol{x}, k)$ *exist as* $\rho \to 0$. *Write*

$$\lim_{\rho \to 0} \rho V^\rho(\boldsymbol{x}, k) = \widehat{\lambda} \quad \text{and} \quad \lim_{\rho \to 0} \widetilde{V}^\rho(\boldsymbol{x}, k) = V(\boldsymbol{x}, k). \tag{3.92}$$

Furthermore, $(\widehat{\lambda}, V(\boldsymbol{x}, k))$ *is a viscosity solution of the* HJB *equation* (3.90).

Proof. The proof is similar to that of Corollary 3.4 and Theorem 4.1, and the details are omitted here. □

Theorem 7.6. *Let Assumptions* (A3) *and* (A6)–(A7) *hold.* $(\widehat{\lambda}, V(\boldsymbol{x},k))$ *defined in Theorem* 7.5 *is a solution of the* HJBDD *equation* (3.91).

Proof. The convexity of $V(\boldsymbol{x},k)$ directly follows from the convexity of $V^\rho(\boldsymbol{x},k)$. Using (i) of Lemma C.3, we know that the relative interior of a nonempty convex set, denoted by $\mathrm{ri}(D^-V(\boldsymbol{x},k))$, is nonempty, where $D^-V(\boldsymbol{x},k)$ is defined by (3.34) in Section 3.4.

First we show that, for $\boldsymbol{r} \in D^-V(\boldsymbol{x},k)$,

$$\widehat{\lambda} = \inf_{\boldsymbol{u}\in\mathcal{U}(k)} \{\langle \boldsymbol{r}, \boldsymbol{u}-\boldsymbol{z}\rangle + g(\boldsymbol{x},\boldsymbol{u})\} + QV(\boldsymbol{x},\cdot)(k). \tag{3.93}$$

Let

$$\Gamma(\boldsymbol{x}) = \left\{ \lim_{n\to\infty} \frac{\partial V(\boldsymbol{x}_n,k)}{\partial \boldsymbol{x}} : V(\boldsymbol{x},k) \text{ is differentiable at } \boldsymbol{x}_n, \right. \\ \left. \text{and } \lim_{n\to\infty} \boldsymbol{x}_n = \boldsymbol{x} \right\},$$

let $\mathrm{co}(\Gamma(\boldsymbol{x}))$ be the convex hull of $\Gamma(\boldsymbol{x})$, and let $\overline{\mathrm{co}}(\Gamma(\boldsymbol{x}))$ be the closure of $\mathrm{co}(\Gamma(\boldsymbol{x}))$. From Lemma C.2 we have

$$\overline{\mathrm{co}}(\Gamma(\boldsymbol{x})) = D^-V(\boldsymbol{x},k). \tag{3.94}$$

Since $V(\boldsymbol{x},k)$ is a continuous convex function (see Lemma C.1), it is differentiable a.e. For any $\boldsymbol{x}$, we can take a sequence $\boldsymbol{x}_n \to \boldsymbol{x}$ such that $V(\cdot,k)$ is differentiable at $\boldsymbol{x}_n$ for each n. Thus, from Theorem 7.5, we have

$$\begin{aligned}\widehat{\lambda} &= \inf_{\boldsymbol{u}\in\mathcal{U}(k)} \left\{ \left\langle \frac{\partial V(\boldsymbol{x}_n,k)}{\partial \boldsymbol{x}}, \boldsymbol{u}-\boldsymbol{z} \right\rangle + g(\boldsymbol{x}_n,\boldsymbol{u}) \right\} + QV(\boldsymbol{x}_n,\cdot)(k) \\ &\le \left\langle \frac{\partial V(\boldsymbol{x}_n,k)}{\partial \boldsymbol{x}}, \boldsymbol{u}-\boldsymbol{z} \right\rangle + g(\boldsymbol{x}_n,\boldsymbol{u}) + QV(\boldsymbol{x}_n,\cdot)(k),\end{aligned}$$

for any $\boldsymbol{u} \in \mathcal{U}(k)$. Taking $\boldsymbol{x}_n \to \boldsymbol{x}$ as $n \to \infty$, we obtain from the continuity of $V(\cdot,k)$ and $g(\cdot,\boldsymbol{u})$,

$$\widehat{\lambda} \le \langle \boldsymbol{r}, \boldsymbol{u}-\boldsymbol{z}\rangle + g(\boldsymbol{x},\boldsymbol{u}) + QV(\boldsymbol{x},\cdot)(k), \quad \boldsymbol{r} \in \Gamma(\boldsymbol{x}). \tag{3.95}$$

From the linearity of the right-hand side of (3.95) in $\boldsymbol{r}$, we can extend the inequality to $\mathrm{co}(\Gamma(\boldsymbol{x}))$. Taking $\boldsymbol{r}_n \in \mathrm{co}(\Gamma(\boldsymbol{x}))$ such that $\boldsymbol{r}_n \to \boldsymbol{r}$, the extension can be made to $\overline{\mathrm{co}}(\Gamma(\boldsymbol{x}))$. By (3.94) and (3.95), we have shown that, for $\boldsymbol{r} \in D^-V(\boldsymbol{x},k)$,

$$\widehat{\lambda} \le \inf_{\boldsymbol{u}\in\mathcal{U}(k)} \{\langle \boldsymbol{r}, \boldsymbol{u}-\boldsymbol{z}\rangle + g(\boldsymbol{x},\boldsymbol{u})\} + QV(\boldsymbol{x},\cdot)(k).$$

On the other hand, the viscosity solution property implies the opposite inequality, and hence we get (3.93).

We can now show that for any $\widehat{\boldsymbol{y}} \in \mathrm{ri}(D^-V(\boldsymbol{x},k))$, if we let $\boldsymbol{u}^*$ be such that

$$\widehat{\lambda} = \langle \widehat{\boldsymbol{y}}, \boldsymbol{u}^* - \boldsymbol{z} \rangle + g(\boldsymbol{x}, \boldsymbol{u}^*) + QV(\boldsymbol{x},\cdot)(k), \tag{3.96}$$

then (3.96) still holds when we replace $\widehat{\boldsymbol{y}}$ by any $\boldsymbol{y} \in D^-V(\boldsymbol{x},k)$. In fact, by (ii) of Lemma C.3, for any $\boldsymbol{y} \in D^-V(\boldsymbol{x},k), \boldsymbol{y} \neq \widehat{\boldsymbol{y}}$, there exists a $\widetilde{\boldsymbol{y}} \in D^-V(\boldsymbol{x},k)$ such that

$$\widehat{\boldsymbol{y}} = \delta \boldsymbol{y} + (1-\delta)\widetilde{\boldsymbol{y}}, \quad 0 < \delta < 1.$$

From (3.96) we can write

$$\begin{aligned}\widehat{\lambda} &= \delta[\langle \boldsymbol{y}, \boldsymbol{u}^* - \boldsymbol{z} \rangle + g(\boldsymbol{x}, \boldsymbol{u}^*) + QV(\boldsymbol{x},\cdot)(k)] \\ &\quad + (1-\delta)[\langle \widetilde{\boldsymbol{y}}, \boldsymbol{u}^* - \boldsymbol{z} \rangle + g(\boldsymbol{x}, \boldsymbol{u}^*) + QV(\boldsymbol{x},\cdot)(k)].\end{aligned}$$

From Theorem 7.5, we have

$$\widehat{\lambda} \le \langle \boldsymbol{y}, \boldsymbol{u}^* - \boldsymbol{z} \rangle + g(\boldsymbol{x}, \boldsymbol{u}^*) + QV(\boldsymbol{x},\cdot)(k).$$

If

$$\widehat{\lambda} < \langle \boldsymbol{y}, \boldsymbol{u}^* - \boldsymbol{z} \rangle + g(\boldsymbol{x}, \boldsymbol{u}^*) + QV(\boldsymbol{x},\cdot)(k),$$

then we must have

$$\widehat{\lambda} > \langle \widetilde{\boldsymbol{y}}, \boldsymbol{u}^* - \boldsymbol{z} \rangle + g(\boldsymbol{x}, \boldsymbol{u}^*) + QV(\boldsymbol{x},\cdot)(k),$$

which contradicts (3.93). That is,

$$\widehat{\lambda} = \inf_{\boldsymbol{u} \in \mathcal{U}(k)} \{\langle \widetilde{\boldsymbol{y}}, \boldsymbol{u} - \boldsymbol{z} \rangle + g(\boldsymbol{x}, \boldsymbol{u})\} + QV(\boldsymbol{x},\cdot)(k),$$

because $\widetilde{\boldsymbol{y}} \in D^-V(\boldsymbol{x},k)$. From this and (3.93), we can conclude that

$$\begin{aligned}\widehat{\lambda} &= \inf_{\boldsymbol{u} \in \mathcal{U}(k)} \{\langle \boldsymbol{y}, \boldsymbol{u} - \boldsymbol{z} \rangle + g(\boldsymbol{x}, \boldsymbol{u})\} + QV(\boldsymbol{x},\cdot)(k) \\ &= \langle \boldsymbol{y}, \boldsymbol{u}^* - \boldsymbol{z} \rangle + g(\boldsymbol{x}, \boldsymbol{u}^*) + QV(\boldsymbol{x},\cdot)(k), \quad \boldsymbol{y} \in D^-V(\boldsymbol{x},k).\end{aligned} \tag{3.97}$$

Apply again Lemma C.2 to conclude that

$$\widehat{\lambda} = \partial_{\boldsymbol{u}^* - \boldsymbol{z}} V(\boldsymbol{x},k) + g(\boldsymbol{x}, \boldsymbol{u}^*) + QV(\boldsymbol{x},\cdot)(k). \tag{3.98}$$

It follows directly from (3.98) that

$$\widehat{\lambda} \ge \inf_{\boldsymbol{u} \in \mathcal{U}(k)} \{\partial_{\boldsymbol{u} - \boldsymbol{z}} V(\boldsymbol{x},k) + g(\boldsymbol{x}, \boldsymbol{u})\} + QV(\boldsymbol{x},\cdot)(k). \tag{3.99}$$

On the other hand, Lemma C.2 allows us, for every $\boldsymbol{u}$, to find a $\boldsymbol{y}_{\boldsymbol{u}} \in D^-V(\boldsymbol{x},k)$ such that $\langle \boldsymbol{y}_{\boldsymbol{u}}, \boldsymbol{u} - \boldsymbol{z} \rangle = \partial_{\boldsymbol{u}-\boldsymbol{z}} V(\boldsymbol{x},k)$. Using this fact and (3.97), we obtain the opposite inequality for $\boldsymbol{u} \in \mathcal{U}(k)$,

$$\begin{aligned} &\langle \boldsymbol{y_u}, \boldsymbol{u} - \boldsymbol{z} \rangle + g(\boldsymbol{x}, \boldsymbol{u}) + QV(\boldsymbol{x}, \cdot)(k) \\ &\quad = \partial_{\boldsymbol{u}-\boldsymbol{z}} V(\boldsymbol{x}, k) + g(\boldsymbol{x}, \boldsymbol{u}) + QV(\boldsymbol{x}, \cdot)(k) \\ &\quad \geq \langle \boldsymbol{y_u}, \boldsymbol{u}^* - \boldsymbol{z} \rangle + g(\boldsymbol{x}, \boldsymbol{u}^*) + QV(\boldsymbol{x}, \cdot)(k) \\ &\quad = \widehat{\lambda}. \end{aligned} \tag{3.100}$$

Thus, combining (3.99) and (3.100), we have shown that

$$\widehat{\lambda} = \inf_{\boldsymbol{u} \in \mathcal{U}(k)} \{\partial_{\boldsymbol{u}-\boldsymbol{z}} V(\boldsymbol{x}, k) + g(\boldsymbol{x}, \boldsymbol{u})\} + QV(\boldsymbol{x}, \cdot)(k),$$

i.e., $(\widehat{\lambda}, V(\boldsymbol{x}, k))$ is a solution of the HJBDD equation (3.91). □

The following verification theorem can now be proved.

Theorem 7.7. *Let Assumptions* (A3) *and* (A6)–(A7) *hold, and let* $(\lambda, W(\boldsymbol{x}, k))$ *be a solution of the* HJBDD *equation* (3.91). *Then:*

(i) *If there is a control* $\boldsymbol{u}^*(\cdot) \in \mathcal{A}(k)$ *such that*

$$\begin{aligned} &\inf_{\boldsymbol{u} \in \mathcal{U}(k(t))} \{\partial_{\boldsymbol{u}-\boldsymbol{z}} W(\boldsymbol{x}^*(t), k(t)) + g(\boldsymbol{x}^*(t), \boldsymbol{u})\} \\ &\qquad = \partial_{\boldsymbol{u}^*(t)-\boldsymbol{z}} W(\boldsymbol{x}^*(t), k(t)) + g(\boldsymbol{x}^*(t), \boldsymbol{u}^*(t)) \end{aligned} \tag{3.101}$$

for a.e. $t \geq 0$ *with probability* 1, *where* $\boldsymbol{x}^*(\cdot)$ *is the surplus process corresponding to the control* $\boldsymbol{u}^*(\cdot)$ *and the initial condition* $\boldsymbol{x}^*(0) = \boldsymbol{x}$, *and*

$$\lim_{T \to \infty} \frac{E\left[W(\boldsymbol{x}^*(T), k(T))\right]}{T} = 0, \tag{3.102}$$

then

$$\lambda = J(\boldsymbol{x}, k, \boldsymbol{u}^*(\cdot)).$$

(ii) *For any* $\boldsymbol{u}(\cdot) \in \mathcal{A}(k)$, *we have* $\lambda \leq J(\boldsymbol{x}, k, \boldsymbol{u}(\cdot))$, *i.e.,*

$$\limsup_{T \to \infty} \frac{1}{T} E \int_0^T g(\boldsymbol{x}(t), \boldsymbol{u}(t))\, dt \geq \lambda.$$

(iii) *Furthermore, for any (stable) control* $\boldsymbol{u}(\cdot) \in \mathcal{S}(k)$, *we have*

$$\liminf_{T \to \infty} \frac{1}{T} E \int_0^T g(\boldsymbol{x}(t), \boldsymbol{u}(t))\, dt \geq \lambda. \tag{3.103}$$

Proof. The proof can be obtained along the lines of the proof of Theorem 4.3. □

Remark 7.2. From Theorems 7.6 and 7.7, we know that $\widehat{\lambda}$ defined by (3.92) is the optimal cost, i.e., $\widehat{\lambda} = \lambda^*$, and $V(\boldsymbol{x}, k)$ defined by (3.92) is a potential function. □

In this chapter we have assumed a constant demand. Nevertheless, the results obtained in this chapter, except the monotonicity of the turnpike sets in Section 6.1, still hold when the demand rate is modeled as a Markov chain.

3.8 Notes

This chapter is based on Sethi, Suo, Taksar, and Zhang [113], Sethi, Yan, Zhang, and Zhang [116], and Sethi, Suo, Taksar, and Yan [112].

Theorems 4.1–4.3, which concern the properties of the potential functions and the optimal controls for the single product case, are due to Sethi, Suo, Taksar, and Zhang [113]. For the multiproduct case, Theorems 7.4–7.7 are from Sethi, Suo, Taksar, and Yan [112]. Theorem 6.1 concerning the monotonicity property of the turnpike sets obtained by Sethi, Yan, Zhang and Zhang [116]. Similar results have been obtained by Hu and Xiang [71]. The monotonicity results are extended in Hu and Xiang [72] and Liberopoulos and Hu [90] to models with non-Markovian processes.

Eleftheriu [44] and Zhang and Yin [152] consider the problem with a finite horizon. Eleftheriu [44] uses a stochastic maximum principle of Sworder [134] to analyze the undiscounted version of the problem. Zhang and Yin [152] show that the turnpike sets of the problem become turnpike curves in the finite horizon case. They derive explicit solutions and show that the optimal controls can be written in terms of the turnpike curves under appropriate "traceability" conditions.

Bielecki and Kumar [20] study the problem in which there is no production cost, the machine has two states (up and down), and the production cost is linear. For this problem, they are able to explicitly obtain a threshold-type policy. Furthermore, with the same cost structure and the production machine property, when the demand process follows a Poisson process, Feng and Yan [51] also establish the explicit solution for the problem. Sharifnia [128] deals with an extension of their work with more than two-machine states. Liberopoulos and Caramanis [89] show that Sharifnia's method for evaluating hedging point policies applies even when the transition rates of the machine states depend on the production rate. Caramanis and Sharifnia [28] consider an average-cost problem with multiple part-types. They decompose the problem to many analytically tractable single-product problems in order to obtain near-optimal hedging points for the problem.

Other multiproduct systems are considered by Presman, Sethi, Zhang, and Zhang [104], Sethi, Zhang, and Zhang [121], Srivatsan [131] and Srivatsan and Dallery [133] for a two-product problem. Caramanis and Sharifnia [28] decompose a multiproduct problem into analytically tractable single-product problems in order to obtain approximately optimal hedging points for the problem. Other existing results for multiproduct systems involve either approximations or numerical solutions (Sethi and Zhang [118], Veatch [140], Yan, Yin and Lou [147], Wein [142], and Liberopoulos [88]).

Manufacturing systems involving preventive maintenance are studied by Boukas [21], Boukas and Haurie [22], Boukas, Zhang, and Zhu [25], and Boukas, Zhu, and Zhang [26]. The maintenance activity involves lubrication, routine adjustments, etc., which reduce the machine failure rate. The objective in these systems is to choose the rate of maintenance and the rate of production in order to minimize the total discounted cost of surplus, production, and maintenance.

4

Optimal Control of Dynamic Flowshops

4.1 Introduction

In this chapter we consider planning of production in a stochastic m-machine flowshop described in Chapter 2. The goal is to choose rates of production at each machine to meet the demand for a single product facing the system at a minimum long-run average cost. The machine capacities process is assumed to be a finite state Markov chain. The control variables are the input rates to these machines. We take the inventory (number of parts) in the buffer of the first $(m-1)$ machines and the surplus at the last machine to be the state variables. Since the number of parts in the internal buffers between any two machines needs to be nonnegative, the problem is inherently a state-constrained problem. Our objective is to choose admissible input rates to various machines in order to minimize the expected long-run average surplus and production costs.

In Chapter 3, we studied average-cost systems with parallel machines using Hamilton-Jacobi-Bellman equations in terms of directional derivatives (HJBDD). However, the flowshop problem is more complicated because of the presence of internal buffers and of the resulting nonnegativity state constraints. Certain boundary conditions need to be taken into account for the associated Hamilton-Jacobi-Bellman (HJB) equation. Optimal control policy can no longer be described simply in terms of some threshold levels (or turnpike sets).

To rigorously deal with the general flowshop problem under consideration, we use the HJBDD equation to characterize the system dynamics.

The equation involves an infimum taken over *admissible directions.* If a convex function is differentiable at an inner point of the state space, then the HJBDD equation at this point coincides with the usual HJB equation. If the restriction of the function on some face of the boundary of the state space is differentiable at an inner point of this face, then the m-dimensional gradient can be defined at this point, and the HJBDD equation at this point gives the corresponding boundary condition.

Under the framework of HJBDD equation, we follow the vanishing discount approach to prove the existence of the minimum average cost and the potential function. Furthermore, we prove a verification theorem associated with our dynamic programming formulation.

The plan of this chapter is as follows. In Section 4.2 we formulate the problem under consideration. In Section 4.3 we recapitulate some properties of the value function of the corresponding discounted-cost problem, and construct stable policies which will be used to analyze the long-run average-cost problem. In Section 4.4 we discuss the HJBDD equation and the verification theorem. In Section 4.5, a two-machine flowshop with bounded internal buffer is studied. The chapter is concluded in Section 4.6.

4.2 Problem Formulation

Let $\mathcal{M} = \{\boldsymbol{k}^1, \ldots, \boldsymbol{k}^p\}$ for a given integer $p \geq 1$, where $\boldsymbol{k}^i = (k_1^i, \ldots, k_m^i)$ with k_j^i denoting the capacity of the jth machine, $j = 1, \ldots, m$. Let $\boldsymbol{k}(\cdot) = (k_1(\cdot), \ldots, k_m(\cdot))$ denote a Markov chain with the state space $\mathcal{M}$. We use $u_j(t)$ to denote the input rate to the jth machine, $j = 1, \ldots, m$, and $x_j(t)$ to denote the number of parts in the buffer between the jth and $(j+1)$th machines, $j = 1, \ldots, m-1$. Finally, the surplus is denoted by $x_m(t)$. The dynamics of the system can then be written as follows:

$$\frac{d}{dt}\boldsymbol{x}(t) = A\boldsymbol{u}(t) + Bz, \quad \boldsymbol{x}(0) = \boldsymbol{x}, \tag{4.1}$$

where

$$A = \begin{pmatrix} 1 & -1 & 0 & \cdots & 0 \\ 0 & 1 & -1 & \cdots & 0 \\ \vdots & \vdots & \vdots & \cdots & \vdots \\ 0 & 0 & 0 & \cdots & 1 \end{pmatrix} \quad \text{and} \quad B = \begin{pmatrix} 0 \\ 0 \\ \vdots \\ 0 \\ -1 \end{pmatrix}.$$

Since the number of parts in the internal buffers cannot be negative, we have to impose the state constraints $x_j(t) \geq 0, j = 1, \ldots, m-1$. To formulate the problem precisely, let $\mathcal{X} = \Re_+^{m-1} \times \Re \subset \Re^m$ denote the state constraint domain. For $\boldsymbol{k} = (k_1, \ldots, k_m)$, $k_j \geq 0$, $j = 1, \ldots, m$, let

$$\mathcal{U}(\boldsymbol{k}) = \{\boldsymbol{u} = (u_1, \ldots, u_m)' : 0 \leq u_j \leq k_j, \ j = 1, \ldots, m\}, \tag{4.2}$$

and for $\boldsymbol{x} \in \mathcal{X}$, let

$$\begin{aligned}\mathcal{U}(\boldsymbol{x},\boldsymbol{k}) = \{&\boldsymbol{u} : \boldsymbol{u} \in \mathcal{U}(\boldsymbol{k}) \text{ and} \\ &x_j = 0 \Rightarrow u_j - u_{j+1} \geq 0,\ j = 1,\ldots,m-1\}.\end{aligned} \tag{4.3}$$

Finally, let the σ-algebra $\mathcal{F}_t = \sigma\{\boldsymbol{k}(s) : 0 \leq s \leq t\}$.

Definition 2.1. We say that a control $\boldsymbol{u}(\cdot)$ is *admissible* with respect to the initial values $\boldsymbol{x} \in \mathcal{X}$ and $\boldsymbol{k} \in \mathcal{M}$, if: (i) $\boldsymbol{u}(\cdot)$ is an $\{\mathcal{F}_t\}$-adapted process; (ii) $\boldsymbol{u}(t) \in \mathcal{U}(\boldsymbol{k}(t))$ with $\boldsymbol{k}(0) = \boldsymbol{k}$ for all $t \geq 0$; and (iii) the corresponding state process $\boldsymbol{x}(t) = (x_1(t),\ldots,x_m(t))' \in \mathcal{X}$ with $\boldsymbol{x}(0) = \boldsymbol{x}$ for all $t \geq 0$. □

Let $\mathcal{A}(\boldsymbol{x},\boldsymbol{k})$ denote the set of admissible controls with the initial conditions $\boldsymbol{x}(0) = \boldsymbol{x}$ and $\boldsymbol{k}(0) = \boldsymbol{k}$.

Remark 2.1. Condition (iii) of Definition 2.1 is equivalent to $\boldsymbol{u}(t) \in \mathcal{U}(\boldsymbol{x}(t),\boldsymbol{k}(t))$ for all $t \geq 0$. □

Definition 2.2. A function $\boldsymbol{u}(\cdot,\cdot)$ defined on $\mathcal{X} \times \mathcal{M}$ is called an *admissible feedback control*, or simply *feedback control*, if: (i) for any given initial surplus level $\boldsymbol{x}$ and machine capacity $\boldsymbol{k}$, the equation

$$\frac{d}{dt}\boldsymbol{x}(t) = A\boldsymbol{u}(\boldsymbol{x}(t),\boldsymbol{k}(t)) + Bz$$

has a unique solution; and (ii) $\boldsymbol{u}(\cdot) = \{\boldsymbol{u}(t) = \boldsymbol{u}(\boldsymbol{x}(t),\boldsymbol{k}(t)),\ t \geq 0\} \in \mathcal{A}(\boldsymbol{x},\boldsymbol{k})$. □

The objective of the problem is to find an admissible control $\boldsymbol{u}(\cdot)$ that minimizes the long-run average cost

$$J(\boldsymbol{x},\boldsymbol{k},\boldsymbol{u}(\cdot)) = \limsup_{T\to\infty} \frac{1}{T} E \int_0^T g(\boldsymbol{x}(t),\boldsymbol{u}(t))\,dt, \tag{4.4}$$

where $\boldsymbol{x}(0) = \boldsymbol{x}$, $\boldsymbol{k}(0) = \boldsymbol{k}$, and $g(\cdot,\cdot)$ is the cost of surplus and production defined on $\mathcal{X} \times \Re^m_+$.

The minimum long-run average cost is defined as

$$\lambda^*(\boldsymbol{x},\boldsymbol{k}) = \inf_{\boldsymbol{u}(\cdot)\in\mathcal{A}(\boldsymbol{x},\boldsymbol{k})} J(\boldsymbol{x},\boldsymbol{k},\boldsymbol{u}(\cdot)). \tag{4.5}$$

We make the following assumptions on the Markov chain

$$\boldsymbol{k}(\cdot) = (k_1(\cdot),\ldots,k_m(\cdot))$$

and the cost function $g(\boldsymbol{x},\boldsymbol{u})$.

(A1) $g(\boldsymbol{x},\boldsymbol{u})$ is a nonnegative convex function with $g(0,0) = 0$. There are positive constants C_{g1}, C_{g2}, and $\beta_{g1} \geq 1$ such that, for any fixed $\boldsymbol{u}$,

$$g(\boldsymbol{x},\boldsymbol{u}) \geq C_{g1}|\boldsymbol{x}|^{\beta_{g1}} - C_{g2}, \quad \boldsymbol{x} \in \mathcal{X}.$$

Moreover, there are constants $C_{g3} > 0$ and $\beta_{g2} \geq \beta_{g1}$ such that, for $\boldsymbol{u}, \widehat{\boldsymbol{u}} \in \Re^m_+$, and $\boldsymbol{x}, \widehat{\boldsymbol{x}} \in \mathcal{X}$,

$$\begin{aligned}&|g(\boldsymbol{x}, \boldsymbol{u}) - g(\widehat{\boldsymbol{x}}, \widehat{\boldsymbol{u}})| \\ &\leq C_{g3}\left[\left(1 + |\boldsymbol{x}|^{\beta_{g2}-1} + |\widehat{\boldsymbol{x}}|^{\beta_{g2}-1}\right)|\boldsymbol{x} - \widehat{\boldsymbol{x}}| + |\boldsymbol{u} - \widehat{\boldsymbol{u}}|\right].\end{aligned}$$

(A2) The capacity process $\boldsymbol{k}(t) \in \mathcal{M}$, $t \geq 0$, is a finite state Markov chain with the infinitesimal generator $Q = (q_{\boldsymbol{k}^i \boldsymbol{k}^j})_{p\times p}$ with $q_{\boldsymbol{k}^i\boldsymbol{k}^j} \geq 0$, $i \neq j$, $q_{\boldsymbol{k}^i\boldsymbol{k}^i} = -\sum_{j\neq i} q_{\boldsymbol{k}^i\boldsymbol{k}^j}$, and

$$Q\varphi(\cdot)(\boldsymbol{k}^j) = \sum_{i\neq j} q_{\boldsymbol{k}^j\boldsymbol{k}^i}[\varphi(\boldsymbol{k}^i) - \varphi(\boldsymbol{k}^j)], \tag{4.6}$$

for any function $\varphi(\cdot)$ defined on $\mathcal{M}$. Moreover, the Markov chain is strongly irreducible, that is, the equations

$$(\nu_{\boldsymbol{k}^1}, ..., \nu_{\boldsymbol{k}^p})Q = 0 \quad \text{and} \quad \sum_{i=1}^{p} \nu_{\boldsymbol{k}^i} = 1$$

have a unique solution with $\nu_{\boldsymbol{k}^i} > 0$, $i = 1, \ldots, p$.

(A3) With $p_j := \sum_{i=1}^{p} \nu_{\boldsymbol{k}^i} k^i_j$, $\min_{1\leq j\leq m} p_j > z$.

Without requiring differentiability, it is convenient to write an HJBDD equation for the average-cost problem. Let $\mathcal{O}^m \subset \Re^m$ be a convex subset of $\Re^m$. A vector $\boldsymbol{y}$ at boundary point $\boldsymbol{x}$ of $\mathcal{O}^m$ is called an admissible direction if there exists a $\delta > 0$ such that $\boldsymbol{x} + \delta\boldsymbol{y} \in \mathcal{O}^m$. Note that $\{A\boldsymbol{u} + Bz : \boldsymbol{u} \in \mathcal{U}(\boldsymbol{x}, \boldsymbol{k})\}$ is the set of admissible directions at $\boldsymbol{x}$ when $\mathcal{O}^m = \mathcal{X}$. Note that a continuous convex function defined on $\mathcal{O}^m$ is differentiable a.e., and it has a directional derivative both along any direction at any inner point of $\mathcal{O}^m$ and along any admissible direction at any boundary point of $\mathcal{O}^m$ (see Appendix C).

Typically, under Assumptions (A1)–(A3), $\lambda^*(\boldsymbol{x}, \boldsymbol{k})$ does not depend on the initial states $\boldsymbol{x}$ and $\boldsymbol{k}$ (Theorem 4.2). We can formally write the associated HJBDD equation as follows:

$$\lambda = \inf_{\boldsymbol{u}\in\mathcal{U}(\boldsymbol{x},\boldsymbol{k})} \{\partial_{(A\boldsymbol{u}+Bz)}\phi(\boldsymbol{x}, \boldsymbol{k}) + g(\boldsymbol{x}, \boldsymbol{u})\} + Q\phi(\boldsymbol{x}, \cdot)(\boldsymbol{k}), \tag{4.7}$$

where $\phi(\cdot, \cdot)$ is defined on $\mathcal{X} \times \mathcal{M}$. Before defining a solution to the equation (4.7), we first introduce some notation. Let $\mathcal{G}$ denote the family of real-valued functions $G(\cdot, \cdot)$ defined on $\mathcal{X} \times \mathcal{M}$ such that:

(i) $G(\cdot, \boldsymbol{k})$ is convex for each $\boldsymbol{k} \in \mathcal{M}$;

(ii) $G(\cdot, \cdot)$ has polynomial growth, i.e., there exists a constant $C > 0$ such that

$$|G(\boldsymbol{x}, \boldsymbol{k})| \leq C(1 + |\boldsymbol{x}|^{\beta_{g2}+1}), \quad \boldsymbol{x} \in \mathcal{X} \text{ and } \boldsymbol{k} \in \mathcal{M},$$

where β_{g2} is defined in Assumption (A1).

A *solution* to the HJBDD equation (4.7) is a pair $(\lambda, W(\cdot,\cdot))$ with λ a constant and $W(\cdot,\cdot) \in \mathcal{G}$. The function $W(\cdot,\cdot)$ is called the *potential function* for the control problem, if λ is the minimum long-run average cost defined by (4.5).

Definition 2.3. A control $\boldsymbol{u}(\cdot) \in \mathcal{A}(\boldsymbol{x}, \boldsymbol{k})$ is *stable* if it satisfies the condition

$$\lim_{T\to\infty} \frac{E|\boldsymbol{x}(T)|^{\beta_{g2}+1}}{T} = 0,$$

where $\boldsymbol{x}(\cdot)$ is the surplus process corresponding to the control $\boldsymbol{u}(\cdot)$ with the initial condition $(\boldsymbol{x}(0), \boldsymbol{k}(0)) = (\boldsymbol{x}, \boldsymbol{k})$ and β_{g2} is as defined in Assumption (A1). □

Let $\mathcal{S}(\boldsymbol{x}, \boldsymbol{k})$ denote the class of stable controls with the initial conditions $\boldsymbol{x}(0) = \boldsymbol{x}$ and $\boldsymbol{k}(0) = \boldsymbol{k}$. Of course, $\mathcal{S}(\boldsymbol{x}, \boldsymbol{k}) \subset \mathcal{A}(\boldsymbol{x}, \boldsymbol{k})$. It will be seen in the next section that the set of stable admissible controls $\mathcal{S}(\boldsymbol{x}, \boldsymbol{k})$ is nonempty.

4.3 Estimates for Discounted Cost Value Functions

In order to study the existence of a solution to the HJBDD equation (4.7), we introduce the corresponding discounted cost problem with the discount rate $\rho > 0$. For $\boldsymbol{u}(\cdot) \in \mathcal{A}(\boldsymbol{x}, \boldsymbol{k})$, we define the expected discounted cost as

$$J^\rho(\boldsymbol{x}, \boldsymbol{k}, \boldsymbol{u}(\cdot)) = E \int_0^\infty e^{-\rho t} g(\boldsymbol{x}(t), \boldsymbol{u}(t))\, dt, \tag{4.8}$$

where $\boldsymbol{x}(\cdot)$ is the surplus process corresponding to the control $\boldsymbol{u}(\cdot)$ with the initial surplus level $\boldsymbol{x}(0) = \boldsymbol{x}$, and $\boldsymbol{k}$ is the initial value of the process $\boldsymbol{k}(\cdot)$. The value function is defined as

$$V^\rho(\boldsymbol{x}, \boldsymbol{k}) = \inf_{\boldsymbol{u}(\cdot)\in\mathcal{A}(\boldsymbol{x},\boldsymbol{k})} J^\rho(\boldsymbol{x}, \boldsymbol{k}, \boldsymbol{u}(\cdot)). \tag{4.9}$$

The HJBDD equation associated with this discounted cost problem is

$$\rho\phi^\rho(\boldsymbol{x}, \boldsymbol{k}) = \inf_{\boldsymbol{u}\in\mathcal{U}(\boldsymbol{x},\boldsymbol{k})} \{\partial_{(A\boldsymbol{u}+B\boldsymbol{z})}\phi^\rho(\boldsymbol{x}, \boldsymbol{k}) + g(\boldsymbol{x}, \boldsymbol{u})\} + Q\phi^\rho(\boldsymbol{x}, \cdot)(\boldsymbol{k}), \tag{4.10}$$

where $\phi^\rho(\boldsymbol{x}, \boldsymbol{k})$ is defined on $\mathcal{X} \times \mathcal{M}$. Then we have the following result which can be proved as Theorem 3.1 in Chapter 4 of Sethi and Zhang [125].

Theorem 3.1. *Under Assumptions* (A1) *and* (A2), *the value function* $V^\rho(\boldsymbol{x}, \boldsymbol{k})$ *has the following properties:*

(i) *for each* $\boldsymbol{k} \in \mathcal{M}$, $V^\rho(\boldsymbol{x}, \boldsymbol{k})$ *is convex, continuous on* $\mathcal{X}$, *and satisfies the condition*

$$|V^\rho(\widehat{\boldsymbol{x}}, \boldsymbol{k}) - V^\rho(\widetilde{\boldsymbol{x}}, \boldsymbol{k})| \le C\left(1 + |\widehat{\boldsymbol{x}}|^{\beta_{g2}-1} + |\widetilde{\boldsymbol{x}}|^{\beta_{g2}-1}\right) |\widehat{\boldsymbol{x}} - \widetilde{\boldsymbol{x}}|, \tag{4.11}$$

for some constant C and all $\widehat{\boldsymbol{x}}, \widetilde{\boldsymbol{x}} \in \mathcal{X}$, where β_{g2} is as given in Assumption (A1);

(ii) $V^\rho(\boldsymbol{x}, \boldsymbol{k})$ *satisfies the* HJBDD *equation* (4.10).

In order to study the long-run average-cost control problem using the vanishing discount approach, we must first obtain some estimates for the value function $V^\rho(\boldsymbol{x}, \boldsymbol{k})$ for small values of ρ. To do this, we establish the following theorem which concerns the stable control policies.

Theorem 3.2. *Let Assumptions* (A2) *and* (A3) *hold. For any $r \geq 1$, $(\widehat{\boldsymbol{x}}, \widehat{\boldsymbol{k}}) \in \mathcal{X} \times \mathcal{M}$, and $(\overline{\boldsymbol{x}}, \overline{\boldsymbol{k}}) \in \mathcal{X} \times \mathcal{M}$, there exist a control $\boldsymbol{u}(\cdot) \in \mathcal{A}(\widehat{\boldsymbol{x}}, \widehat{\boldsymbol{k}})$, and a positive constant C_r, such that*

$$E\big[\tau(\widehat{\boldsymbol{x}}, \widehat{\boldsymbol{k}}, \overline{\boldsymbol{x}}, \overline{\boldsymbol{k}})\big]^r \leq C_r \left(1 + \sum_{j=1}^{m} |\widehat{x}_j - \overline{x}_j|^r\right), \tag{4.12}$$

where

$$\tau(\widehat{\boldsymbol{x}}, \widehat{\boldsymbol{k}}, \overline{\boldsymbol{x}}, \overline{\boldsymbol{k}}) = \inf\left\{t \geq 0 : \boldsymbol{x}(t) = \overline{\boldsymbol{x}},\ \boldsymbol{k}(t) = \overline{\boldsymbol{k}}\right\},$$

and $\boldsymbol{x}(t), t \geq 0$, is the surplus process corresponding to the control $\boldsymbol{u}(\cdot)$ and the initial condition $(\boldsymbol{x}(0), \boldsymbol{k}(0)) = (\widehat{\boldsymbol{x}}, \widehat{\boldsymbol{k}})$.

Proof. The proof of the theorem is divided into four steps involving some intermediate results stated as lemmas. Here is an outline of the proof. We begin by modifying the process $\boldsymbol{k}(\cdot)$ in such a way that the modified average capacity of any machine is larger than the modified average capacity of the machine that follows it, and the modified average capacity of the last machine is larger than z. Then we alternate between the following two policies. In the first policy, the production rate at each machine is the maximum admissible modified capacity. In the second policy, we stop producing at the first machine and have the maximum possible production rate at other machines under the restriction that the content of each buffer j, $1 \leq j \leq m-1$, is not less than $\overline{x}_j$ where $\overline{\boldsymbol{x}} = (\overline{x}_1, \ldots, \overline{x}_m)'$. The first policy is used until such time when the content of the first buffer exceeds the value $\overline{x}_1$ and the content of each buffer j, $2 \leq j \leq m$, exceeds the value $M + \overline{x}_j$ for some $M > 0$. At that time we switch to the second policy. We use the second policy until such time when the content of the last buffer drops to the level $\overline{x}_m$. After that we revert to the first policy, and so on. Using this alternating procedure, it is possible to specify $\tau(\widehat{\boldsymbol{x}}, \widehat{\boldsymbol{k}}, \overline{\boldsymbol{x}}, \overline{\boldsymbol{k}})$ and provide an estimate for it.

Step 1. We construct an auxiliary Markov chain $\widetilde{\boldsymbol{k}}(\cdot)$ from the Markov chain $\boldsymbol{k}(\cdot)$ specified by Assumptions (A2) and (A3). It follows from Assumption (A3) that we can select vectors $\widetilde{\boldsymbol{k}}^i = (\widetilde{k}_1^i, \ldots, \widetilde{k}_m^i)$, $i = 1, \ldots, p$, such that $\widetilde{k}_1^i = k_1^i$, $\widetilde{k}_j^i \le k_j^i$, $i = 1, \ldots, p$, $j = 2, \ldots, m$, and

$$\widetilde{p}_j = \sum_{i=1}^{p} \nu_{\boldsymbol{k}^i} \widetilde{k}_j^i > \widetilde{p}_{j+1} = \sum_{i=1}^{p} \nu_{\boldsymbol{k}^i} \widetilde{k}_{j+1}^i > z, \quad j = 1, \ldots, m-1. \tag{4.13}$$

Let us define the process $\widetilde{\boldsymbol{k}}(\cdot)$ as follows:

$$\widetilde{\boldsymbol{k}}(t) = \widetilde{\boldsymbol{k}}^i \quad \text{when} \quad \boldsymbol{k}(t) = \boldsymbol{k}^i,\ t \ge 0.$$

Let $\widetilde{\mathcal{M}} = \{\widetilde{\boldsymbol{k}}^1, \ldots, \widetilde{\boldsymbol{k}}^p\}$. We know that $\widetilde{\boldsymbol{k}}(\cdot)$ is still a Markov chain with the state space $\widetilde{\mathcal{M}}$. Furthermore, $\widetilde{\boldsymbol{k}}(\cdot)$ is strongly irreducible and has the stationary distribution $\nu_{\widetilde{\boldsymbol{k}}^i}$ with $\nu_{\widetilde{\boldsymbol{k}}^i} = \nu_{\boldsymbol{k}^i}$, $i = 1, \ldots, p$. Let

$$\widetilde{p}_j = \sum_{i=1}^{p} \widetilde{k}_j^i \nu_{\boldsymbol{k}^i}. \tag{4.14}$$

Then, (4.13) gives

$$p_1 = \widetilde{p}_1 > \widetilde{p}_2 > \cdots > \widetilde{p}_m > z. \tag{4.15}$$

Step 2. We construct a family of auxiliary processes $\boldsymbol{x}^0(t|s, \boldsymbol{x})$ ($t \ge s \ge 0$ and $\boldsymbol{x} \in \mathcal{X}$), which will be used to construct the control policy $\boldsymbol{u}(t)$, $t \ge 0$. To do this, consider the function $\boldsymbol{u}^0(\boldsymbol{x}, \widetilde{\boldsymbol{k}}) = (u_1^0(\boldsymbol{x}, \widetilde{\boldsymbol{k}}), \ldots, u_m^0(\boldsymbol{x}, \widetilde{\boldsymbol{k}}))$ with $u_1^0(\boldsymbol{x}, \widetilde{\boldsymbol{k}}) = \widetilde{k}_1$ and

$$u_j^0(\boldsymbol{x}, \widetilde{\boldsymbol{k}}) = \begin{cases} \widetilde{k}_j, & \text{if } x_{j-1} > 0, \\ \widetilde{k}_j \wedge u_{j-1}^0(\boldsymbol{x}, \widetilde{\boldsymbol{k}}), & \text{if } x_{j-1} = 0, \end{cases}$$

$j = 2, \ldots, m$. We define $\boldsymbol{x}^0(t|s, \boldsymbol{x})$, $t \ge 0$, as the process which satisfies the equation

$$\frac{d}{dt}\boldsymbol{x}^0(t|s, \boldsymbol{x}) = A\boldsymbol{u}^0(\boldsymbol{x}^0(t|s, \boldsymbol{x}), \widetilde{\boldsymbol{k}}(t)) + Bz, \quad \boldsymbol{x}^0(s|s, \boldsymbol{x}) = \boldsymbol{x}.$$

Clearly $\boldsymbol{x}^0(t|s, \boldsymbol{x}) \in \mathcal{X}$ for all $t \ge s$. For a fixed s, $\boldsymbol{x}^0(t|s, \boldsymbol{x})$ is the state of the system with the production rate obtained by using the maximum admissible modified capacity at each machine.

Now define the Markov time with respect to the process $\widetilde{\boldsymbol{k}}(\cdot)$:

$$\begin{aligned} \theta(s, \boldsymbol{x}, \overline{\boldsymbol{x}}) = \inf \big\{ t \ge s : {} & x_1^0(t|s, \boldsymbol{x}) \ge \overline{x}_1 \text{ and} \\ & x_j^0(t|s, \boldsymbol{x}) \ge M + \overline{x}_j,\ j = 2, \ldots, m \big\}, \end{aligned} \tag{4.16}$$

where $M > 0$ is a constant specified later. By definition, $\theta(s, \boldsymbol{x}, \overline{\boldsymbol{x}})$ is the first time when the state process $\boldsymbol{x}^0(t|s, \boldsymbol{x})$ exceeds $(\overline{x}_1, M + \overline{x}_2, \ldots, M + \overline{x}_m)'$

under the production rate $\boldsymbol{u}^0(\boldsymbol{x}^0(t|s,\boldsymbol{x}),\widetilde{\boldsymbol{k}}(t))$. Since each machine's modified average capacity is larger than the modified average capacity of the machine that follows it, and since the last machine's modified average capacity is larger than the demand rate z (see (4.13)), intuitively, the following lemma should hold. Its proof is given in Appendix B.

Lemma 3.1. *Let Assumptions* (A2) *and* (A3) *hold. Then there exists a constant* $\widehat{C}_r$ *such that*

$$E\left[\theta(s,\boldsymbol{x},\overline{\boldsymbol{x}})-s\right]^{2r}<\widehat{C}_r\left(1+\sum_{j=1}^{m}\left((\overline{x}_j-x_j)^+\right)^r\right)^2.$$

Step 3. In this step we construct a family of auxiliary processes $\boldsymbol{x}^1(t|s,\boldsymbol{x})$ ($t\geq s\geq 0$, and $\boldsymbol{x}\in\mathcal{X}$) that will be used to construct the control policy $\boldsymbol{u}(t)$. Consider the following function $\boldsymbol{u}^1(\boldsymbol{x},\widetilde{\boldsymbol{k}})=(u_1^1(\boldsymbol{x},\widetilde{\boldsymbol{k}}),\ldots,u_m^1(\boldsymbol{x},\widetilde{\boldsymbol{k}}))$ with $u_1^1(\boldsymbol{x},\widetilde{\boldsymbol{k}})=0$ and

$$u_j^1(\boldsymbol{x},\widetilde{\boldsymbol{k}})=\begin{cases}\widetilde{k}_j, & \text{if } x_{j-1}>\overline{x}_{j-1},\\ \widetilde{k}_j\wedge u_{j-1}^1(\boldsymbol{x},\widetilde{\boldsymbol{k}}), & \text{if } x_{j-1}=\overline{x}_{j-1},\end{cases}$$

for $j=2,\ldots,m$. We define $\boldsymbol{x}^1(t|s,\boldsymbol{x})$, $t\geq 0$, as a continuous process which coincides with $\boldsymbol{x}^0(t|s,\boldsymbol{x})$ for $s\leq t\leq\theta(s,\boldsymbol{x},\overline{\boldsymbol{x}})$, and for $t\geq\theta(s,\boldsymbol{x},\overline{\boldsymbol{x}})$,

$$\frac{d}{dt}\boldsymbol{x}^1(t|s,\boldsymbol{x})=A\boldsymbol{u}^1(\boldsymbol{x}^1(t|s,\boldsymbol{x}),\widetilde{\boldsymbol{k}}(t))+Bz.$$

Clearly, $\boldsymbol{x}^1(t|s,\boldsymbol{x})\in\mathcal{X}$ for all $t\geq s$. Furthermore, for $j=1,\ldots,m-1$,

$$x_j^1(t|s,\boldsymbol{x})\geq\overline{x}_j,\quad t\geq\theta(s,\boldsymbol{x},\overline{\boldsymbol{x}}). \tag{4.17}$$

This process corresponds to a policy in which after the time $\theta(s,\boldsymbol{x},\overline{\boldsymbol{x}})$, we stop producing at the first machine and have the maximum possible production rate at other machines under the restriction that the content of each buffer j ($1\leq j\leq m-1$) is not less than $\overline{x}_j$.

Now define a Markov time

$$\widehat{\theta}(s,\boldsymbol{x},\overline{\boldsymbol{x}})=\inf\left\{t\geq\theta(s,\boldsymbol{x},\overline{\boldsymbol{x}}):x_m^1(t|s,\boldsymbol{x})=\overline{x}_m\right\}. \tag{4.18}$$

Then we have the following result which will be proved in Appendix B.

Lemma 3.2. *Let Assumptions* (A2) *and* (A3) *hold. Then:*

(i) *For a given* $q\in(0,\ 1)$, *a positive constant* M *can be chosen in such a way that, for all* $s\geq 0$ *and* $\boldsymbol{x}\in\mathcal{X}$,

$$P\left((\boldsymbol{x}^1(\widehat{\theta}(s,\boldsymbol{x},\overline{\boldsymbol{x}})|s,\boldsymbol{x}),\ \boldsymbol{k}(\widehat{\theta}(s,\boldsymbol{x},\overline{\boldsymbol{x}})))=(\overline{\boldsymbol{x}},\widetilde{\boldsymbol{k}})\right)\geq 1-q>0.$$

(ii) *There exists a constant* C *such that*

$$\frac{M}{z} \le \widehat{\theta}(s, \boldsymbol{x}, \overline{\boldsymbol{x}}) - s \le \frac{1}{z}\left(\sum_{j=1}^{m}(x_j - \overline{x}_j) + C[\theta(s, \boldsymbol{x}, \overline{\boldsymbol{x}}) - s]\right) \tag{4.19}$$

and

$$\sum_{j=1}^{m} x_j^1(\widehat{\theta}(s, \boldsymbol{x}, \overline{\boldsymbol{x}})|s, \boldsymbol{x}) \le \sum_{j=1}^{m} x_j + C[\theta(s, \boldsymbol{x}, \overline{\boldsymbol{x}}) - s]. \tag{4.20}$$

Step 4. Now we construct a process $\boldsymbol{x}(t)$, $t \ge 0$, and the corresponding control policy $\boldsymbol{u}(t)$ which satisfy the statement of Theorem 3.2.

Define a sequence of Markov times $\{\widehat{\theta}_\ell\}_{\ell=0}^{\infty}$ with respect to $\widetilde{\boldsymbol{k}}(\cdot)$ and the process $\boldsymbol{x}(t)$ for $\widehat{\theta}_\ell \le t < \widehat{\theta}_{\ell+1}$ $(\ell = 1, 2, \ldots)$ as follows:

$$\widehat{\theta}_0 = 0, \ \widehat{\theta}_1 = \widehat{\theta}(0, \widehat{\boldsymbol{x}}, \overline{\boldsymbol{x}})$$

and

$$\boldsymbol{x}(t) = \boldsymbol{x}^1(t|0, \widehat{\boldsymbol{x}}) \quad \text{with } 0 \le t < \widehat{\theta}_1.$$

If $\widehat{\theta}_\ell$ is defined for $\ell \ge 1$ and $\boldsymbol{x}(t)$ is defined for $0 \le t < \widehat{\theta}_\ell$, then we let

$$\widehat{\theta}_{\ell+1} = \widehat{\theta}(\widehat{\theta}_\ell, \boldsymbol{x}(\widehat{\theta}_\ell), \overline{\boldsymbol{x}}) \quad \text{and} \quad \boldsymbol{x}(t) = \boldsymbol{x}^1(t|\widehat{\theta}_\ell, \boldsymbol{x}(\widehat{\theta}_\ell)) \quad \text{with} \quad \widehat{\theta}_\ell \le t < \widehat{\theta}_{\ell+1}.$$

According to the left inequality in (4.19), $\boldsymbol{x}(t)$ is defined for all $t \ge 0$. Let

$$\theta_1 = \theta(0, \widehat{\boldsymbol{x}}, \overline{\boldsymbol{x}}) \quad \text{and} \quad \theta_{\ell+1} = \theta(\widehat{\theta}_\ell, \boldsymbol{x}(\widehat{\theta}_\ell), \overline{\boldsymbol{x}}), \quad \ell = 1, 2, \ldots.$$

Define the control $\boldsymbol{u}(\cdot)$ as

$$\boldsymbol{u}(t) = \begin{cases} \boldsymbol{u}^0(\boldsymbol{x}(t), \widetilde{\boldsymbol{k}}(t)), & \text{if } \widehat{\theta}_{\ell-1} \le t < \theta_\ell, \\ \boldsymbol{u}^1(\boldsymbol{x}(t), \widetilde{\boldsymbol{k}}(t)), & \text{if } \theta_\ell \le t < \widehat{\theta}_\ell, \end{cases} \tag{4.21}$$

for $\ell = 1, 2, \ldots$. Let $\boldsymbol{x}(\cdot)$ be the corresponding state process. It is clear that $\boldsymbol{u}(\cdot) \in \mathcal{A}(\widehat{\boldsymbol{x}}, \widehat{\boldsymbol{k}})$.

For the process $\boldsymbol{x}(\cdot)$, a Markov time is defined as

$$\tau(\widehat{\boldsymbol{x}}, \widehat{\boldsymbol{k}}, \overline{\boldsymbol{x}}, \overline{\boldsymbol{k}}) = \inf\left\{t \ge 0 \ : \ (\boldsymbol{x}(t), \boldsymbol{k}(t)) = (\overline{\boldsymbol{x}}, \overline{\boldsymbol{k}})\right\}.$$

Let

$$\mathcal{B}_\ell = \{\omega : (\boldsymbol{x}(\widehat{\theta}_\ell)(\omega), \boldsymbol{k}(\widehat{\theta}_\ell)(\omega)) = (\overline{\boldsymbol{x}}, \overline{\boldsymbol{k}})\}.$$

Using conditional probabilities (see (3.15)), we have from Lemma 3.2(i) that

$$P\left(\bigcap_{\ell_1=1}^{\ell} \mathcal{B}_{\ell_1}^c\right) \le q^\ell, \quad \ell = 1, 2, \ldots, \tag{4.22}$$

where $\mathcal{B}_{\ell_1}^c$ is the complement set of $\mathcal{B}_{\ell_1}$. Using (4.21) and the definition of $\boldsymbol{x}(t)$ we get that

$$\tau(\widehat{\boldsymbol{x}}, \widehat{\boldsymbol{k}}, \overline{\boldsymbol{x}}, \overline{\boldsymbol{k}}) = \sum_{\ell=1}^{\infty} \widehat{\theta}_\ell I_{\{\cap_{\ell_1=0}^{\ell-1} \mathcal{B}_{\ell_1}^c \cap \mathcal{B}_\ell\}},$$

which implies

$$(\tau(\widehat{\boldsymbol{x}}, \widehat{\boldsymbol{k}}, \overline{\boldsymbol{x}}, \overline{\boldsymbol{k}}))^r = \sum_{\ell=1}^{\infty} (\widehat{\theta}_\ell)^r I_{\{\cap_{\ell_1=0}^{\ell-1} \mathcal{B}_{\ell_1}^c \cap \mathcal{B}_\ell\}}, \tag{4.23}$$

where $\mathcal{B}_0^c = \Omega$. Using (4.19) and (4.20) we have that, for $\ell = 1, 2, \ldots$,

$$\widehat{\theta}_\ell - \widehat{\theta}_{\ell-1} \le \frac{1}{z} \left(\sum_{j=1}^{m} x_j(\widehat{\theta}_{\ell-1}) - \sum_{j=1}^{m} \overline{x}_j + C[\theta_\ell - \widehat{\theta}_{\ell-1}] \right) \tag{4.24}$$

and

$$\sum_{j=1}^{m} x_j(\widehat{\theta}_\ell) \le \sum_{j=1}^{m} x_j(\widehat{\theta}_{\ell-1}) + C[\theta_\ell - \widehat{\theta}_{\ell-1}]. \tag{4.25}$$

Using (4.24) and (4.25), there exists a constant $C_1 > 0$ independent of $\widehat{\boldsymbol{x}}$ and $\overline{\boldsymbol{x}}$ such that, for $\ell = 1, 2, \ldots$,

$$\widehat{\theta}_\ell \le \frac{C_1(\ell+1)}{z} \left[\sum_{j=1}^{m} (\widehat{x}_j - \overline{x}_j)^+ + \sum_{\ell_1=1}^{\ell} (\theta_{\ell_1} - \widehat{\theta}_{\ell_1-1}) \right].$$

This implies that

$$(\widehat{\theta}_\ell)^r \le C_2(\ell+1)^{2r} \left[\sum_{j=1}^{m} \left((\widehat{x}_j - \overline{x}_j)^+ \right)^r + \sum_{\ell_1=1}^{\ell} (\theta_{\ell_1} - \widehat{\theta}_{\ell_1-1})^r \right], \tag{4.26}$$

for some $C_2 > 0$. By (4.17) and (4.18), we know that $\boldsymbol{x}(\widehat{\theta}_\ell) \ge \overline{\boldsymbol{x}}$ for $\ell = 1, 2, \ldots$. Using the Schwarz inequality (Corollary 3 on page 104 of Chow and Teicher [33]), we get from (4.22) and Lemma 3.1 that there exists a positive constant $\overline{C}_r$ dependent on r such that

$$\begin{aligned} E\left(\theta_1^r I_{\{\cap_{\ell_1=1}^{\ell-1} \mathcal{B}_{\ell_1}^c \cap \mathcal{B}_\ell\}}\right) &\le q^{(\ell-1)/2} \left(E[\theta_1^{2r}]\right)^{1/2} \\ &\le \overline{C}_r q^{(\ell-1)/2} \left(1 + \sum_{j=1}^{m} \left((\overline{x}_j - \widehat{x}_j)^+ \right)^r \right), \qquad (4.27) \\ &\qquad \ell = 1, 2, \ldots, \end{aligned}$$

and

$$
\begin{aligned}
E&\left((\theta_{\ell_1}-\widehat{\theta}_{\ell_1-1})^r I_{\{\cap_{\ell_2=1}^{\ell-1}\mathcal{B}_{\ell_2}^c\cap\mathcal{B}_\ell\}}\right)\\
&\le q^{(\ell-1)/2}\left(E(\theta_{\ell_1}-\widehat{\theta}_{\ell_1-1})^{2r}\right)^{1/2}\\
&\le \overline{C}_r q^{(\ell-1)/2}\left(1+\sum_{j=1}^m((\overline{x}_j-\widehat{x}_j)^+)^r\right),\\
&\qquad 2\le \ell_1\le \ell=2,3,\ldots.
\end{aligned}
\tag{4.28}
$$

Substituting (4.26) into (4.23), taking expectation, and using (4.27) and (4.28), we get (4.12). This completes the proof. □

4.4 Verification Theorem and HJBDD Equations

Our goal is to construct a pair $(\lambda, W(\cdot,\cdot))$ which satisfies (4.7). To get this pair, we use the vanishing discount approach. First, based on Theorem 3.2, we have the following convergence results.

Theorem 4.1. *Let Assumptions* (A1)–(A3) *hold. There exists a sequence* $\{\rho_\ell : \ell \ge 1\}$ *with* $\rho_\ell \to 0$ *as* $\ell \to \infty$ *such that for* $(\boldsymbol{x},\boldsymbol{k}) \in \mathcal{X}\times\mathcal{M}$, *the limits of* $\rho_\ell V^{\rho_\ell}(\boldsymbol{x},\boldsymbol{k})$ *and* $[V^{\rho_\ell}(\boldsymbol{x},\boldsymbol{k}) - V^{\rho_\ell}(0,\boldsymbol{k})]$ *exist as* $\ell\to\infty$. *Write*

$$
\widehat{\lambda} = \lim_{\ell\to\infty} \rho_\ell V^{\rho_\ell}(\boldsymbol{x},\boldsymbol{k}), \tag{4.29}
$$

$$
V(\boldsymbol{x},\boldsymbol{k}) = \lim_{\ell\to\infty}[V^{\rho_\ell}(\boldsymbol{x},\boldsymbol{k}) - V^{\rho_\ell}(0,\boldsymbol{k})]. \tag{4.30}
$$

The limit $V(\boldsymbol{x},\boldsymbol{k})$ *is convex for any fixed* $\boldsymbol{k}\in\mathcal{M}$.

Proof. For the value function $V^\rho(\boldsymbol{x},\boldsymbol{k})$ of the discounted cost problem, we define the difference

$$
\widetilde{V}^\rho(\boldsymbol{x},\boldsymbol{k}) = V^\rho(\boldsymbol{x},\boldsymbol{k}) - V^\rho(0,\boldsymbol{k}).
$$

By Theorem 3.2 we know that there exists a control $\boldsymbol{u}(\cdot)\in\mathcal{A}(\boldsymbol{x},\boldsymbol{k})$ such that, for each $r\ge 1$,

$$
E[\tau(\boldsymbol{x},\boldsymbol{k},\boldsymbol{x},\boldsymbol{k})]^r \le C_1, \tag{4.31}
$$

where $C_1 > 0$ is a constant (which depends on r) and

$$
\tau(\boldsymbol{x},\boldsymbol{k},\boldsymbol{x},\boldsymbol{k}) = \inf\{t>0 : (\boldsymbol{x}(t),\boldsymbol{k}(t)) = (\boldsymbol{x},\boldsymbol{k})\},
$$

with $\boldsymbol{x}(\cdot)$ being the surplus process corresponding to the control $\boldsymbol{u}(\cdot)$ and the initial condition $(\boldsymbol{x}(0),\boldsymbol{k}(0)) = (\boldsymbol{x},\boldsymbol{k})$. For notational simplicity, in what follows we write $\tau(\boldsymbol{x},\boldsymbol{k},\boldsymbol{x},\boldsymbol{k})$ as τ.

By the optimality principle, we have

$$\begin{aligned}V^\rho(\boldsymbol{x},\boldsymbol{k}) &\le E\left\{\int_0^\tau e^{-\rho t} g(\boldsymbol{x}(t),\boldsymbol{u}(t))\,dt + e^{-\rho\tau}V^\rho(\boldsymbol{x}(\tau),\boldsymbol{k}(\tau))\right\}\\ &= E\left\{\int_0^\tau e^{-\rho t} g(\boldsymbol{x}(t),\boldsymbol{u}(t))\,dt + e^{-\rho\tau}V^\rho(\boldsymbol{x},\boldsymbol{k})\right\}.\end{aligned} \tag{4.32}$$

Note that

$$|\boldsymbol{x}(t)| \le \sum_{j=1}^m |x_j| + \left(\sum_{j=1}^m \max_{1\le i\le p} k_j^i + z\right) t, \quad \text{for } 0 \le t \le \tau.$$

Thus, by Assumption (A1),

$$g(\boldsymbol{x}(t),\boldsymbol{u}(t)) \le C_2\left(1 + t + t^{\beta_{g2}}\right),$$

where C_2 is a positive constant dependent on $\boldsymbol{x}$. It follows from (4.31) and (4.32) that

$$[1 - Ee^{-\rho\tau}]V^\rho(\boldsymbol{x},\boldsymbol{k}) \le E\int_0^\tau C_2\left(1+t+t^{\beta_{g2}}\right)dt \le C_3, \tag{4.33}$$

for some positive constant C_3 (independent of ρ). Now using the inequality $1 - e^{-\rho\tau} \ge \rho\tau - \rho^2\tau^2/2$, we can get

$$\left[1 - Ee^{-\rho\tau}\right]V^\rho(\boldsymbol{x},\boldsymbol{k}) \ge \rho\left(E\tau - \rho E[\tau^2]/2\right)V^\rho(\boldsymbol{x},\boldsymbol{k}). \tag{4.34}$$

From the definition of the stopping time $\tau(\boldsymbol{x},\boldsymbol{k},\boldsymbol{x},\boldsymbol{k})$, we have

$$0 < E\tau < \infty \quad \text{and} \quad 0 < E\tau^2 < \infty.$$

Take $\rho_0 = E\tau/E\tau^2$. By (4.33) and (4.34), we have that, for $0 < \rho \le \rho_0$,

$$\rho V^\rho(\boldsymbol{x},\boldsymbol{k}) \le \frac{C_3}{E\tau - \rho_0 E[\tau^2]/2} = \frac{2C_3}{E\tau} < \infty.$$

Consequently, there exists a sequence $\{\rho_\ell : \ell \ge 1\}$ with $\rho_\ell \to 0$ as $\ell \to \infty$ such that, for $(\boldsymbol{x},\boldsymbol{k}) \in \mathcal{X}\times\mathcal{M}$,

$$\lim_{\ell\to\infty} \rho_\ell V^{\rho_\ell}(\boldsymbol{x},\boldsymbol{k}) = \widehat{\lambda}. \tag{4.35}$$

Now we prove (4.30). To do this we first show that there is a constant $C_4 > 0$ such that

$$|\widetilde{V}^\rho(\boldsymbol{x},\boldsymbol{k})| \le C_4\left(1 + |\boldsymbol{x}|^{\beta_{g2}+1}\right), \tag{4.36}$$

for all $(\boldsymbol{x},\boldsymbol{k}) \in \mathcal{X}\times\mathcal{M}$ and $\rho > 0$. Without loss of generality, we suppose that $V^\rho(\boldsymbol{x},\boldsymbol{k}) \ge V^\rho(0,\boldsymbol{k})$ (the case $V^\rho(\boldsymbol{x},\boldsymbol{k}) \le V^\rho(0,\boldsymbol{k})$ is treated in the same way).

By Theorem 3.2 there exists a control process $\boldsymbol{u}(\cdot)$ such that, for any $r \geq 1$,

$$E(\tau_1)^r \leq C_r \left(1 + \sum_{j=1}^{m} |x_j|^r \right), \tag{4.37}$$

where

$$\tau_1 = \inf\{t > 0 : (\boldsymbol{x}(t), \boldsymbol{k}(t)) = (0, \boldsymbol{k})\},$$

and $\boldsymbol{x}(\cdot)$ is the state process corresponding to $\boldsymbol{u}(\cdot)$ with the initial condition $(\boldsymbol{x}(0), \boldsymbol{k}(0)) = (\boldsymbol{x}, \boldsymbol{k})$. Using the optimality principle, we have

$$V^\rho(\boldsymbol{x}, \boldsymbol{k}) \leq E\left\{ \int_0^{\tau_1} e^{-\rho t} g(\boldsymbol{x}(t), \boldsymbol{u}(t))\, dt + e^{-\rho\tau_1} V^\rho(0, \boldsymbol{k}) \right\}.$$

Therefore,

$$\begin{aligned} |\widetilde{V}^\rho(\boldsymbol{x}, \boldsymbol{k})| &= |V^\rho(\boldsymbol{x}, \boldsymbol{k}) - V^\rho(0, \boldsymbol{k})| \\ &\leq E\left[\int_0^{\tau_1} e^{-\rho t} g(\boldsymbol{x}(t), \boldsymbol{u}(t))\, dt \right]. \end{aligned} \tag{4.38}$$

By Assumption (A1), there exists a constant $C_5 > 0$ such that

$$g(\boldsymbol{x}(t), \boldsymbol{u}(t)) \leq C_5 \left(1 + |\boldsymbol{x}|^{\beta_{g2}} + t + t^{\beta_{g2}} \right), \tag{4.39}$$

where we use the fact that $\boldsymbol{u}(t)$ is bounded. Therefore, (4.37) and (4.39) imply that

$$\begin{aligned} E &\int_0^{\tau_1} e^{-\rho t} g(\boldsymbol{x}(t), \boldsymbol{u}(t))\, dt \\ &\leq E \int_0^{\tau_1} C_5 (1 + |\boldsymbol{x}|^{\beta_{g2}} + t + t^{\beta_{g2}})\, dt \\ &\leq C_6 \left(1 + \sum_{j=1}^{m} |x_j|^{\beta_{g2}+1} \right) \end{aligned}$$

for some $C_6 > 0$. Thus (4.38) gives (4.36).

For $\delta \in (0,\ 1)$, let

$$\mathcal{O}_\delta = \left[\delta,\ \frac{1}{\delta} \right]^{m-1} \times \left[-\frac{1}{\delta},\ \frac{1}{\delta} \right].$$

Based on (4.36) it follows from Lemma C.4 that there is a $C(\delta)$ such that, for $\boldsymbol{x}, \widehat{\boldsymbol{x}} \in \mathcal{O}_\delta$,

$$|\widetilde{V}^\rho(\boldsymbol{x}, \boldsymbol{k}) - \widetilde{V}^\rho(\widehat{\boldsymbol{x}}, \boldsymbol{k})| \leq C_\delta |\boldsymbol{x} - \widehat{\boldsymbol{x}}|. \tag{4.40}$$

Without loss of generality, we assume that C_δ is decreasing in δ. For $1 \leq \hat{n} \leq m-1$ and $1 \leq j_1 < \cdots < j_{\hat{n}} \leq m-1$, let

$$\mathcal{O}^{j_1 \cdots j_{\hat{n}}} = \{\boldsymbol{x} \in \mathcal{X} : x_{j_\ell} = 0 \text{ for } \ell = 1, ..., \hat{n}\},$$

and

$$\mathrm{ri}(\mathcal{O}^{j_1\cdots j_{\hat{n}}}) = \{\boldsymbol{x} \in \mathcal{O}^{j_1\cdots j_{\hat{n}}} : x_j > 0,\ j \neq j_\ell \text{ and } 1 \leq \ell \leq \hat{n}\}.$$

That is, $\mathrm{ri}(\mathcal{O}^{j_1\cdots j_{\hat{n}}})$ is the *relative interior* of $\mathcal{O}^{j_1\cdots j_{\hat{n}}}$ relative to $[0,\infty)^{m-\hat{n}-1}\times\{0\}^{\hat{n}} \times (-\infty,\ +\infty)$. Note that the function $V^\rho(\boldsymbol{x},\boldsymbol{k})$ is convex also on $\mathcal{O}^{j_1\cdots j_{\hat{n}}}$. Let

$$\mathcal{O}_\delta^{j_1\cdots j_{\hat{n}}} = \prod_{\ell=1}^{m-1} \Upsilon_\ell^\delta \times \left[-\frac{1}{\delta},\ \frac{1}{\delta}\right]$$

with

$$\Upsilon_\ell^\delta = \begin{cases} \{0\}, & \text{if } \ell \in \{j_1, ..., j_{\hat{n}}\}, \\ [\delta,\ 1/\delta], & \text{if } \ell \notin \{j_1, ..., j_{\hat{n}}\}. \end{cases}$$

Using once again Lemma C.4, in view of (4.36), there is a $C_\delta^{j_1\cdots j_{\hat{n}}} > 0$ such that, for $\boldsymbol{x}, \widehat{\boldsymbol{x}} \in \mathcal{O}_\delta^{j_1\cdots j_{\hat{n}}}$,

$$|\widetilde{V}^\rho(\boldsymbol{x},\boldsymbol{k}) - \widetilde{V}^\rho(\widehat{\boldsymbol{x}},\boldsymbol{k})| \leq C_\delta^{j_1\cdots j_{\hat{n}}} |\boldsymbol{x} - \widehat{\boldsymbol{x}}|. \tag{4.41}$$

Also we assume that $C_\delta^{j_1\cdots j_{\hat{n}}}$ is a decreasing function of δ. From the arbitrariness of δ and (4.40)–(4.41), there exist $V(\boldsymbol{x},\boldsymbol{k})$ and a sequence of $\{\rho_\ell : \ell \geq 1\}$ with $\rho_\ell \to 0$ as $\ell \to \infty$ such that, for $(\boldsymbol{x},\boldsymbol{k}) \in \mathcal{X} \times \mathcal{M}$,

$$\lim_{\ell\to\infty} [V^{\rho_\ell}(\boldsymbol{x},\boldsymbol{k}) - V^{\rho_\ell}(0,\boldsymbol{k})] = V(\boldsymbol{x},\boldsymbol{k}). \tag{4.42}$$

Moreover, it follows from the convexity of $V^{\rho_\ell}(\boldsymbol{x},\boldsymbol{k})$ that the limit function $V(\boldsymbol{x},\boldsymbol{k})$ is also convex on $\mathcal{X} \times \mathcal{M}$. Therefore, the proof of the theorem is completed. □

Let $\partial V^{\rho_\ell}(\boldsymbol{x},\boldsymbol{k})/\partial \boldsymbol{x}$ be the derivative of $V^{\rho_\ell}(\boldsymbol{x},\boldsymbol{k})$ at the point $\boldsymbol{x}$ where the derivative exists.

Theorem 4.2. *Let Assumptions* (A1)–(A3) *hold. Then:*

(i) $\lambda^*(\boldsymbol{x},\boldsymbol{k})$ *defined by* (4.5) *does not depend on* $(\boldsymbol{x},\boldsymbol{k})$.

(ii) *The pair* $(\lambda, V(\boldsymbol{x},\boldsymbol{k}))$ *defined in Theorem* 4.1 *satisfies the* HJBDD *equation* (4.7) *in the interior of* $\mathcal{X}$.

(iii) *If there exists an open subset* $\widehat{\mathcal{X}}$ *of* $\mathcal{X}$ *such that* $b(\mathcal{X}) \subseteq b(\widehat{\mathcal{X}})$, *where* $b(\widehat{\mathcal{X}})$ *and* $b(\mathcal{X})$ *are the boundaries of* $\widehat{\mathcal{X}}$ *and* $\mathcal{X}$, *respectively, and* $\{\partial V^{\rho_\ell}(\boldsymbol{x},\boldsymbol{k})/\partial \boldsymbol{x} : \ell \geq 1\}$ *is uniformly equi-Lipschitz continuous on* $\widehat{\mathcal{X}}$, *then the pair* $(\widehat{\lambda}, V(\cdot,\cdot))$ *defined in Theorem* 4.1 *is a solution of the* HJBDD *equation* (4.7).

Proof. First we consider (i). Suppose to the contrary that there exist $(\widetilde{\boldsymbol{x}}, \widetilde{\boldsymbol{k}}) \in \mathcal{X} \times \mathcal{M}$ and $(\widehat{\boldsymbol{x}}, \widehat{\boldsymbol{k}}) \in \mathcal{X} \times \mathcal{M}$ such that

$$\lambda^*(\widetilde{\boldsymbol{x}}, \widetilde{\boldsymbol{k}}) < \lambda^*(\widehat{\boldsymbol{x}}, \widehat{\boldsymbol{k}}). \tag{4.43}$$

We choose $\delta > 0$ and a control $\widetilde{\boldsymbol{u}}(\cdot) \in \mathcal{A}(\widetilde{\boldsymbol{x}}, \widetilde{\boldsymbol{k}})$ such that

$$\lambda^*(\widetilde{\boldsymbol{x}}, \widetilde{\boldsymbol{k}}) + \delta < \lambda^*(\widehat{\boldsymbol{x}}, \widehat{\boldsymbol{k}}) \tag{4.44}$$

and

$$\limsup_{T\to\infty} \frac{1}{T} E \int_0^T g(\widetilde{\boldsymbol{x}}(t), \widetilde{\boldsymbol{u}}(t))\, dt \le \lambda^*(\widetilde{\boldsymbol{x}}, \widetilde{\boldsymbol{k}}) + \delta, \tag{4.45}$$

where

$$\frac{d}{dt}\widetilde{\boldsymbol{x}}(t) = A\widetilde{\boldsymbol{u}}(t) + Bz, \quad \widetilde{\boldsymbol{x}}(0) = \widetilde{\boldsymbol{x}}.$$

Let $\boldsymbol{u}(t)$, $t \ge 0$, be the one given in Theorem 3.2, and let

$$\tau(\widehat{\boldsymbol{x}}, \widehat{\boldsymbol{k}}, \widetilde{\boldsymbol{x}}, \widetilde{\boldsymbol{k}}) = \inf\left\{t \ge 0 : (\boldsymbol{x}(t), \boldsymbol{k}(t)) = (\widetilde{\boldsymbol{x}}, \widetilde{\boldsymbol{k}})\right\},$$

where

$$\frac{d}{dt}\boldsymbol{x}(t) = A\boldsymbol{u}(t) + Bz, \quad \boldsymbol{x}(0) = \widehat{\boldsymbol{x}}.$$

We define

$$\widehat{\boldsymbol{u}}(t) = \begin{cases} \boldsymbol{u}(t), & \text{if } t \le \tau(\widehat{\boldsymbol{x}}, \widehat{\boldsymbol{k}}, \widetilde{\boldsymbol{x}}, \widetilde{\boldsymbol{k}}), \\ \widetilde{\boldsymbol{u}}(t - \tau(\widehat{\boldsymbol{x}}, \widehat{\boldsymbol{k}}, \widetilde{\boldsymbol{x}}, \widetilde{\boldsymbol{k}})), & \text{if } t > \tau(\widehat{\boldsymbol{x}}, \widehat{\boldsymbol{k}}, \widetilde{\boldsymbol{x}}, \widetilde{\boldsymbol{k}}). \end{cases}$$

Let $\widehat{\boldsymbol{x}}(\cdot)$ be the state process corresponding to the control $\widehat{\boldsymbol{u}}(\cdot)$. It directly follows from Theorem 3.2 that

$$\limsup_{T\to\infty} \frac{1}{T} E \int_0^T g(\widehat{\boldsymbol{x}}(t), \widehat{\boldsymbol{u}}(t))\, dt = \limsup_{T\to\infty} \frac{1}{T} E \int_0^T g(\widetilde{\boldsymbol{x}}(t), \widetilde{\boldsymbol{u}}(t))\, dt. \tag{4.46}$$

On the other hand,

$$\limsup_{T\to\infty} \frac{1}{T} E \int_0^T g(\widehat{\boldsymbol{x}}(t), \widehat{\boldsymbol{u}}(t))\, dt \ge \lambda^*(\widehat{\boldsymbol{x}}, \widehat{\boldsymbol{k}}). \tag{4.47}$$

Thus from (4.46), (4.47) contradicts (4.45). Consequently, (i) is proved.

Now we prove (ii). Let $\mathcal{O}^m$ be the set of all points in the interior of $\mathcal{X}$ on which $V(\boldsymbol{x}, \boldsymbol{k})$ is differentiable. From the convexity of $V(\boldsymbol{x}, \boldsymbol{k})$ we know that $\mathcal{O}^m$ is dense in $\mathcal{X}$. It follows from the properties of convex functions that, for $\boldsymbol{x} \in \mathcal{O}^m$ and for any $\boldsymbol{p}$,

$$\lim_{\ell\to\infty} \partial_{\boldsymbol{p}} \widetilde{V}^{\rho_\ell}(\boldsymbol{x}, \boldsymbol{k}) = \partial_{\boldsymbol{p}} V(\boldsymbol{x}, \boldsymbol{k}). \tag{4.48}$$

Using Theorem 3.1, we have

$$\begin{aligned}\rho_\ell V^{\rho_\ell}(\boldsymbol{x},\boldsymbol{k}) = & \inf_{\boldsymbol{u}\in\mathcal{U}(\boldsymbol{x},\boldsymbol{k})}\{\partial_{A\boldsymbol{u}+Bz}V^{\rho_\ell}(\boldsymbol{x},\boldsymbol{k})+g(\boldsymbol{x},\boldsymbol{u})\}\\ & +QV^{\rho_\ell}(\boldsymbol{x},\cdot)(\boldsymbol{k}).\end{aligned}$$

This implies that

$$\begin{aligned}\rho_\ell V^{\rho_\ell}(\boldsymbol{x},\boldsymbol{k}) = & \inf_{\boldsymbol{u}\in\mathcal{U}(\boldsymbol{x},\boldsymbol{k})}\left\{\partial_{A\boldsymbol{u}+Bz}\widetilde{V}^{\rho_\ell}(\boldsymbol{x},\boldsymbol{k})+g(\boldsymbol{x},\boldsymbol{u})\right\}\\ & +Q\widetilde{V}^{\rho_\ell}(\boldsymbol{x},\cdot)(\boldsymbol{k}).\end{aligned} \tag{4.49}$$

Taking the limit on both sides, we have that, for $\boldsymbol{x}\in\mathcal{O}^m$,

$$\widehat{\lambda} = \inf_{\boldsymbol{u}\in\mathcal{U}(\boldsymbol{x},\boldsymbol{k})}\{\partial_{A\boldsymbol{u}+Bz}V(\boldsymbol{x},\boldsymbol{k})+g(\boldsymbol{x},\boldsymbol{u})\}+QV(\boldsymbol{x},\cdot)(\boldsymbol{k}). \tag{4.50}$$

If $\boldsymbol{x}\notin\mathcal{O}^m$ but $\boldsymbol{x}$ is an interior point of $\mathcal{X}$, then for any direction $\boldsymbol{p}$ it follows from Lemma C.2 that there exist a sequence $\{\boldsymbol{x}_n\}_{n=1}^\infty$ such that $\boldsymbol{x}_n\in\mathcal{O}^m$ and $\partial_{\boldsymbol{p}}V(\boldsymbol{x}_n,\boldsymbol{k})\to\partial_{\boldsymbol{p}}V(\boldsymbol{x},\boldsymbol{k})$. Consequently, it follows from the continuity of $V(\boldsymbol{x},\boldsymbol{k})$ that (4.50) holds for all $\boldsymbol{x}$ in the interior part of $\mathcal{X}$. Hence we get (ii).

Finally we prove (iii). Consider now the boundary $b(\mathcal{X})$ of $\mathcal{X}$. From the uniform equi-Lipschitzian property of $\{\partial V^{\rho_\ell}(\boldsymbol{x},\boldsymbol{k})/\partial\boldsymbol{x} : \ell\geq 1\}$ on $\widehat{\mathcal{X}}$, we know that (4.48) holds for all $\boldsymbol{x}\in b(\mathcal{X})$. Therefore, we have (4.50) in $b(\mathcal{X})$. Thus by (ii), the proof of (iii) is completed. □

Finally we establish the following verification theorem, which explains why equation (4.7) is the HJBDD equation for our problem.

Theorem 4.3. *Let Assumptions* (A1)–(A3) *hold, and let* $(\lambda, W(\boldsymbol{x},\boldsymbol{k}))$ *be a solution to the* HJBDD *equation* (4.7). *Then:*

(i) *For any* $\boldsymbol{u}(\cdot)\in\mathcal{A}(\boldsymbol{x},\boldsymbol{k})$, *we have* $\lambda\leq J(\boldsymbol{x},\boldsymbol{k},\boldsymbol{u}(\cdot))$, *i.e.,*

$$\limsup_{T\to\infty}\frac{1}{T}E\int_0^T g(\boldsymbol{x}(t),\boldsymbol{u}(t))\,dt\geq\lambda.$$

(ii) *Furthermore, for any (stable) control policy* $\boldsymbol{u}(\cdot)\in\mathcal{S}(\boldsymbol{x},\boldsymbol{k})$, *we have*

$$\liminf_{T\to\infty}\frac{1}{T}E\int_0^T g(\boldsymbol{x}(t),\boldsymbol{u}(t))\,dt\geq\lambda. \tag{4.51}$$

(iii) *If there is a control* $\boldsymbol{u}^*(\cdot)\in\mathcal{A}(\boldsymbol{x},\boldsymbol{k})$ *such that*

$$\begin{aligned}& \inf_{\boldsymbol{u}\in\mathcal{U}(\boldsymbol{k}(t))}\{\partial_{A\boldsymbol{u}+Bz}W(\boldsymbol{x}^*(t),\boldsymbol{k}(t))+g(\boldsymbol{x}^*(t),\boldsymbol{u})\}\\ & = \partial_{A\boldsymbol{u}^*(t)+Bz}W(\boldsymbol{x}^*(t),\boldsymbol{k}(t))+g(\boldsymbol{x}^*(t),\boldsymbol{u}^*(t))\end{aligned} \tag{4.52}$$

for a.e. $t \geq 0$ *with probability* 1, *where* $\boldsymbol{x}^*(\cdot)$ *is the surplus process corresponding to the control* $\boldsymbol{u}^*(\cdot)$, *and*

$$\lim_{T\to\infty} \frac{E[W(\boldsymbol{x}^*(T), \boldsymbol{k}(T))]}{T} = 0, \tag{4.53}$$

then

$$\lambda = J(\boldsymbol{x}, \boldsymbol{k}, \boldsymbol{u}^*(\cdot)).$$

Proof. We first prove (iii). Since $(\lambda, W(\cdot,\cdot))$ is a solution of (4.7) and $(\boldsymbol{x}^*(t), \boldsymbol{u}^*(t))$ satisfies condition (4.52), we have

$$\begin{aligned} &\partial_{A\boldsymbol{u}^*(t)+Bz} W(\boldsymbol{x}^*(t), \boldsymbol{k}(t)) + QW(\boldsymbol{x}^*(t), \cdot)(\boldsymbol{k}(t)) \\ &\quad = \lambda - g(\boldsymbol{x}^*(t), \boldsymbol{u}^*(t)). \end{aligned} \tag{4.54}$$

Since $W(\boldsymbol{x}, \boldsymbol{k}) \in \mathcal{G}$, we apply Dynkin's formula and use (4.54) to get

$$\begin{aligned} E&\left[W(\boldsymbol{x}^*(T), \boldsymbol{k}(T))\right] \\ &= W(\boldsymbol{x}, \boldsymbol{k}) + E\int_0^T \left[\partial_{A\boldsymbol{u}^*(t)+Bz} W(\boldsymbol{x}^*(t), \boldsymbol{k}(t))\right. \\ &\qquad\qquad \left. + QW(\boldsymbol{x}^*(t), \cdot)(\boldsymbol{k}(t))\right] dt \\ &= W(\boldsymbol{x}, \boldsymbol{k}) + E\int_0^T [\lambda - g(\boldsymbol{x}^*(t), \boldsymbol{u}^*(t)] \, dt \\ &= W(\boldsymbol{x}, \boldsymbol{k}) + \lambda T - E\int_0^T g(\boldsymbol{x}^*(t), \boldsymbol{u}^*(t)) \, dt. \end{aligned} \tag{4.55}$$

We can rewrite (4.55) as

$$\begin{aligned} \lambda = &\frac{1}{T}\left[E(W(\boldsymbol{x}^*(T), \boldsymbol{k}(T))) - W(\boldsymbol{x}, \boldsymbol{k})\right] \\ &+ \frac{1}{T} E\int_0^T g(\boldsymbol{x}^*(t), \boldsymbol{u}^*(t)) \, dt. \end{aligned} \tag{4.56}$$

Using (4.56) and taking the limit as $T \to \infty$, we get

$$\lambda = \limsup_{T\to\infty} \frac{1}{T} E\int_0^T g(\boldsymbol{x}^*(t), \boldsymbol{u}^*(t)) \, dt.$$

For the proof of part (ii), if $\boldsymbol{u}(\cdot) \in \mathcal{S}(\boldsymbol{x}, \boldsymbol{k})$, then from $W(\boldsymbol{x}, \boldsymbol{k}) \in \mathcal{G}$, we know that

$$\lim_{T\to\infty} \frac{E[W(\boldsymbol{x}(T), \boldsymbol{k}(T))]}{T} = 0.$$

Moreover, from the HJBDD equation (4.7) we have

$$\begin{aligned} &\partial_{(A\boldsymbol{u}(t)+Bz)} W(\boldsymbol{x}(t), \boldsymbol{k}(t)) + QW(\boldsymbol{x}(t), \cdot)(\boldsymbol{k}(t)) \\ &\quad \geq \lambda - g(\boldsymbol{x}(t), \boldsymbol{u}(t)). \end{aligned}$$

Now (4.51) can be proved similarly as before.

Finally, we apply Lemma F.4 to show part (i). Let $\boldsymbol{u}(\cdot) \in \mathcal{A}(\boldsymbol{x}, \boldsymbol{k})$ be any control and let $\boldsymbol{x}(\cdot)$ be the corresponding surplus process. Suppose that

$$J(\boldsymbol{x}, \boldsymbol{k}, \boldsymbol{u}(\cdot)) < \lambda. \tag{4.57}$$

We can apply Lemma F.4 and (4.57) to obtain

$$\limsup_{\ell \to \infty} \rho_\ell J^{\rho_\ell}(\boldsymbol{x}, \boldsymbol{k}, \boldsymbol{u}(\cdot)) < \lambda. \tag{4.58}$$

On the other hand, we know from Theorem 4.1 that

$$\lim_{\ell \to \infty} \rho_\ell V^{\rho_\ell}(\boldsymbol{x}, \boldsymbol{k}) = \widehat{\lambda} = \lambda.$$

This equation and (4.58) imply the existence of a $\rho > 0$ such that

$$\rho J^{\rho}(\boldsymbol{x}, \boldsymbol{k}, \boldsymbol{u}(\cdot)) < \rho V^{\rho}(\boldsymbol{x}, \boldsymbol{k}),$$

which contradicts the definition of $V^{\rho}(\boldsymbol{x}, \boldsymbol{k})$. Thus (i) is proved. □

4.5 Two-Machine Flowshop with Finite Internal Buffer

In this section, we consider the two-machine flowshop shown in Figure 4.1 with a finite internal buffer. This imposes an upper bound constraint on the work-in-process. Each machine has a finite number of states, resulting in a system involving a finite state Markov chain. We use $k_j(t)$ to denote the state of machine j at time t, $j = 1, 2$. We denote the number of parts in the buffer between the first and second machines, called work-in-process, as $x_1(t)$, and the difference of the real and planned cumulative productions, called surplus at the second machine, as $x_2(t)$. Since the number of parts in the internal buffer cannot be negative and buffers usually have limited storage capacities, we impose the state constraints $0 \leq x_1(t) \leq H$, $0 < H < \infty$, where H represents the upper bound on the work-in-process. If $x_2(t) > 0$, we have finished goods inventories, and if $x_2(t) < 0$, we have backlogs. The input rates to the machines are denoted by $u_j(t)$, $j = 1, 2$. Owing to the capacity constraints, $u_j(t)$ can vary from 0 to $k_j(t)$, $j = 1, 2$. We assume a constant demand rate z.

Let $\mathcal{X}_H = [0, H] \times \Re^1$ denote the state constraint domain. The dynamics and constraints of the system can then be written as follows:

$$\begin{cases} \dfrac{d}{dt} x_1(t) = u_1(t) - u_2(t), \\ \dfrac{d}{dt} x_2(t) = u_2(t) - z, \end{cases} \tag{4.59}$$

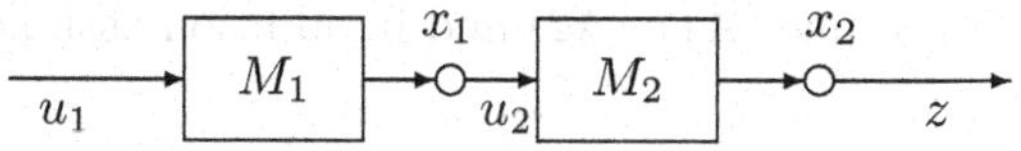

Figure 4.1. A Manufacturing System with a Two-Machine Flowshop.

where $\boldsymbol{x}(t) = (x_1(t), x_2(t))' \in \mathcal{X}_H$ and $0 \le u_j(t) \le k_j(t)$, $\boldsymbol{x}(0) = \boldsymbol{x}$, $k_j(0) = k_j$ for $j = 1, 2$, $t \ge 0$. For $\boldsymbol{k} = (k_1, k_2), k_j \ge 0, j = 1, 2$, let

$$\mathcal{U}(\boldsymbol{k}) = \{\boldsymbol{u} = (u_1, u_2)' \;:\; 0 \le u_j \le k_j, j = 1, 2\},$$

and for $\boldsymbol{x} \in \mathcal{X}_H$, let

$$\begin{aligned}\mathcal{U}(\boldsymbol{x}, \boldsymbol{k}) = \{\boldsymbol{u} \in \mathcal{U}(\boldsymbol{k}) : x_1 = 0 \Rightarrow u_1 - u_2 \ge 0,\\ x_1 = H \Rightarrow u_1 - u_2 \le 0\}.\end{aligned}$$

Furthermore, let

$$A = \begin{pmatrix} 1 & -1 \\ 0 & 1 \end{pmatrix} \quad \text{and} \quad B = \begin{pmatrix} 0 \\ -1 \end{pmatrix}.$$

Then (4.59) can be written as

$$\frac{d}{dt}\boldsymbol{x}(t) = A\boldsymbol{u}(t) + Bz, \quad \boldsymbol{x}(0) = \boldsymbol{x}. \tag{4.60}$$

Let $\mathcal{F}_t$ denote the σ-algebra generated by the process $\boldsymbol{k}(\cdot) = (k_1(\cdot), k_2(\cdot))$, i.e., $\mathcal{F}_t = \sigma\{\boldsymbol{k}(s), 0 \le s \le t\}$. We now define the concept of admissible controls.

Definition 5.1. We say that a control $\boldsymbol{u}(\cdot) = (u_1(\cdot), u_2(\cdot))'$ is *admissible* with respect to the initial state vector $\boldsymbol{x} = (x_1, x_2)' \in \mathcal{X}_H$ if:

(i) $\boldsymbol{u}(\cdot)$ is an $\mathcal{F}_t$-adapted measurable process;

(ii) $\boldsymbol{u}(t) \in \mathcal{U}(\boldsymbol{k}(t))$ for all $t \ge 0$; and

(iii) $\boldsymbol{x}(t) = (x_1 + \int_0^t [u_1(s) - u_2(s)]\, ds, x_2 + \int_0^t [u_2(s) - z]\, ds)' \in \mathcal{X}_H$ for all $t \ge 0$. □

Let $\mathcal{A}(\boldsymbol{x}, \boldsymbol{k})$ denote the set of admissible controls with the initial vector $\boldsymbol{x}$ and $\boldsymbol{k}(0) = \boldsymbol{k}$. Our objective is to obtain an admissible control $\boldsymbol{u}(\cdot) \in \mathcal{A}(\boldsymbol{x}, \boldsymbol{k})$ that minimizes the long-run average cost

$$J(\boldsymbol{x}, \boldsymbol{k}, \boldsymbol{u}(\cdot)) = \limsup_{T \to \infty} \frac{1}{T} E \int_0^T g(\boldsymbol{x}(t), \boldsymbol{u}(t))\, dt, \tag{4.61}$$

where $g(\boldsymbol{x}, \boldsymbol{u})$ is the surplus and production cost.

As in Section 4.2, we assume (A1)–(A2) and in addition assume:

(A4) Let $p_j = \displaystyle\sum_{i=1}^{p} \nu_{\boldsymbol{k}^i} k_j^i$ and $\min_{1\le j\le 2} p_j > z$.

Remark 5.1. Assumption (A4) is necessary for the long-run average cost to be finite. In addition, the buffer size H should not be too small in some cases. For example, if $H = 0$, we choose the admissible control $\boldsymbol{u}(\cdot)$ with $u_1(t) = u_2(t) = k_1(t) \wedge k_2(t)$, $t \ge 0$. If $\sum_{i=1}^{p}(k_1^i \wedge k_2^i)\nu_{\boldsymbol{k}^i} < z$, then the backlog would build up over time. Consequently, the corresponding long-run average cost would be ∞ in the case when $g(\boldsymbol{x}, \cdot)$ tends to ∞ as $x_2 \to -\infty$. This is also true when H is small. Later in this section, we will show that Assumption (A4) is sufficient for having finite average cost when H is suitably large. □

Let $\lambda^*(\boldsymbol{x}, \boldsymbol{k})$ denote the minimal long-run average cost, i.e.,

$$\lambda^*(\boldsymbol{x}, \boldsymbol{k}) = \inf_{\boldsymbol{u}\in\mathcal{A}(\boldsymbol{x},\boldsymbol{k})} J(\boldsymbol{x}, \boldsymbol{k}, \boldsymbol{u}(\cdot)). \tag{4.62}$$

From Theorem 5.2 given in the following, we know that $\lambda^*(\boldsymbol{x}, \boldsymbol{k})$ does not depend on the initial states $\boldsymbol{x}$ and $\boldsymbol{k}$. We can formally write the HJBDD equation for the problem as

$$\lambda = \inf_{\boldsymbol{u}\in\mathcal{U}(\boldsymbol{x},\boldsymbol{k})} \{\partial_{(A\boldsymbol{u}+Bz)}\phi(\boldsymbol{x}, \boldsymbol{k}) + g(\boldsymbol{x}, \boldsymbol{u})\} + Q\phi(\boldsymbol{x}, \cdot)(\boldsymbol{k}), \tag{4.63}$$

where $\phi(\cdot, \cdot)$ is defined on $\mathcal{X}_H \times \mathcal{M}$.

Let $\mathcal{S}(\boldsymbol{x}, \boldsymbol{k})$ denote the class of stable controls with the initial condition $\boldsymbol{k}(0) = \boldsymbol{k}$ (see Definition 2.3). Of course, $\mathcal{S}(\boldsymbol{x}, \boldsymbol{k}) \subset \mathcal{A}(\boldsymbol{x}, \boldsymbol{k})$. It will be seen in the next section that the set of stable admissible controls $\mathcal{S}(\boldsymbol{x}, \boldsymbol{k})$ is nonempty.

In order to study the existence of a solution to the HJBDD equation (4.63), we introduce a corresponding control problem with the cost discounted at a rate $\rho > 0$. For $\boldsymbol{u}(\cdot) \in \mathcal{A}(\boldsymbol{x}, \boldsymbol{k})$, we define the expected discounted cost as

$$J^\rho(\boldsymbol{x}, \boldsymbol{k}, \boldsymbol{u}(\cdot)) = E\int_0^\infty e^{-\rho t} g(\boldsymbol{x}(t), \boldsymbol{u}(t))\, dt, \tag{4.64}$$

where $\boldsymbol{x}(\cdot)$ is the state process corresponding the control $\boldsymbol{u}(\cdot)$ with the initial surplus level $\boldsymbol{x}(0) = \boldsymbol{x}$, and $\boldsymbol{k}$ is the initial value of the process $\boldsymbol{k}(\cdot)$. The value function is defined as

$$V^\rho(\boldsymbol{x}, \boldsymbol{k}) = \inf_{\boldsymbol{u}(\cdot)\in\mathcal{A}(\boldsymbol{x},\boldsymbol{k})} J^\rho(\boldsymbol{x}, \boldsymbol{k}, \boldsymbol{u}(\cdot)). \tag{4.65}$$

The HJBDD equation associated with this discounted cost problem is

$$\begin{aligned}\rho\phi^\rho(\boldsymbol{x},\boldsymbol{k}) = &\inf_{\boldsymbol{u}\in\mathcal{U}(\boldsymbol{x},\boldsymbol{k})}\{\partial_{(A\boldsymbol{u}+Bz)}\phi^\rho(\boldsymbol{x},\boldsymbol{k}) + g(\boldsymbol{x},\boldsymbol{u})\}\\ &+Q\phi^\rho(\boldsymbol{x},\cdot)(\boldsymbol{k}),\end{aligned} \tag{4.66}$$

where $\phi^\rho(\boldsymbol{x},\boldsymbol{k})$ is defined on $\mathcal{X}_H \times \mathcal{M}$. Then we have the following result as in Theorem 3.1 in Chapter 4 of Sethi and Zhang [125].

Theorem 5.1. *Let Assumptions* (A1) *and* (A2) *hold. Then the value function* $V^\rho(\boldsymbol{x},\boldsymbol{k})$ *has the following properties:*

(i) *For each* $\boldsymbol{k}\in\mathcal{M}$, *the value function* $V^\rho(\boldsymbol{x},\boldsymbol{k})$ *is convex and continuous on* $\mathcal{X}_H$, *and satisfies the condition*

$$\begin{aligned}&|V^\rho(\widehat{\boldsymbol{x}},\boldsymbol{k}) - V^\rho(\widetilde{\boldsymbol{x}},\boldsymbol{k})|\\ &\quad\le C\left(1+|\widehat{\boldsymbol{x}}|^{\beta_{g2}-1}+|\widetilde{\boldsymbol{x}}|^{\beta_{g2}-1}\right)|\widehat{\boldsymbol{x}}-\widetilde{\boldsymbol{x}}|\end{aligned} \tag{4.67}$$

for some positive constant C *and all* $\widehat{\boldsymbol{x}},\widetilde{\boldsymbol{x}}\in\mathcal{X}_H$, *where* β_{g2} *is as given in Assumption* (A1);

(ii) $V^\rho(\boldsymbol{x},\boldsymbol{k})$ *satisfies the* HJBDD *equation* (4.66).

In order to study the long-run average-cost control problem using the vanishing discount approach, we must obtain some estimates for the value function $V^\rho(\boldsymbol{x},\boldsymbol{k})$ for small values of ρ. To do this, we establish the following theorem which relates to stable controls.

Theorem 5.2. *Under Assumptions* (A2) *and* (A4), *for any* $r\ge 1$, $(\widehat{\boldsymbol{x}},\widehat{\boldsymbol{k}})\in\mathcal{X}_H\times\mathcal{M}$, *and* $(\overline{\boldsymbol{x}},\overline{\boldsymbol{k}})\in\mathcal{X}_H\times\mathcal{M}$, *there exist a control* $\boldsymbol{u}(t)$, $t\ge 0$, *and a positive constant* C_r *such that*

$$E[\tau(\widehat{\boldsymbol{x}},\widehat{\boldsymbol{k}},\overline{\boldsymbol{x}},\overline{\boldsymbol{k}})]^r \le C_r\left(1+\sum_{j=1}^{2}|\widehat{x}_j-\overline{x}_j|^r\right), \tag{4.68}$$

where

$$\tau(\widehat{\boldsymbol{x}},\widehat{\boldsymbol{k}},\overline{\boldsymbol{x}},\overline{\boldsymbol{k}}) = \inf\left\{t\ge 0 : \boldsymbol{x}(t)=\overline{\boldsymbol{x}},\ \boldsymbol{k}(t)=\overline{\boldsymbol{k}}\right\},$$

and $\boldsymbol{x}(t), t\ge 0$, *is the surplus process corresponding to the control* $\boldsymbol{u}(\cdot)$ *and the initial condition* $(\boldsymbol{x}(0),\boldsymbol{k}(0)) = (\widehat{\boldsymbol{x}},\widehat{\boldsymbol{k}})$.

Proof. First we provide an outline of the proof. We begin by modifying the process $\boldsymbol{k}(\cdot)$ in such a way that the modified average capacity of machine 1 is larger than the modified average capacity of machine 2, and that the modified average capacity of machine 2 is larger than z. Then, we alternatively use the two policies described below. In the first policy, we use the maximum admissible production rate corresponding to the modified process. In the second policy, we use a zero production rate for machine 1 and the maximum possible production rate for machine 2 under the restriction that the work-in-process is larger than $\overline{x}_1$. The first policy is used until the time the work-in-process exceeds the value $\overline{x}_1$ and the surplus process exceeds the value $M + \overline{x}_2$ for some $M > 0$. At this time we switch to the second policy. We use the second policy until such time when the surplus process drops to the level $\overline{x}_2$. After that we revert to the first policy, and so on. Using this alternating procedure, it is possible to specify $\tau(\widehat{\boldsymbol{x}}, \widehat{\boldsymbol{k}}, \overline{\boldsymbol{x}}, \overline{\boldsymbol{k}})$ and provide an estimate for it. The proof is divided into several steps.

Step 1. We construct an auxiliary Markov chain $\widetilde{\boldsymbol{k}}(\cdot)$. It follows from (A4) that we can select $\widetilde{\boldsymbol{k}}^i = (\widetilde{k}_1^i, \widetilde{k}_2^i)$, $i = 1, \ldots, p$, such that $\widetilde{k}_1^i \leq k_1^i$, $\widetilde{k}_2^i \leq k_2^i$, $i = 1, \ldots, p$, $\min_{1 \leq i \leq p} \widetilde{k}_1^i < \max_{1 \leq i \leq p} \widetilde{k}_2^i$, and

$$\widetilde{p}_1 = \sum_{i=1}^{p} \nu_{\boldsymbol{k}^i} \widetilde{k}_1^i > \widetilde{p}_2 = \sum_{i=1}^{p} \nu_{\boldsymbol{k}^i} \widetilde{k}_2^i > z. \tag{4.69}$$

Let us define the process $\widetilde{\boldsymbol{k}}(t) = (\widetilde{k}_1(t), \widetilde{k}_2(t))$ as follows:

$$\widetilde{\boldsymbol{k}}(t) = \widetilde{\boldsymbol{k}}^i \quad \text{when} \quad \boldsymbol{k}(t) = \boldsymbol{k}^i.$$

Let $\widetilde{\mathcal{M}} = \{\widetilde{\boldsymbol{k}}^1, \ldots, \widetilde{\boldsymbol{k}}^p\}$. We know that $\widetilde{\boldsymbol{k}}(t) \in \widetilde{\mathcal{M}}$, $t \geq 0$, is also strongly irreducible and has the stationary distribution $\nu_{\widetilde{\boldsymbol{k}}^i}$ $(= \nu_{\boldsymbol{k}^i})$. Thus, $(\widetilde{p}_1, \widetilde{p}_2)$ corresponds to its stationary expectation, and (4.69) gives

$$\widetilde{p}_1 > \widetilde{p}_2 > z. \tag{4.70}$$

Step 2. We construct a family of auxiliary processes $\boldsymbol{x}^0(t|s, \boldsymbol{x})$, $t \geq s \geq 0$, and $\boldsymbol{x} \in \mathcal{X}_H$. For $\boldsymbol{x} = (x_1, x_2)'$ and $\widetilde{\boldsymbol{k}} = (\widetilde{k}_1, \widetilde{k}_2)$, consider the following function

$$\boldsymbol{u}^0(\boldsymbol{x}, \widetilde{\boldsymbol{k}}) = (u_1^0(\boldsymbol{x}, \widetilde{\boldsymbol{k}}), u_2^0(\boldsymbol{x}, \widetilde{\boldsymbol{k}}))'$$

given by

$$\boldsymbol{u}^0(\boldsymbol{x}, \widetilde{\boldsymbol{k}}) = \begin{cases} (\widetilde{k}_1, \widetilde{k}_2)', & \text{if } 0 < x_1 < H, \\ (\widetilde{k}_1, \widetilde{k}_1 \wedge \widetilde{k}_2)', & \text{if } x_1 = 0, \\ (\widetilde{k}_1 \wedge \widetilde{k}_2, \widetilde{k}_2)', & \text{if } x_1 = H. \end{cases} \tag{4.71}$$

We define $\boldsymbol{x}^0(t|s, \boldsymbol{x})$ as the process which satisfies the following equation (see (4.59)):

$$\frac{d}{dt}\boldsymbol{x}^0(t|s, \boldsymbol{x}) = A\boldsymbol{u}^0(\boldsymbol{x}^0(t|s, \boldsymbol{x}), \widetilde{\boldsymbol{k}}(t)) + Bz, \quad \boldsymbol{x}^0(s|s, \boldsymbol{x}) = \boldsymbol{x}.$$

Clearly, $\boldsymbol{x}^0(t|s,\boldsymbol{x}) \in \mathcal{X}_H$ for all $t \geq s$. For a fixed s, $\boldsymbol{x}^0(t|s,\boldsymbol{x})$ is the state of the system with the production rate which is obtained by using the maximum admissible modified capacity for both machines.

Define now the Markov time

$$\theta(s,\boldsymbol{x},\overline{\boldsymbol{x}}) = \inf\{t \geq s : x_1^0(t|s,\boldsymbol{x}) \geq \overline{x}_1,\ x_2^0(t|s,\boldsymbol{x}) \geq M + \overline{x}_2\}, \tag{4.72}$$

where $M > 0$ is a constant specified later. It follows from this definition that $\theta(s,\boldsymbol{x},\overline{\boldsymbol{x}})$ is the first time when the state process $\boldsymbol{x}^0(t|s,\boldsymbol{x})$ exceeds $(\overline{x}_1, M+\overline{x}_2)'$ under the production rate $\boldsymbol{u}^0(\boldsymbol{x}^0(t|s,\boldsymbol{x}),\widetilde{\boldsymbol{k}}(t))$. Since the modified average capacity of machine 1 is larger than the modified average capacity of machine 2, since the modified average capacity of machine 2 is larger than the required rate z (see (4.69)), and since H is suitably large, we have the following lemma.

Lemma 5.1. *Let Assumptions* (A2) *and* (A4) *hold. If H is suitably large, then there exists a constant $\widehat{C}_r$ such that*

$$E\left[\theta(s,\boldsymbol{x},\overline{\boldsymbol{x}}) - s\right]^r < \widehat{C}_r\left[1 + \left(\sum_{j=1}^{2}(\overline{x}_j - x_j)^+\right)^r\right].$$

Proof. See Appendix B. □

Step 3. We construct a family of auxiliary processes $\boldsymbol{x}^1(t|s,\boldsymbol{x})$, $t \geq s \geq 0$, and $\boldsymbol{x} \in \mathcal{X}_H$. Consider the function

$$\begin{aligned}\boldsymbol{u}^1(\boldsymbol{x},\widetilde{\boldsymbol{k}}) &= (u_1^1(\boldsymbol{x},\widetilde{\boldsymbol{k}}), u_2^1(\boldsymbol{x},\widetilde{\boldsymbol{k}}))\\ &= \begin{cases}(0,\widetilde{k}_2)', & \text{if } x_1 > \overline{x}_1,\\ (0,0)', & \text{if } x_1 = \overline{x}_1.\end{cases}\end{aligned} \tag{4.73}$$

defined for $\boldsymbol{x}$ such that $x_1 \geq \overline{x}_1$.

We define $\boldsymbol{x}^1(t|s,\boldsymbol{x})$ as a continuous process which coincides with $\boldsymbol{x}^0(t|s,\boldsymbol{x})$ for $s \leq t \leq \theta(s,\boldsymbol{x},\overline{\boldsymbol{x}})$, and satisfies the following equation (see (4.59)):

$$\frac{d}{dt}\boldsymbol{x}^1(t|s,\boldsymbol{x}) = A\boldsymbol{u}^1(\boldsymbol{x}^1(t|s,\boldsymbol{x}),\widetilde{\boldsymbol{k}}(t)) + Bz, \quad t \geq \theta(s,\boldsymbol{x},\overline{\boldsymbol{x}}).$$

Clearly, $\boldsymbol{x}^1(t|s,\boldsymbol{x}) \in \mathcal{X}_H$ for all $t \geq s$, and $x_1^1(t|s,\boldsymbol{x}) \geq \overline{x}_1$ for $t \geq \theta(s,\boldsymbol{x},\overline{\boldsymbol{x}})$. This process corresponds to the policy in which after $\theta(s,\boldsymbol{x},\overline{\boldsymbol{x}})$ we stop production at machine 1 and have the maximum possible production rate at machine 2 under the restriction that the work-in-process is larger than $\overline{x}_1$.

We now define a Markov time

$$\widehat{\theta}(s,\boldsymbol{x},\overline{\boldsymbol{x}}) = \inf\{t \geq \theta(s,\boldsymbol{x},\overline{\boldsymbol{x}})\ :\ x_2^1(t|s,\boldsymbol{x}) = \overline{x}_2\}. \tag{4.74}$$

Lemma 5.2. *Let Assumptions* (A2) *and* (A4) *hold. If H is suitably large, then:*

(i) *for a given* $q \in (0,\ 1)$, *a positive constant* M *can be chosen in such a way that, for all* s *and* $\boldsymbol{x} \in \mathcal{X}_H$,

$$P(\boldsymbol{x}^1(\widehat{\theta}(s,\boldsymbol{x},\overline{\boldsymbol{x}})|s,\boldsymbol{x}) = \overline{\boldsymbol{x}},\ \boldsymbol{k}(\widehat{\theta}(s,\boldsymbol{x},\overline{\boldsymbol{x}})) = \overline{\boldsymbol{k}}) \geq 1 - q > 0.$$

(ii) *there exists a constant* C *such that*

$$\frac{M}{z} \leq \widehat{\theta}(s,\boldsymbol{x},\overline{\boldsymbol{x}}) - s \leq \frac{1}{z}\left(\sum_{j=1}^{2}(x_j - \overline{x}_j) + C[\theta(s,\boldsymbol{x},\overline{\boldsymbol{x}}) - s]\right) \quad (4.75)$$

and

$$\sum_{j=1}^{2} x_j^1(\widehat{\theta}(s,\boldsymbol{x},\overline{\boldsymbol{x}})|s,\boldsymbol{x}) \leq \sum_{j=1}^{2} x_j + C[\theta(s,\boldsymbol{x},\overline{\boldsymbol{x}}) - s]. \quad (4.76)$$

Proof. See Appendix B. □

Step 4. We construct a process $\boldsymbol{x}(t)$, $t \geq 0$, and the corresponding control $\boldsymbol{u}(t)$, $t \geq 0$, which satisfies the statement of Theorem 5.2.

Define a sequence of Markov times $\{\widehat{\theta}_\ell\}_{\ell=0}^{\infty}$ with respect to $\widetilde{\boldsymbol{k}}(\cdot)$ and the process $\boldsymbol{x}(t)$ for $\widehat{\theta}_\ell \leq t < \widehat{\theta}_{\ell+1}$ $(\ell = 0, 1, \ldots)$ as follows:

$$\widehat{\theta}_0 = 0, \quad \widehat{\theta}_1 = \widehat{\theta}(0, \widehat{\boldsymbol{x}}, \overline{\boldsymbol{x}}),$$

and

$$\boldsymbol{x}(t) = \boldsymbol{x}^1(t|0,\widehat{\boldsymbol{x}}), \quad 0 \leq t < \widehat{\theta}_1.$$

If $\widehat{\theta}_\ell$ is defined for $\ell \geq 1$ and $\boldsymbol{x}(t)$ is defined for $0 \leq t < \widehat{\theta}_\ell$, then we let

$$\widehat{\theta}_{\ell+1} = \widehat{\theta}(\widehat{\theta}_\ell, \boldsymbol{x}(\widehat{\theta}_\ell), \overline{\boldsymbol{x}}) \quad \text{and} \quad \boldsymbol{x}(t) = \boldsymbol{x}^1(t|\widehat{\theta}_\ell, \boldsymbol{x}(\widehat{\theta}_\ell)), \quad \widehat{\theta}_\ell \leq t < \widehat{\theta}_{\ell+1}.$$

According to the left inequality in (4.75), $\boldsymbol{x}(t)$ is defined for all $t \geq 0$. Let

$$\theta_1 = \theta(0, \widehat{\boldsymbol{x}}, \overline{\boldsymbol{x}}) \quad \text{and} \quad \theta_{\ell+1} = \theta(\widehat{\theta}_\ell, \boldsymbol{x}(\widehat{\theta}_\ell), \overline{\boldsymbol{x}}), \quad \ell = 1, 2, \ldots.$$

The control corresponding to the process $\boldsymbol{x}(\cdot)$ is given by

$$\boldsymbol{u}(t) = \begin{cases} \boldsymbol{u}^0(\boldsymbol{x}(t), \widetilde{\boldsymbol{k}}(t)), & \text{if } \widehat{\theta}_{\ell-1} \leq t < \theta_\ell, \\ \boldsymbol{u}^1(\boldsymbol{x}(t), \widetilde{\boldsymbol{k}}(t)), & \text{if } \theta_\ell \leq t < \widehat{\theta}_\ell, \end{cases} \quad (4.77)$$

for $\ell = 1, 2, \ldots$. It is clear that $\boldsymbol{u}(\cdot) \in \mathcal{A}(\widehat{\boldsymbol{x}}, \widehat{\boldsymbol{k}})$.

For the process $\boldsymbol{x}(\cdot)$, a Markov time is defined as

$$\tau(\widehat{\boldsymbol{x}}, \widehat{\boldsymbol{k}}, \overline{\boldsymbol{x}}, \overline{\boldsymbol{k}}) = \inf\left\{t \geq 0 :\ (\boldsymbol{x}(t), \boldsymbol{k}(t)) = (\overline{\boldsymbol{x}}, \overline{\boldsymbol{k}})\right\}.$$

Let

$$\mathcal{B}_\ell = \{\omega : (\boldsymbol{x}(\widehat{\theta}_\ell)(\omega), \boldsymbol{k}(\widehat{\theta}_\ell)(\omega)) = (\overline{\boldsymbol{x}}, \overline{\boldsymbol{k}})\}.$$

Using conditional probabilities (see (3.15)), we have from Lemma 5.2(i) that

$$P\left(\bigcap_{\ell_1=1}^{\ell} \mathcal{B}_{\ell_1}^c\right) \le q^\ell, \quad \ell = 1, 2, \ldots, \tag{4.78}$$

where $\mathcal{B}_{\ell_1}^c$ is the complement set of $\mathcal{B}_{\ell_1}$. Using (4.77) and the definition of $\boldsymbol{x}(t)$ we get that

$$\tau(\widehat{\boldsymbol{x}}, \widehat{\boldsymbol{k}}, \overline{\boldsymbol{x}}, \overline{\boldsymbol{k}}) = \sum_{\ell=1}^{\infty} \widehat{\theta}_\ell I_{\{\cap_{\ell_1=0}^{\ell-1} \mathcal{B}_{\ell_1}^c \cap \mathcal{B}_\ell\}},$$

which implies

$$[\tau(\widehat{\boldsymbol{x}}, \widehat{\boldsymbol{k}}, \overline{\boldsymbol{x}}, \overline{\boldsymbol{k}})]^r = \sum_{\ell=1}^{\infty} [\widehat{\theta}_\ell]^r I_{\{\cap_{\ell_1=0}^{\ell-1} \mathcal{B}_{\ell_1}^c \cap \mathcal{B}_\ell\}}, \tag{4.79}$$

where $\mathcal{B}_0^c = \Omega$. Using (4.75) and (4.76), we have, for $\ell = 1, 2, \ldots,$

$$\widehat{\theta}_\ell - \widehat{\theta}_{\ell-1} \le \frac{1}{z}\left(\sum_{j=1}^{2} x_j(\widehat{\theta}_{\ell-1}) - \sum_{j=1}^{2} \overline{x}_j + C[\theta_\ell - \widehat{\theta}_{\ell-1}]\right) \tag{4.80}$$

and

$$\sum_{j=1}^{2} x_j(\widehat{\theta}_\ell) \le \sum_{j=1}^{2} x_j(\widehat{\theta}_{\ell-1}) + C[\theta_\ell - \widehat{\theta}_{\ell-1}]. \tag{4.81}$$

Using (4.80) and (4.81), there exists a constant $C_1 > 0$ independent of $\widehat{\boldsymbol{x}}$ and $\overline{\boldsymbol{x}}$ such that, for $\ell = 1, 2, \ldots,$

$$\widehat{\theta}_\ell \le \frac{C_1(\ell+1)}{z}\left(\sum_{j=1}^{2} (\widehat{x}_j - \overline{x}_j)^+ + \sum_{\ell_1=1}^{\ell} (\theta_{\ell_1} - \widehat{\theta}_{\ell_1-1})\right),$$

which implies

$$(\widehat{\theta}_\ell)^r \le C_2(\ell+1)^{2r}\left(\sum_{j=1}^{2} ((\widehat{x}_j - \overline{x}_j)^+)^r + \sum_{\ell_1=1}^{\ell} (\theta_{\ell_1} - \widehat{\theta}_{\ell_1-1})^r\right) \tag{4.82}$$

for some $C_2 > 0$. Note that $\boldsymbol{x}(\widehat{\theta}_\ell) \ge \overline{\boldsymbol{x}}$ for $\ell = 1, 2, \ldots$. Using the Schwarz inequality (Corollary 3 on page 104 of Chow and Teicher [33]), we get from

(4.78) and Lemma 5.1 that there exists a positive constant $\overline{C}_r$ dependent on r such that

$$E\left(\theta_1^r I_{\{\cap_{\ell_1=1}^{\ell-1}\mathcal{B}_{\ell_1}^c\cap\mathcal{B}_\ell\}}\right) \le q^{(\ell-1)/2}\left(E(\theta_1^{2r})\right)^{1/2} \le \overline{C}_r q^{(\ell-1)/2}\left(1+\sum_{j=1}^{2}\left[(\overline{x}_j-\widehat{x}_j)^+\right]^r\right), \quad \ell=1,2,\ldots, \tag{4.83}$$

and

$$\begin{aligned} &E\left[(\theta_{\ell_1}-\widehat{\theta}_{\ell_1-1})^r I_{\{\cap_{\ell_2=1}^{\ell-1}\mathcal{B}_{\ell_2}^c\cap\mathcal{B}_\ell\}}\right] \\ &\le q^{(\ell-1)/2}\left(E(\theta_{\ell_1}-\widehat{\theta}_{\ell_1-1})^{2r}\right)^{1/2} \\ &\le \overline{C}_r q^{(\ell-1)/2}\left(1+\sum_{j=1}^{2}\left[(\overline{x}_j-\widehat{x}_j)^+\right]^r\right), \\ &\qquad 2\le \ell_1\le \ell=2,3,\ldots. \end{aligned} \tag{4.84}$$

Substituting (4.82) into (4.79), taking expectation, and using (4.83) and (4.84), we get (4.68). □

Theorem 5.3. *Let Assumptions* (A1), (A2), *and* (A4) *hold. Then there exist a sequence* $\{\rho_\ell : \ell \ge 1\}$ *with* $\rho_\ell \to 0$ *as* $\ell \to \infty$, *a constant* $\widehat{\lambda}$, *and a convex function* $V(\boldsymbol{x},\boldsymbol{k})$, *such that, for* $(\boldsymbol{x},\boldsymbol{k}) \in \mathcal{X}_H \times \mathcal{M}$,

$$\lim_{\ell\to\infty} \rho_\ell V^{\rho_\ell}(\boldsymbol{x},\boldsymbol{k}) = \widehat{\lambda},$$
$$\lim_{\ell\to\infty} \left[V^{\rho_\ell}(\boldsymbol{x},\boldsymbol{k}) - V^{\rho_\ell}(\boldsymbol{0},\boldsymbol{k})\right] = V(\boldsymbol{x},\boldsymbol{k}).$$

Proof. The proof is similar to the proof of Theorem 4.1 and, therefore, the details are omitted. □

Theorem 5.4. *Let Assumptions* (A1), (A2), *and* (A4) *hold. Then the following assertions hold:*

(i) $\lambda^*(\boldsymbol{x},\boldsymbol{k})$ *is independent of* $(\boldsymbol{x},\boldsymbol{k})$, *i.e.*, $\lambda^*(\boldsymbol{x},\boldsymbol{k}) = \lambda^*$;

(ii) *the pair* $(\widehat{\lambda}, V(\cdot,\cdot))$ *defined in Theorem* 5.3 *is a solution of* (4.63) *for* $0 < x_1 < H$.

Remark 5.2. For $x_1 = 0$ and/or $x_1 = H$, whether the pair $(\lambda, V(\cdot,\cdot))$ satisfies equation (4.63) or not, is a question to be answered. We believe that the answer would be in the affirmative. □

Proof of Theorem 5.4. The proof of (i) is similar to the proof of Theorem 4.2(i), and hence it is omitted here. Now we prove (ii). Let $\mathcal{O}_H^2$ be the set of all points in the interior of $\mathcal{X}_H$ on which $V(\boldsymbol{x}, \boldsymbol{k})$ is differentiable. ¿From the convexity of $V(\boldsymbol{x}, \boldsymbol{k})$ we know that $\mathcal{O}_H^2$ is dense in $\mathcal{X}_H$. It follows from the properties of convex functions that, for $\boldsymbol{x} \in \mathcal{O}_H^2$ and any $\boldsymbol{p}$,

$$\lim_{\ell\to\infty} \partial_{\boldsymbol{p}} \left[V^{\rho_\ell}(\boldsymbol{x}, \boldsymbol{k}) - V^{\rho_\ell}(\boldsymbol{0}, \boldsymbol{k})\right] = \partial_{\boldsymbol{p}} V(\boldsymbol{x}, \boldsymbol{k}). \tag{4.85}$$

It can be seen in Lemma E.4 that for any $\boldsymbol{x} \in \mathcal{X}_H$, the value function $V^{\rho_\ell}(\boldsymbol{x}, \boldsymbol{k})$ of the discounted cost problem satisfies

$$\rho_\ell V^{\rho_\ell}(\boldsymbol{x}, \boldsymbol{k}) = \inf_{\boldsymbol{u}\in\mathcal{U}(\boldsymbol{x},\boldsymbol{k})} \{\partial_{A\boldsymbol{u}+Bz} V^{\rho_\ell}(\boldsymbol{x}, \boldsymbol{k}) + g(\boldsymbol{x}, \boldsymbol{u})\} + QV^{\rho_\ell}(\boldsymbol{x}, \boldsymbol{k}).$$

This implies that

$$\rho_\ell V^{\rho_\ell}(\boldsymbol{x}, \boldsymbol{k}) = \inf_{\boldsymbol{u}\in\mathcal{U}(\boldsymbol{x},\boldsymbol{k})} \{\partial_{A\boldsymbol{u}+Bz} \widetilde{V}^{\rho_\ell}(\boldsymbol{x}, \boldsymbol{k}) + g(\boldsymbol{x}, \boldsymbol{u})\} + Q\widetilde{V}^{\rho_\ell}(\boldsymbol{x}, \boldsymbol{k}), \tag{4.86}$$

where $\widetilde{V}^{\rho_\ell}(\boldsymbol{x}, \boldsymbol{k}) = V^{\rho_\ell}(\boldsymbol{x}, \boldsymbol{k}) - V^{\rho_\ell}(\boldsymbol{0}, \boldsymbol{k})$. Taking the limit on both sides, we have that, for $\boldsymbol{x} \in \mathcal{O}_H^2$,

$$\widehat{\lambda} = \inf_{\boldsymbol{u}\in\mathcal{U}(\boldsymbol{x},\boldsymbol{k})} \{\partial_{A\boldsymbol{u}+Bz} V(\boldsymbol{x},\ \boldsymbol{k}) + g(\boldsymbol{x}, \boldsymbol{u})\} + QV(\boldsymbol{x}, \boldsymbol{k}). \tag{4.87}$$

Let $\boldsymbol{x}$ be any interior point of $\mathcal{X}_H$. If $\boldsymbol{x} \notin \mathcal{O}_H^2$, then for any direction $\boldsymbol{p}$ there exist a sequence $\{\boldsymbol{x}_n\}_{n=1}^{\infty}$ such that $\boldsymbol{x}_n \in \mathcal{O}_H^2$ and $\partial_{\boldsymbol{p}} V(\boldsymbol{x}_n, \boldsymbol{k}) \to \partial_{\boldsymbol{p}} V(\boldsymbol{x}, \boldsymbol{k})$. From this fact and from the continuity of $V(\boldsymbol{x}, \boldsymbol{k})$, it follows that (4.87) holds for all $\boldsymbol{x}$ in the interior of $\mathcal{X}_H$. □

Theorem 5.5. *Under Assumptions* (A1), (A2), *and* (A4), *let* $(\lambda, W(\boldsymbol{x}, \boldsymbol{k}))$ *be a solution to the* HJBDD *equation* (4.63). *Then:*

(i) *for any* $\boldsymbol{u}(\cdot) \in \mathcal{A}(\boldsymbol{x}, \boldsymbol{k})$, *we have* $\lambda \le J(\boldsymbol{x}, \boldsymbol{k}, \boldsymbol{u}(\cdot))$, *i.e.,*

$$\limsup_{T\to\infty} \frac{1}{T} E \int_0^T g(\boldsymbol{x}(t), \boldsymbol{u}(t))\, dt \ge \lambda;$$

(ii) *for any (stable) control policy* $\boldsymbol{u}(\cdot) \in \mathcal{S}(\boldsymbol{x}, \boldsymbol{k})$, *we have*

$$\liminf_{T\to\infty} \frac{1}{T} E \int_0^T g(\boldsymbol{x}(t), \boldsymbol{u}(t))\, dt \ge \lambda;$$

(iii) *if there is a control* $\boldsymbol{u}^*(\cdot) \in \mathcal{A}(\boldsymbol{x}, \boldsymbol{k})$ *such that*

$$\begin{aligned} &\inf_{\boldsymbol{u}\in\mathcal{U}(\boldsymbol{k}(t))} \{\partial_{A\boldsymbol{u}+Bz} W(\boldsymbol{x}^*(t), \boldsymbol{k}(t)) + g(\boldsymbol{x}^*(t), \boldsymbol{u})\} \\ &\quad = \partial_{A\boldsymbol{u}^*(t)+Bz} W(\boldsymbol{x}^*(t), \boldsymbol{k}(t)) + g(\boldsymbol{x}^*(t), \boldsymbol{u}^*(t)) \end{aligned} \tag{4.88}$$

for a.e. $t \geq 0$ *with probability* 1, *where* $\boldsymbol{x}^*(\cdot)$ *is the surplus process corresponding to the control* $\boldsymbol{u}^*(\cdot)$, *and*

$$\lim_{T\to\infty} \frac{E[W(\boldsymbol{x}^*(T), \boldsymbol{k}(T))]}{T} = 0,$$

then

$$\lambda = J(\boldsymbol{x}, \boldsymbol{k}, \boldsymbol{u}^*(\cdot)) = \lambda^*.$$

Proof. The proof is similar to the proof of Theorem 4.3, and therefore the details are omitted. □

Remark 5.3. In this section, we have developed a theory of dynamic programming in terms of directional derivatives for a stochastic two-machine flowshop with a limited internal buffer, convex cost, and the long-run average-cost minimization criterion. Claiming the existence of a solution to the dynamic programming equation, a verification theorem has been established only when the upper bound on work-in-process is sufficiently large. The above results would not hold for a small upper bound on the internal buffer. □

4.6 Notes

This chapter is based on Presman, Sethi, and Suo [100], Presman, Sethi, Zhang, and Zhang [103, 105, 106] and Presman, Sethi, Zhang, and Bisi [102]. Sections 4.2–4.4 are from Presman, Sethi, Zhang, and Zhang [105]. Theorem 5.2, on the two-machine flowshop with a limited internal buffer, is derived in Presman, Sethi, Zhang, and Bisi [102]. Establishment of these results for an m-machine flowshop with limited buffers remains an open problem for $m > 2$.

Eleftheriu [44], Eleftheriu and Desrochers [45], Bai [8], and Bai and Gershwin [11] consider a two-machine flowshop with the objective of minimizing the long-run average surplus cost; the cost of production is assumed to be zero. They recognize the difficulty of solving the problem analytically, and use heuristic arguments to obtain suboptimal controls. In addition to the natural nonnegativity constraint on the inventory level in the internal buffer, they also consider the buffer to be of limited size, which imposes an upper bound on the inventory level in the buffer. Furthermore, Srivatsan, Bai, and Gershwin [132] apply their results to semiconductor manufacturing.

A number of papers deal with the problem of optimal control of a flowshop with a discounted cost objective. Van Ryzin, Lou, and Gershwin [139] treated a two-machine flowshop. Presman, Sethi, and Zhang [107] studied an m-machine flowshop and carried out rigorous analysis on optimal control. Van Ryzin, Lou, and Gershwin [139] and Lou and Van Ryzin [93] developed simple control rules for the two-machine flowshop, termed two-boundary controls. Lou and Kager [92] constructed a robust production control policy based on the results obtained in Van Ryzin, Lou, and Gershwin [139], and apply it to a VLSI wafer fabrication facility. Samaratunga, Sethi, and Zhou [110] repeat the calculations of Van Ryzin, Lou, and Gershwin [139] using more refined numerical methods of solving dynamic programming equations. Lou and Van Ryzin [93] also analyze a three-machine flowshop and extend the two-boundary control to an m-machine flowshop.

5

Optimal Controls of Dynamic Jobshops

5.1 Introduction

In this chapter, we consider a manufacturing system producing a variety of products in a general network configuration, which generalizes both the parallel and the tandem machine models. Each product follows a given process plan or recipe that specifies the sequence of machines it must visit and the operations performed by them. A recipe may call for multiple visits to a given machine, as is the case in semiconductor manufacturing (Lou and Kager [92], Srivatsan, Bai, and Gershwin [132], and Uzsoy, Lee, and Martin-Vega [136, 137]). The machines are failure-prone and they break down and must be repaired from time to time. A manufacturing system so described will be termed a *dynamic jobshop*. The term will be made mathematically precise in the next section.

The problem that we are interested in is how to obtain rates of production of intermediate parts and finished products in a network of failure-prone machines. The objective is to meet demand for finished products at the minimum possible long-run average cost of production, inventories, and backlogs.

In order to carry out the analysis of such a general network system as a jobshop, one needs to address a number of major difficulties. Foremost among these is the specification of the general dynamics. Note that a system with m tandem machines along with their capacities, where m is a positive integer, immediately defines a general flowshop. This is no longer the case with jobshops. A jobshop with n machines can have many different configurations. Moreover, specifying other parameters, such as the number of buffers, the number of processing steps, etc., still cannot uniquely define a network system. Therefore, our first task is to establish a mathematical framework for a dynamic jobshop that appropriately describes and uniquely determines its system dynamics along with the state and control constraints.

Once the mathematical representation for a jobshop is in place, the next task is to construct a control policy which takes a given system state to any other state in a time whose rth moment has a finite expectation, this will be the key to implement the vanishing discount approach. This plays an essential role in analyzing our problem. In the flowshop case (Chapter 4), these tasks were accomplished by constructive proofs carried out in a sequential manner. Indeed, in the case of a flowshop, there is an obvious natural sequence, and the sequential step is easy to imagine: if one removes the first machine and the first internal buffer from an m-machine flowshop, the remaining system is an $(m-1)$-machine flowshop having exactly the same structure as its "parent." Let us call this property the "inheritance" property.

In a jobshop, however, one does not have a natural sequence, nor can one easily conceive of an inheritance property. The main reasons for this are that a machine may feed into many buffers and that there are reentrant flows. Yet, as we shall see, we are able to define an appropriate sequence and obtain a needed inheritance property for jobshops.

We present a graph-theoretic framework in which a dynamic jobshop is modeled as a directed graph (digraph) with a "placement of machines" that reflects system dynamics and the control constraints. The conceptual framework, it should be noted, is based on the intuitive notion of what constitutes a jobshop or a general manufacturing system with a network of machines. We introduce a labeling procedure that defines a *sequence* appropriate for constructing the control required in the proof of the finiteness of the rth moment of the time required to go from a given initial system state to any other system state. More importantly, we use graph theory to isolate some key properties of the dynamics of a general network that will be *inherited* by the subsystem obtained by removing the first internal buffer (in the labeled sequence) from the system. It is interesting to observe that this inheritance property, borne out of intuitive considerations in the modeling of a dynamic jobshop, is exactly the property needed to establish finiteness of the rth moment mentioned above.

The chapter is organized as follows. In Section 5.2 we present a graph-theoretic framework for manufacturing systems with machines in a network, analyze its structure, and give the system dynamics equations. At the end of the section, we formulate the optimization problem for dynamic jobshops with unreliable machines under the long-run average-cost criterion. Section 5.3 is devoted to analysis of the corresponding discounted cost problem. Section 5.4 analyses the original problem. Section 5.5 concludes the chapter.

5.2 A Graph-Theoretic Framework and Problem Formulation

The purpose of this section is to develop a graph-theoretic framework for general dynamic manufacturing systems. The framework allows us to describe them precisely as well as to analyze them asymptotically.

We begin with the example of a simple jobshop described in Section 2.5. For convenience in exposition, we reproduce the associated figure, the system dynamics, and the constraints below.

System dynamics:

$$\begin{aligned}
&\frac{d}{dt}x_1(t) = u_{01}(t) - u_{12}(t) - u_{14}(t), && \frac{d}{dt}x_4(t) = u_{14}(t) + u_{34}(t) - z_4,\\
&\frac{d}{dt}x_2(t) = u_{12}(t) - u_{23}(t), && \frac{d}{dt}x_5(t) = u_{05}(t) - z_5.\\
&\frac{d}{dt}x_3(t) = u_{23}(t) - u_{34}(t),
\end{aligned} \tag{5.1}$$

Control constraints:

$$\begin{aligned}
&r_{01}u_{01}(t) \le k_1(t), && r_{05}u_{05}(t) + r_{14}u_{14}(t) \le k_4(t),\\
&r_{12}u_{12}(t) + r_{34}u_{34}(t) \le k_3(t), && r_{23}u_{23}(t) \le k_2(t).
\end{aligned} \tag{5.2}$$

State constraints:

$$x_i(t) \ge 0, \quad i = 1, 2, 3. \tag{5.3}$$

Since we are interested in general jobshops of the type shown in Figure 5.1 and described by (5.1), (5.2), and (5.3), we must find a way to generalize the formulation. In order to do so, note immediately that the placement of machines in Figure 5.1 plays no role in describing the dynamics (5.1) and constraints (5.3). Removal of machines from Figure 5.1 leaves us with the digraph shown in Figure 5.2.

The placement of machines supplies us with the control constraints (5.2). Note that a different placement of a possibly different number of machines would result in a different set of control constraints. This gives us the idea that a general dynamic jobshop of interest to us could be obtained by a digraph with a placement of a number of machines. We pursue this idea toward modeling a dynamic jobshop as the appropriate generalization of Figure 5.1.

Next we define a class of digraphs to represent a general system dynamics, analyze its structure from the viewpoint of deriving an analysis for our problem in Sections 5.3 and 5.4, and supplement any digraph in the class with a placement of machines to complete the model of a dynamic jobshop.

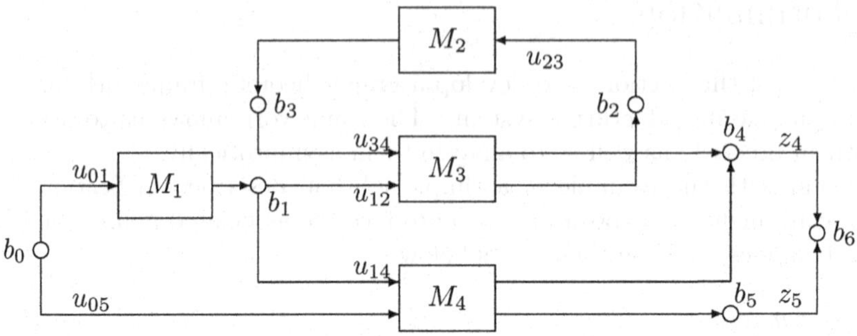

Figure 5.1. A Typical Dynamic Jobshop.

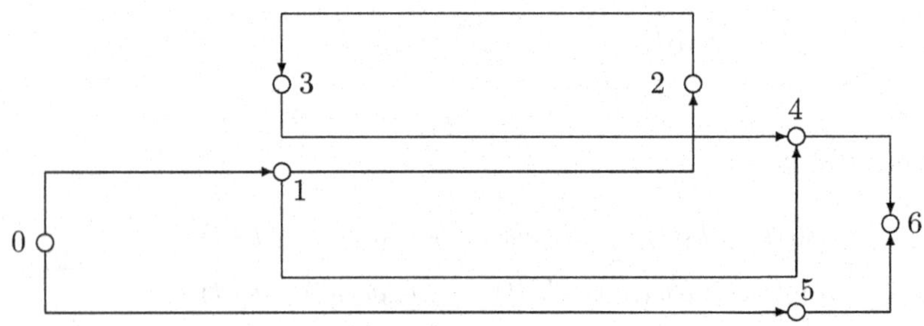

Figure 5.2. The Digraph Corresponding to Figure 5.1.

To do this we first introduce the digraph and some of its properties, for more discussion on digraph theory, see Chartrand and Lesniak [30].

Definition 2.1. A *digraph* denoted by (V, A) is a finite nonempty set V, whose elements are called *vertices*, together with a set A of ordered pairs (b, c) called *arcs*, where b and c are two distinct vertices. □

Definition 2.2. In a digraph the number of arcs beginning at the vertex b is called the *outdegree* of b. The number of arcs ending at the vertex b is called the *indegree* of b. A vertex is a *source* if its indegree is zero and its outdegree is nonzero. A vertex is a *sink* if its outdegree is zero and its indegree is nonzero. A vertex is *isolated* if both its indegree and outdegree are zero. □

Definition 2.3. A *walk* $\mathcal{W}$ in a digraph is a finite sequence which consists of vertices and arcs alternately and begins and ends with vertices. Furthermore, in the walk

$$\mathcal{W} = b_1, (b_1, b_2), b_2, (b_2, b_3), \ldots, (b_{n-1}, b_n), b_n,$$

b_1 is called its *first vertex* and b_n its *last vertex*. We shall also say that $\mathcal{W}$ is a walk from b_1 to b_n and simply express $\mathcal{W}$ as $\langle b_1 b_2 \cdots b_n \rangle$. □

Definition 2.4. A walk whose first and last vertices coincide is called a *cycle*. A walk, all of whose vertices are distinct, is called a *path*. A path whose first vertex is a source and last vertex is a sink is called a *complete path*. □

We define a class of digraphs that are of interest, in the sense that any digraph in the class would correspond to the dynamics of a manufacturing system such as the one described by (5.1) and (5.2).

Definition 2.5. A *manufacturing digraph* is a digraph (V, A) satisfying the following properties:

(i) there are two nonempty sets, one containing sources and the other containing a sink in the digraph;

(ii) no vertex in the digraph is isolated;

(iii) the digraph does not contain any cycle. □

Definition 2.6. In a manufacturing digraph, a vertex is called a *supply node* if it is a source, the *customer node* if it is a sink, vertices immediately proceeding the sink are called *external buffers*, and the others are called an *internal buffer*. □

Remark 2.1. Condition (ii) in Definition 2.5 is not an essential restriction. Inclusion of isolated vertices is merely a nuisance. This is because an isolated vertex is like a warehouse that can only ship out parts of a particular type to meet their demand. Since no machine (or production) is involved, its inclusion or exclusion does not affect the optimization problem under consideration. Condition (iii) in Definition 2.5 is imposed to rule out the following two trivial situations: (a) a part of type i in buffer i gets processed on a machine without any transformation and returns to buffer i; and (b) a part of type i is processed and converted into a part of type j, $j \neq i$, and is then processed further on a number of machines to be converted back into a part of type i. Moreover, if we had included any cycle in our manufacturing system, the flow of parts that leave buffer i, *only* to return to buffer i, would be zero in any optimal solution. It is unnecessary, therefore, to complicate the problem by including cycles. □

In order to obtain the system dynamics from a given manufacturing digraph and to get an appropriate "sequence" as mentioned in Section 5.1, a systematic procedure is required to label the state and control variables. To present such a procedure, we need an additional definition.

Definition 2.7. In a manufacturing digraph, let b be a vertex contained in a complete path $\mathcal{W} = \langle b_1 b_2 \cdots b \cdots b_l \rangle$. The *depth of* b *with respect to* $\mathcal{W}$ is

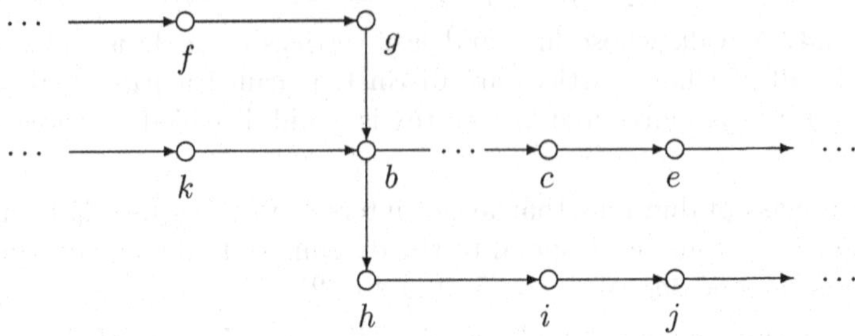

Figure 5.3. Illustration of the Proof of Theorem 2.1.

the length of the walk $\langle b_1 b_2 \cdots b \rangle$. The *depth* of a vertex b, denoted by $d(b)$, is the maximum of the depths of b with respect to all possible complete paths that contain b. □

Remark 2.2. The depth of any source is zero. □

Theorem 2.1. *In a manufacturing digraph, the depths of all the buffers along any complete path are in a strictly increasing order.*

Proof. If not, then let us suppose that there are two buffers b and c with depths $d(b)$ and $d(c)$, respectively, in a complete path $\langle \cdots kb \cdots ce \cdots \rangle$ (see Figure 5.3), while $d(b) \geq d(c)$. By definition, there is a complete path $\langle \cdots fgbhij \cdots \rangle$ with respect to which the depth of buffer b is just $d(b)$. We see that the depth of buffer c with respect to the complete path $\langle \cdots fgb \cdots ce \cdots \rangle$ is at least $d(b)+1$. This implies that the depth of buffer c is strictly greater than $d(c)$, which is a contradiction that proves the result. □

Let us now suppose that a given manufacturing digraph contains a total of (n_0+n+1) vertices including n_0 sources, the sink, m internal buffers, and $(n-m)$ external buffers for some integers m and n with $0 \leq m \leq n-1$ and $n \geq 1$.

Theorem 2.2. *We can label all the vertices from* $-n_0+1$ *to* $n+1$ *in a way so that the label numbers of the vertices along every complete path are in a strictly increasing order. In particular, the* n_0 *sources can be labeled as* $\{-n_0+1, \ldots, 0\}$*, and the external buffers (vertices immediately preceding the sink) can be labeled as* $\{m+1, m+2, \ldots, n\}$.

Proof. We label all the vertices by the following procedure:

Step 1. Label the n_0 sources arbitrarily using the numbers $\{-n_0+1, -n_0+2, \ldots, 0\}$, label the sink as $n+1$, and label the $n-m$ vertices immediately preceding the sink arbitrarily using the numbers $\{m+1, m+2, \ldots, n\}$.

Step 2. Classify the remaining m vertices, which are neither sources nor the sink nor the vertices that immediately precedes the sink, according to their depths. Suppose that there are m_1 vertices with depth 1, m_2 vertices with depth 2, ..., and m_ℓ vertices with depth ℓ, where $m_1 + m_2 + \cdots + m_\ell = m$.

Step 3. Label the m_1 depth-1 vertices arbitrarily from the numbers $\{1, 2, \ldots, m_1\}$, then the m_2 depth-2 vertices arbitrarily from the numbers $\{m_1 + 1, m_1 + 2, \ldots, m_1 + m_2\}$, ..., and finally the m_ℓ depth-ℓ vertices arbitrarily from the numbers $\{m - m_\ell + 1, \ldots, m\}$.

By virtue of Theorem 2.1, it is easily seen that our labeling procedure meets the requirement of the theorem. □

With the help of Theorem 2.2, one is able to formally write the dynamics and the state constraints associated with a given manufacturing digraph; at the same time, one can prove some important properties which are crucial for the analysis later in the chapter. First let us give a few definitions.

Definition 2.8. For each arc (i,j), $j \neq n+1$, in a manufacturing digraph, the rate at which parts in buffer i are converted to parts in buffer j is labeled as *control* $u_{i,j}$. Moreover, the control $u_{i,j}$ associated with the arc (i,j) is called an *output* of i and an *input* to j. In particular, outputs of the source $(i = -n_0 + 1, \ldots, 0)$ are called *primary controls* of the digraph. For each arc $(i, n+1)$, $i = m+1, \ldots, n$, the demand for products in buffer i is denoted by z_i. □

Remark 2.3. There is a control associated with each arc (i,j) $(j \neq n+1)$ in a manufacturing digraph, and there is an arc (i,j) $(j \neq n+1)$ associated with each control. In other words, the controls and the arcs have a one-to-one correspondence. □

Remark 2.4. Later in the section, we will associate an appropriate machine with each arc (i,j) $(j \neq n+1)$ that is capable of converting parts in i to parts in j. □

In what follows, we shall also set

$$u_{i,j} = 0, \quad \text{for } (i,j) \notin A, \ -n_0 + 1 \le i \le m, \ 1 \le j \le n,$$

for a unified notation suggested in Presman, Sethi, and Suo [99]. In this way, we can consider the controls as an $(n_0 + m) \times n$ matrix $(u_{i,j})$ of the following form

$$\begin{pmatrix} u_{-n_0+1,1} & u_{-n_0+1,2} & \cdots & u_{-n_0+1,m+1} & \cdots & u_{-n_0+1,n} \\ \vdots & \vdots & \vdots & \vdots & \vdots & \vdots \\ u_{0,1} & \cdots & \cdots & u_{0,m+1} & \cdots & u_{0,n} \\ 0 & u_{1,2} & \cdots & u_{1,m+1} & \cdots & u_{1,n} \\ \vdots & \vdots & \vdots & \vdots & \vdots & \vdots \\ 0 & 0 & \cdots & u_{m,m+1} & \cdots & u_{m,n} \end{pmatrix}.$$

The set of all such controls is written as $\mathcal{U}$, i.e.,

$$\begin{aligned} \mathcal{U} = \{(u_{i,j}) : -n_0+1 \le i \le m,\ 1 \le j \le n, \\ u_{i,j} = 0 \text{ for } (i,j) \notin A\}. \end{aligned} \tag{5.4}$$

Now we shall write the dynamics and the state constraints corresponding to a manufacturing digraph (V, A) containing $(n_0 + n + 1)$ vertices consisting of n_0 sources, a sink, m internal buffers, and $(n - m)$ external buffers associated with the $(n - m)$ distinct final products to be manufactured. We label all the vertices according to Theorem 2.2. For simplicity in the sequel, we shall call the buffer whose label is j as buffer j $(j = 1, 2, ..., n)$. We denote the surplus at time t in buffer j by $x_j(t)$. Write it in vector form

$$\boldsymbol{x}(t) = (x_1(t), \ldots, x_n(t))'.$$

Clearly,

$$\{1, \ldots, n\} = V \setminus \{-n_0+1, -n_0+2, \ldots, 0, n+1\}.$$

The control at time t associated with arc (i, j) by $u_{i,j}(t)$, $(i, j) \in A$ and $j \neq n+1$.

The dynamics of the system are therefore

$$\begin{aligned} \frac{d}{dt}x_j(t) &= \sum_{\ell=-n_0+1}^{j-1} u_{\ell,j}(t) - \sum_{\ell=j+1}^{n} u_{j,\ell}(t), \quad 1 \le j \le m, \\ \frac{d}{dt}x_j(t) &= \sum_{\ell=-n_0+1}^{m} u_{\ell,j}(t) - z_j, \quad m+1 \le j \le n, \end{aligned} \tag{5.5}$$

with $\boldsymbol{x}(0) = (x_1(0), ..., x_n(0))' = (x_1, ..., x_n)' = \boldsymbol{x}$. The state constraints are

$$\begin{aligned} &x_j(t) \ge 0, \quad t \ge 0, \quad j = 1, ..., m, \\ &-\infty < x_j(t) < +\infty, \quad t \ge 0, \quad j = m+1, ..., n. \end{aligned} \tag{5.6}$$

Note that if $x_j(t) > 0$, $j = 1, ..., n$, we have an inventory in buffer j, and if $x_j(t) < 0$, $j = m+1, ..., n$, we have a shortage of the finished product j.

It is convenient in the following discussion to write the control in vector form rather than in a matrix form given earlier. To do this, for $\ell = -n_0 + 1, \ldots, 0$, and $j = 1, \ldots, m$, let

$$\boldsymbol{u}_\ell(t) = \begin{pmatrix} u_{\ell,1}(t) \\ \vdots \\ u_{\ell,n}(t) \end{pmatrix}, \quad \boldsymbol{u}_j(t) = \begin{pmatrix} u_{j,j+1}(t) \\ \vdots \\ u_{j,n}(t) \end{pmatrix}, \tag{5.7}$$

$$\boldsymbol{u}(t) = \begin{pmatrix} \boldsymbol{u}_{-n_0+1}(t) \\ \vdots \\ \boldsymbol{u}_m(t) \end{pmatrix}, \quad \text{and} \quad \boldsymbol{z}(t) = \begin{pmatrix} z_{m+1} \\ \vdots \\ z_n \end{pmatrix}. \tag{5.8}$$

According to the definition of vector $\boldsymbol{u}(\cdot)$, there is a one-to-one mapping between the control matrix $(u_{i,j}(\cdot))_{(n_0+m)\times n}$ and the vector $\boldsymbol{u}(t)$ given by (5.8). Thus, hereafter, we use $\boldsymbol{u}(\cdot)$ to represent our control. Furthermore, for $i = 1, \ldots, m-1$, let O_i be an $i \times (n-i-1)$ zero matrix, O_m is an $m \times (n-m)$ zero matrix, and for $i = 1, \ldots, m$,

$$B_i = \begin{pmatrix} -1 & -1 & \cdots & -1 \\ 1 & 0 & \cdots & 0 \\ 0 & 1 & \cdots & 0 \\ \vdots & \vdots & \cdots & \vdots \\ 0 & 0 & \cdots & 1 \end{pmatrix}_{(n-i+1)\times(n-i)},$$

and

$$B_{m+1} = \begin{pmatrix} -1 & 0 & & & \\ 0 & -1 & 0 & & \\ & \ddots & \ddots & \ddots & \\ & & 0 & -1 & 0 \\ & & & 0 & -1 \end{pmatrix}_{(n-m)\times(n-m)}.$$

Let

$$A_{-n_0+1} = \cdots = A_0 = \begin{pmatrix} 1 & 0 & & & \\ 0 & 1 & 0 & & \\ & \ddots & \ddots & \ddots & \\ & & 0 & 1 & 0 \\ & & & 0 & 1 \end{pmatrix}_{n\times n},$$

and

$$A_1 = B_1, \quad A_j = \begin{pmatrix} O_{j-1} \\ B_j \end{pmatrix}, \quad j = 2, \ldots, m+1.$$

Based on the notation introduced above, relation (5.5) can be written in the following vector form:

$$\frac{d}{dt}\boldsymbol{x}(t) = (\, A_{-n_0+1}, \ldots, A_{m+1}\,) \begin{pmatrix} \boldsymbol{u}(t) \\ \boldsymbol{z} \end{pmatrix}, \quad \boldsymbol{x}(0) = \boldsymbol{x}. \tag{5.9}$$

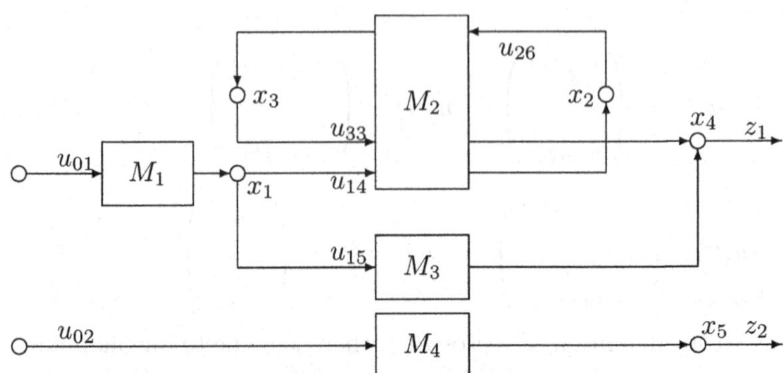

Figure 5.4. Another System Corresponding to the Digraph in Figure 5.2.

Note that

$$(A_{-n_0+1}, \ldots, A_{m+1})$$

is an $n \times \widetilde{n}$ matrix with $\widetilde{n} = n_0 n + (n-m) + \sum_{\ell=1}^{m}(n-\ell)$. Let $\mathcal{Y} = \Re_+^m \times \Re^{n-m}$. Furthermore, we introduce a linear operator $\mathcal{L}$ from $\Re_+^{\widetilde{n}}$ to $\mathcal{Y}$:

$$\mathcal{L}(\boldsymbol{u}(t), \boldsymbol{z}) = (A_{-n_0+1}, \ldots, A_{m+1}) \begin{pmatrix} \boldsymbol{u}(t) \\ \boldsymbol{z} \end{pmatrix}.$$

Then, (5.9) can be simply written as

$$\frac{d}{dt}\boldsymbol{x}(t) = \mathcal{L}(\boldsymbol{u}(t), \boldsymbol{z}), \quad \boldsymbol{x}(0) = \boldsymbol{x}.$$

Now we introduce the control constraints. As mentioned earlier, the control constraints (5.2) for Figure 5.1 depend on the placement of the machines on the digraph in Figure 5.2, and that different placements on the same digraph will give rise to different jobshops. In other words, a jobshop corresponds to a unique digraph, whereas a digraph may correspond to many different jobshops. For example, the system depicted in Figure 5.4 also corresponds to the digraph in Figure 5.2. Therefore, to uniquely characterize a jobshop using graph theory, we need to introduce the concept of a "placement of machines," or simply a "placement." Let m_c denote the number of machines to be placed. Then, $m_c \leq \#A - n + m$, where $\#A$ denotes the total number of arcs in A.

Definition 2.9. In a manufacturing digraph (V, A), a set $\mathcal{K} = \{K_1, K_2, ..., K_{m_c}\}$ is called a *placement of machines* $1, 2, ..., m_c$, if $\mathcal{K}$ is a partition of $B = \{(i,j) \in A : j \leq m\}$, namely, $\emptyset \neq K_j \subset B$, $K_j \cap K_\ell = \emptyset$ for $j \neq \ell$, and $\bigcup_{j=1}^{m_c} K_j = B$. □

So far, we have been using the term dynamic jobshop loosely to refer to a manufacturing system of the type described in Section 2.5 or by Figure 5.1. We are now ready to specify precisely a general dynamic jobshop.

A *dynamic jobshop*, or simply a *jobshop*, can be uniquely specified by a triple $(V, A, \mathcal{K})$, which denotes a manufacturing system that corresponds to a manufacturing digraph (V, A) along with a placement $\mathcal{K} = (K_1, \ldots, K_{m_c})$ satisfying:

(i) If b is a source, then $(b, c) \in A$ if and only if there is one and only one machine such that raw parts go to this machine for processing in order to be stocked in buffer c; if c is not a sink, then $(b, c) \in A$ if and only if there is one and only one machine such that parts in buffer b go to this machine for processing and are then stocked as parts of type c (in buffer c). In either case, we will say that the machine is *associated* with arc (b, c).

(ii) There are a total of m_c distinct machines, and the jth machine is associated with each of the arcs in K_j, $j = 1, 2, \ldots, m_c$.

(iii) The number of the external buffers equals the number of types of finished products to be produced, and different external buffers store different finished products.

Remark 2.5. Occasionally we will also say that a machine connects vertices b and c if the machine is associated with arc (b, c). □

Remark 2.6. The placements for the jobshop in Figures 5.1 and 5.4 are $K_1 = \{(0,1)\}$, $K_2 = \{(2,3)\}$, $K_3 = \{(3,4),(1,2)\}$, $K_4 = \{(0,5),(1,4)\}$), and $K_1 = \{(0,1)\}$, $K_2 = \{(2,3),(1,2),(3,4)\}$, $K_3 = \{(1,4)\}$, $K_4 = \{(0,5)\}$, respectively. □

Consider a jobshop $(V, A, \mathcal{K})$, where the dynamics of (V, A) are given in (5.5) and $\mathcal{K} = (K_1, K_2, ..., K_{m_c})$. Suppose we are given a stochastic process $\boldsymbol{k}(t) = (k_1(t), ..., k_{m_c}(t))$, with $k_j(t)$ representing the capacity of the jth machine at time t, $j = 1, ..., m_c$. Therefore, the controls $u_{i,\ell}(t)$ with $(i, \ell) \in K_j$, $j = 1, ..., m_c$, $t \geq 0$, in (5.5) should satisfy the following constraints:

$$0 \leq \sum_{(i,\ell) \in K_j} u_{i,\ell}(t) \leq k_j(t), \quad \text{for all} \quad t \geq 0,\ j = 1, ..., m_c, \tag{5.10}$$

where we have assumed that the required machine capacity $r_{i\ell}$ (for a unit production rate of type ℓ from part type i) equals 1, for convenience in exposition. Moreover, the corresponding surplus process satisfies the constraints (5.6). The analysis in this chapter can be readily extended to the case when the required machine capacity for the unit production rate of part j from part i is any given positive constant.

We are now in a position to formulate our stochastic optimal control problem for the jobshop defined by (5.5), (5.6), and (5.10).

For $\boldsymbol{k} = (k_1, ..., k_{m_c})$, let

$$\mathcal{U}(\boldsymbol{k}) = \left\{ (u_{i,\ell}) : (u_{i,\ell}) \in \mathcal{U},\ 0 \leq \sum_{(i,\ell)\in K_j} u_{i,\ell} \leq k_j,\ 1 \leq j \leq m_c \right\}.$$

By (5.7) and (5.8), for each $(u_{i,j}) \in \mathcal{U}(\boldsymbol{k})$, we can generate a unique nonnegative $\widetilde{n}$-dimensional vector $\boldsymbol{u}$ ($\widetilde{n} = n_0 + (n-m) + \sum_{\ell=1}^{m}(n-\ell)$). In the rest of this chapter, we use $\boldsymbol{u}$ and $(u_{i,j})$ ($\in \mathcal{U}(\boldsymbol{k})$) interchangeably.

For $\boldsymbol{x} \in \mathcal{Y}$ and $\boldsymbol{k}$,

$$\mathcal{U}(\boldsymbol{x}, \boldsymbol{k}) = \left\{ \boldsymbol{u} : \boldsymbol{u} \in \mathcal{U}(\boldsymbol{k}) \text{ and } x_j = 0 \right.$$
$$\left. \Rightarrow \sum_{i=-n_0+1}^{j-1} u_{i,j} - \sum_{i=j+1}^{n} u_{j,i} \geq 0,\ j = 1, ..., m \right\}.$$

Let the stochastic process $\boldsymbol{k}(\cdot) = (k_1(\cdot), ..., k_{m_c}(\cdot))$, defined on a probability space $(\Omega, \mathcal{F}, P)$, denote the machine capacity process.

Definition 2.10. We say that a control $\boldsymbol{u}(\cdot)$ is *admissible* with respect to the initial state vector $\boldsymbol{x} = (x_1, \ldots, x_n)' \in \mathcal{Y}$ and $\boldsymbol{k}(0) = \boldsymbol{k}$ if: (i) $\boldsymbol{u}(\cdot)$ is an $\mathcal{F}_t$-adapted measurable process with $\mathcal{F}_t = \sigma\{\boldsymbol{k}(s) : 0 \leq s \leq t\}$; (ii) $\boldsymbol{u}(t) \in \mathcal{U}(\boldsymbol{k}(t))$ for all $t \geq 0$; and (iii) the corresponding state process $\boldsymbol{x}(t) = (x_1(t), \ldots, x_n(t))' \in \mathcal{Y}$ for all $t \geq 0$. □

Remark 2.7. Condition (iii) in the above definition is equivalent to $\boldsymbol{u}(t) \in \mathcal{U}(\boldsymbol{x}(t), \boldsymbol{k}(t))$, for $t \geq 0$. □

We use $\mathcal{A}(\boldsymbol{x}, \boldsymbol{k})$ to denote the set of all admissible controls with respect to $\boldsymbol{x} \in \mathcal{Y}$ and $\boldsymbol{k}(0) = \boldsymbol{k}$. The problem is to find an admissible control $\boldsymbol{u}(\cdot) \in \mathcal{A}(\boldsymbol{x}, \boldsymbol{k})$ that minimizes the long-run average-cost function

$$J(\boldsymbol{x}, \boldsymbol{k}, \boldsymbol{u}(\cdot)) = \limsup_{T\to\infty} \frac{1}{T} E \int_0^T g(\boldsymbol{x}(t), \boldsymbol{u}(t))\, dt, \tag{5.11}$$

where $g(\boldsymbol{x}, \boldsymbol{u})$ is the cost of surplus and production and $\boldsymbol{k}$ is the initial value of $\boldsymbol{k}(t)$.

We impose the following assumptions on the process $\boldsymbol{k}(t) = (k_1(t), \ldots, k_{m_c}(t))$ and the cost function $g(\cdot, \cdot)$ throughout this chapter.

(A1) Let $\mathcal{M} = \{\boldsymbol{k}^1, \ldots, \boldsymbol{k}^p\}$ for some positive integer $p \geq 1$, where $\boldsymbol{k}^i = (k_1^i, \ldots, k_{m_c}^i)$, with k_j^i, $j = 1, \ldots, m_c$, denoting the capacity of the jth machine, $i = 1, \ldots, p$. The capacity process $\boldsymbol{k}(t) \in \mathcal{M}$ is a finite state Markov chain with the following infinitesimal generator Q:

$$Q\phi(\boldsymbol{k}^i) = \sum_{j\neq i} q_{\boldsymbol{k}^i\boldsymbol{k}^j}[\phi(\boldsymbol{k}^j) - \phi(\boldsymbol{k}^i)]$$

for some $q_{\boldsymbol{k}^i\boldsymbol{k}^j} \geq 0$ with $j \neq i$, and any function $\phi(\cdot)$ defined on $\mathcal{M}$. Moreover, the Markov chain is strongly irreducible and has the stationary distribution $\nu_{\boldsymbol{k}^i}$, $i = 1, \ldots, p$. That is, the equations

$$(\nu_{\boldsymbol{k}^1}, \ldots, \nu_{\boldsymbol{k}^p})Q = 0 \quad \text{and} \quad \sum_{i=1}^{p} \nu_{\boldsymbol{k}^i} = 1$$

have a unique solution with $\nu_{\boldsymbol{k}^i} > 0$, $i = 1, \ldots, p$.

(A2) Let $p_j = \sum_{i=1}^{p} k_j^i \nu_{\boldsymbol{k}^i}$ and $c(i,j) = \ell$ if $(i,j) \in K_\ell$ for $(i,j) \in A$ and $j \neq n+1$. Here p_j represents the average capacity of machine j, and $c(i,j)$ is the machine number placed at the arc (i,j). Assume that there exist $\{p_{ij} > 0 : (i,j) \in K_\ell\}$ $(\ell = 1, ..., m_c)$ such that

$$\sum_{(i,j)\in K_\ell} p_{ij} \leq 1, \quad \ell = 1, \ldots, m_c, \tag{5.12}$$

$$\sum_{i=-n_0+1}^{m} p_{ij} p_{c(i,j)} > z_j, \quad j = m+1, ..., n, \tag{5.13}$$

and

$$\sum_{i=-n_0+1}^{j-1} p_{ij} p_{c(i,j)} > \sum_{i=j+1}^{n} p_{ji} p_{c(j,i)}, \quad j = 1, ..., m. \tag{5.14}$$

(A3) $g(\boldsymbol{x}, \boldsymbol{u})$ defined on $\mathcal{Y} \times \Re_+^{\tilde{n}}$ is a nonnegative convex function that is strictly convex in either $\boldsymbol{x}$ or $\boldsymbol{u}$ or both with $g(0,0) = 0$. There are positive constants C_{g1}, C_{g2}, and $\beta_{g1} \geq 1$ such that, for any fixed $\boldsymbol{u}$,

$$g(\boldsymbol{x}, \boldsymbol{u}) \geq C_{g1}|\boldsymbol{x}|^{\beta_{g1}} - C_{g2}, \quad \boldsymbol{x} \in \mathcal{Y};$$

moreover, there are constants $C_{g3} > 0$ and $\beta_{g2} \geq \beta_{g1}$ such that, for $\boldsymbol{u}, \widehat{\boldsymbol{u}} \in \Re_+^{\tilde{n}}$, and $\boldsymbol{x}, \widehat{\boldsymbol{x}} \in \mathcal{Y}$,

$$\begin{aligned} &|g(\boldsymbol{x}, \boldsymbol{u}) - g(\widehat{\boldsymbol{x}}, \widehat{\boldsymbol{u}})| \\ &\quad \leq C_{g3}\left[\left(1 + |\boldsymbol{x}|^{\beta_{g2}-1} + |\widehat{\boldsymbol{x}}|^{\beta_{g2}-1}\right)|\boldsymbol{x} - \widehat{\boldsymbol{x}}| + |\boldsymbol{u} - \widehat{\boldsymbol{u}}|\right]. \end{aligned}$$

Let $\lambda^*(\boldsymbol{x}, \boldsymbol{k})$ denote the minimal expected cost with the initial condition $(\boldsymbol{x}(0), \boldsymbol{k}(0)) = (\boldsymbol{x}, \boldsymbol{k})$, i.e.,

$$\lambda^*(\boldsymbol{x}, \boldsymbol{k}) = \inf_{\boldsymbol{u}(\cdot)\in\mathcal{A}(\boldsymbol{x},\boldsymbol{k})} J(\boldsymbol{x}, \boldsymbol{k}, \boldsymbol{u}(\cdot)). \tag{5.15}$$

The Hamilton-Jacobi-Bellman equation in the directional derivative sense (HJBDD) associated with the long-run average-cost optimal control problem formulated in the above takes the following form:

$$\lambda = \inf_{\boldsymbol{u}\in\mathcal{U}(\boldsymbol{x},\boldsymbol{z})} \left\{\partial_{\mathcal{L}(\boldsymbol{u},\boldsymbol{z})}\phi(\boldsymbol{x}, \boldsymbol{k}) + g(\boldsymbol{x}, \boldsymbol{u})\right\} + Q\phi(\boldsymbol{x}, \cdot)(\boldsymbol{k}), \tag{5.16}$$

where λ is a constant, and $\phi(\cdot,\cdot)$ is a real-valued function defined on $\mathcal{Y}\times\mathcal{M}$.

Before we define a solution to the HJBDD equation (5.16), we recall some notation. As in Chapter 4, let $\mathcal{G}$ denote the family of real-valued convex functions $G(\cdot,\boldsymbol{k})$ defined on $\mathcal{Y}$ for each $\boldsymbol{k}\in\mathcal{M}$ with at most the $(\beta_{g2}+1)$th degree polynomial growth. Similar to Definition 4.2.3, a control $\boldsymbol{u}(\cdot)\in\mathcal{A}(\boldsymbol{x},\boldsymbol{k})$ is stable if its corresponding state process $\boldsymbol{x}(\cdot)$ satisfies $\lim_{T\to\infty} E|\boldsymbol{x}(T)|^{\beta_{g2}+1}/T=0$.

A solution to the HJBDD equation (5.16) is a pair $(\lambda, W(\cdot,\cdot))$ with λ a constant and $W(\cdot,\cdot)\in\mathcal{G}$. The function $W(\cdot,\cdot)$ is called a *potential function* for the control problem, if $(\lambda, W(\cdot,\cdot))$ is a solution of the HJBDD equation (5.16) and λ is the minimum long-run average cost.

Since we will use the vanishing discount approach to study our problem, we provide an analysis of the discounted problem in the next section. Before proceeding to the next section, it may be convenient to reiterate the following system parameters that we have used:

m_c	:	number of machines;
m	:	number of internal buffers;
n	:	number of buffers (internal and external);
$n-m$	:	number of external buffers; and
n_0	:	number of the sources.

For example, in Figure 5.1, $m_c=4$, $m=3$, $n=5$, and $n_0=1$. In the particular case of flowshop, $n_0=1$ and $n-m=1$. We should caution that in Chapter 4 which deals with flowshops, we have denoted the number of machines by m.

5.3 Estimates for Discounted Cost Value Functions

First we introduce the corresponding control problem with the cost discounted at a rate $\rho>0$. For $\boldsymbol{u}(\cdot)\in\mathcal{A}(\boldsymbol{x},\boldsymbol{k})$, we define the expected discounted cost as

$$J^{\rho}(\boldsymbol{x},\boldsymbol{k},\boldsymbol{u}(\cdot)) = E\int_0^\infty e^{-\rho t} g(\boldsymbol{x}(t),\boldsymbol{u}(t))\,dt, \tag{5.17}$$

where $\boldsymbol{x}(\cdot)$ is the system state corresponding to the control $\boldsymbol{u}(\cdot)$, and the initial conditions $\boldsymbol{x}(0)=\boldsymbol{x}$ and $\boldsymbol{k}(0)=\boldsymbol{k}$. Define the value function of the discounted problem as

$$V^{\rho}(\boldsymbol{x},\boldsymbol{k}) = \inf_{\boldsymbol{u}(\cdot)\in\mathcal{A}(\boldsymbol{x},\boldsymbol{k})} J^{\rho}(\boldsymbol{x},\boldsymbol{k},\boldsymbol{u}(\cdot)).$$

The corresponding HJBDD equation can be written as

$$\rho\phi^{\rho}(\boldsymbol{x},\boldsymbol{k}) = \inf_{\boldsymbol{u}\in\mathcal{U}(\boldsymbol{x},\boldsymbol{k})} \left\{\partial_{\mathcal{L}(\boldsymbol{u},\boldsymbol{z})}\phi^{\rho}(\boldsymbol{x},\boldsymbol{k}) + g(\boldsymbol{x},\boldsymbol{u})\right\} + Q\phi^{\rho}(\boldsymbol{x},\cdot)(\boldsymbol{k}), \tag{5.18}$$

where $\phi^\rho(\cdot,\cdot)$ is defined on $\mathcal{Y}\times\mathcal{M}$. Just as in the flowshop case (see Theorem 4.3.1), we have the following result:

Theorem 3.1. *Under Assumptions* (A1) *and* (A3), *the value function* $V^\rho(\boldsymbol{x},\boldsymbol{k})$ *has the following properties:*

(i) *For each* $\boldsymbol{k}\in\mathcal{M}$, *the value function* $V^\rho(\boldsymbol{x},\boldsymbol{k})$ *is convex and continuous on* $\mathcal{Y}$ *and satisfies the condition*

$$|V^\rho(\widehat{\boldsymbol{x}},\boldsymbol{k})-V^\rho(\widetilde{\boldsymbol{x}},\boldsymbol{k})|\le C\left(1+|\widehat{\boldsymbol{x}}|^{\beta_{g2}-1}+|\widetilde{\boldsymbol{x}}|^{\beta_{g2}-1}\right)|\widehat{\boldsymbol{x}}-\widetilde{\boldsymbol{x}}|, \quad (5.19)$$

for some positive constant C *and all* $\widehat{\boldsymbol{x}},\widetilde{\boldsymbol{x}}\in\mathcal{Y}$.

(ii) *The value function* $V^\rho(\boldsymbol{x},\boldsymbol{k})$ *satisfies the* HJBDD *equation* (5.18).

In order to study the long-run average-control problem using the vanishing discount approach, we must first obtain some estimates for the value function $V^\rho(\boldsymbol{x},\boldsymbol{k})$. To get the estimates, we first construct a stable control.

Theorem 3.2. *Let Assumptions* (A1) *and* (A2) *hold, and let* $m=n-1$. *For any* $r\ge 1$, $(\boldsymbol{x},\boldsymbol{k})\in\mathcal{Y}\times\mathcal{M}$ *and* $(\widehat{\boldsymbol{x}},\widehat{\boldsymbol{k}})\in\mathcal{Y}\times\mathcal{M}$, *there exist a positive constant* C_r *independent of* $(\boldsymbol{x},\boldsymbol{k})$ *and* $(\widehat{\boldsymbol{x}},\widehat{\boldsymbol{k}})$, *and a control* $\boldsymbol{u}(t), t\ge 0$, *such that*

$$E\left[\tau(\boldsymbol{x},\boldsymbol{k},\widehat{\boldsymbol{x}},\widehat{\boldsymbol{k}})\right]^r\le C_r\left(1+\sum_{j=1}^{n}|x_j-\widehat{x}_j|^r\right), \quad (5.20)$$

where

$$\tau(\boldsymbol{x},\boldsymbol{k},\widehat{\boldsymbol{x}},\widehat{\boldsymbol{k}})=\inf\{t\ge 0:\boldsymbol{x}(t)=\widehat{\boldsymbol{x}},\ \boldsymbol{k}(t)=\widehat{\boldsymbol{k}}\},$$

and $\boldsymbol{x}(t), t\ge 0$, *is the surplus process corresponding to the control* $\boldsymbol{u}(\cdot)$ *and the initial condition* $(\boldsymbol{x}(0),\boldsymbol{k}(0))=(\boldsymbol{x},\boldsymbol{k})$.

Comparing with the proof of Theorem 4.3.2, the proof of the theorem is more complicated and sophisticated. As in Theorem 4.3.2, in order to find $\tau(\boldsymbol{x},\boldsymbol{k},\widehat{\boldsymbol{x}},\widehat{\boldsymbol{k}})$, we alternate between the two policies described below. In the first policy, the production rate for each input is the maximum admissible capacity. In the second policy, we set $u_{i,j}(t)=0$, for $i=-n_0+1,\dots,0$ and $j=1,\dots,n$, and set the maximum possible production rate for all outputs that are not connected to the external buffers under the restriction that the content of each buffer j, $1\le j\le n-1$, is not less than $\widehat{x}_j$. The first policy is used until such time when the content of the first buffer exceeds the value $\widehat{x}_1$ and the content of each buffer j, $2\le j\le n$, exceeds the value $M+\widehat{x}_j$ for some $M>0$. At that time we switch to the second policy. We use the second policy until such time when the content of the last buffer drops to the level $\widehat{x}_n$. After that we revert to the first policy, and so on.

Using this alternating procedure, it is possible to specify $\tau(\boldsymbol{x}, \boldsymbol{k}, \widehat{\boldsymbol{x}}, \widehat{\boldsymbol{k}})$ and provide an estimation for it.

Proof of Theorem 3.2. We divide the proof into several steps.

Step 1. In this step, for any $s \geq 0$, we construct a family of auxiliary processes $\boldsymbol{x}^0(t|s, \boldsymbol{x})$, $t \geq s \geq 0$, and $\boldsymbol{x} \in \mathcal{Y}$, which will be used to generate the desired policy $\boldsymbol{u}(t)$ with s to be changed into a stopping time with respect to process $\boldsymbol{k}(t)$. The function $\boldsymbol{u}^0(\boldsymbol{x}, \boldsymbol{k}) = \{u^0_{i,j}(\boldsymbol{x}, \boldsymbol{k}) : (i, j) \in A \text{ and } j \neq n+1\}$ is given by

$$u^0_{i,j}(\boldsymbol{x}, \boldsymbol{k}) = p_{ij} k_{c(i,j)}, \quad i = -n_0 + 1, \ldots, 0,$$

$$u^0_{i,j}(\boldsymbol{x}, \boldsymbol{k}) = \begin{cases} p_{ij} k_{c(i,j)}, & \text{if } x_i > 0, \\ \dfrac{\left[\left(\sum\limits_{\ell=-n_0+1}^{i-1} u_{\ell,i}(\boldsymbol{x}, \boldsymbol{k}) \right) \wedge \left(\sum\limits_{\ell=i+1}^{n} p_{i\ell} k_{c(i,\ell)} \right) \right] p_{ij} k_{c(i,j)}}{\sum\limits_{\ell=i+1}^{n} p_{i\ell} k_{c(i,\ell)}}, & \text{if } x_i = 0, \end{cases}$$

$$i = 1, \ldots, n-1.$$

For $(i, j) \notin A$, let $u_{i,j}(\boldsymbol{x}, \boldsymbol{k}) = 0$. We define $\boldsymbol{x}^0(t|s, \boldsymbol{x})$ as the process which satisfies the following equation (see (5.5)):

$$\frac{d}{dt} x^0_j(t|s, \boldsymbol{x}) = \sum_{\ell=-n_0+1}^{j-1} u^0_{\ell,j}(\boldsymbol{x}^0(t|s, \boldsymbol{x}), \boldsymbol{k}(t)) - \sum_{\ell=j+1}^{n} u^0_{j,\ell}(\boldsymbol{x}^0(t|s, \boldsymbol{x}), \boldsymbol{k}(t)),$$
$$1 \leq j \leq n-1,$$

$$\frac{d}{dt} x^0_n(t|s, \boldsymbol{x}) = \sum_{\ell=-n_0+1}^{n-1} u^0_{\ell,n}(\boldsymbol{x}^0(t|s, \boldsymbol{x}), \boldsymbol{k}(t)) - z_n, \quad \boldsymbol{x}^0(s|s, \boldsymbol{x}) = \boldsymbol{x}.$$

Clearly $\boldsymbol{x}^0(t|s, \boldsymbol{x}) \in \mathcal{Y}$ for all $t \geq s$. For fixed s, $\boldsymbol{x}^0(t|s, \boldsymbol{x})$ is the state of the system with the production rate which is obtained by using the maximum admissible modified capacity at each machine.

Define now the stopping time with respect to the Markov chain $\boldsymbol{k}(\cdot)$,

$$\begin{aligned} \theta(s, \boldsymbol{x}, \widehat{\boldsymbol{x}}) = \inf \{ t \geq s : \; & x^0_1(t|s, \boldsymbol{x}) \geq \widehat{x}_1, \text{ and} \\ & x^0_j(t|s, \boldsymbol{x}) \geq M + \widehat{x}_j \text{ for } j = 2, \ldots, n \}, \end{aligned} \tag{5.21}$$

where $M > 0$ is a constant specified in Lemma 3.2. It follows from this definition that $\theta(s, \boldsymbol{x}, \widehat{\boldsymbol{x}})$ is the first time when the state process $\boldsymbol{x}^0(t|s, \boldsymbol{x})$

exceeds $(\widehat{x}_1, M + \widehat{x}_2, \ldots, M + \widehat{x}_n)$ under the control $\boldsymbol{u}^0(\boldsymbol{x}^0(t|s,\boldsymbol{x}), \boldsymbol{k}(t))$. Under Assumption (A.2), the following lemma holds. Its proof will be given in Appendix B.

Lemma 3.1. *Under Assumptions* (A1) *and* (A2), *for any* $r \geq 1$, *there exists a constant* $\widehat{C}_r$ *such that*

$$E\left[\theta(s,\boldsymbol{x},\widehat{\boldsymbol{x}}) - s\right]^{2r} < \widehat{C}_r \left[1 + \sum_{j=1}^{n} \left((\widehat{x}_j - x_j)^+\right)^r\right]^2.$$

Step 2. In this step, for any $s \geq 0$, we construct a family of auxiliary processes $\boldsymbol{x}^1(t|s,\boldsymbol{x})$, $t \geq s \geq 0$ and $\boldsymbol{x} \in \mathcal{Y}$, which will be used to construct the desired policy $\boldsymbol{u}(t)$. To do this, consider the following function $\boldsymbol{u}^1(\boldsymbol{x},\boldsymbol{k}) = \{u^1_{i,j}(\boldsymbol{x},\boldsymbol{k}), (i,j) \in A \text{ and } j \neq n+1\}$.

$$u^1_{i,\ell}(\boldsymbol{x},\boldsymbol{k}) = 0, \quad i = -n_0+1, \ldots, 0,$$

$$u^1_{i,j}(\boldsymbol{x},\boldsymbol{k}) = \begin{cases} p_{ij} k_{c(i,j)}, & \text{if } x_i > \widehat{x}_i, \\ \dfrac{\left[\left(\sum_{\ell=-n_0+1}^{i-1} u_{\ell,i}(\boldsymbol{x},\boldsymbol{k})\right) \wedge \left(\sum_{\ell=i+1}^{n} p_{i\ell} k_{c(i,\ell)}\right)\right] p_{ij} k_{c(i,j)}}{\sum_{\ell=i+1}^{n} p_{i\ell} k_{c(i,\ell)}}, & \text{if } x_i = \widehat{x}_i, \end{cases}$$

$$i = 1, \ldots, n-1.$$

As the policy $\boldsymbol{u}^0(\boldsymbol{x},\boldsymbol{k})$, for $(i,j) \notin A$, let $u^1_{i,j}(\boldsymbol{x},\boldsymbol{k}) = 0$. We define $\boldsymbol{x}^1(t|s,\boldsymbol{x})$ as a continuous process which coincides with $\boldsymbol{x}^0(t|s,\boldsymbol{x})$ for $s \leq t < \theta(s,\boldsymbol{x},\widehat{\boldsymbol{x}})$, however, for $t \geq \theta(s,\boldsymbol{x},\widehat{\boldsymbol{x}})$, define it to be satisfied by the following equation (see (5.5)):

$$\frac{d}{dt} x^1_i(t|s,\boldsymbol{x}) = \sum_{\ell=-n_0+1}^{i-1} u^1_{\ell,i}(\boldsymbol{x}^1(t|s,\boldsymbol{x}), \boldsymbol{k}(t)) - \sum_{\ell=i+1}^{n} u^1_{i,\ell}(\boldsymbol{x}^1(t|s,\boldsymbol{x}), \boldsymbol{k}(t)), \quad 1 \leq i \leq n-1,$$

$$\frac{d}{dt} x^1_n(t|s,\boldsymbol{x}) = \sum_{\ell=-n_0+1}^{n-1} u^1_{\ell,n}(\boldsymbol{x}^1(t|s,\boldsymbol{x}), \boldsymbol{k}(t)) - z_n, \quad \boldsymbol{x}^1(s|s,\boldsymbol{x}) = \boldsymbol{x}^0(s|s,\boldsymbol{x}).$$

Clearly $\boldsymbol{x}^1(t|s,\boldsymbol{x}) \in \mathcal{Y}$ for all $t \geq s$, and $x^1_j(t|s,\boldsymbol{x}) \geq \widehat{x}_j$ $(1 \leq j \leq n-1)$ for $t \geq \theta(s,\boldsymbol{x},\widehat{\boldsymbol{x}})$. This process corresponds to a policy in which, after $\theta(s,\boldsymbol{x},\widehat{\boldsymbol{x}})$, we set $u_{\ell,j}(t) = 0$ for $\ell = -n_0+1, \ldots, 0$ and any j, and set the maximum possible production rate for all other outputs that are not connected to the external buffers, under the restriction that the content of each buffer j $(1 \le j \le n-1)$ is not less than $\widehat{x}_j$.

We define now a stopping time with respect to the Markov chain $\boldsymbol{k}(t)$.

$$\widehat{\theta}(s, \boldsymbol{x}, \widehat{\boldsymbol{x}}) = \inf\{t \ge \theta(s, \boldsymbol{x}, \widehat{\boldsymbol{x}}) : x_n^1(t|s, \boldsymbol{x}) = \widehat{x}_n\}. \tag{5.22}$$

Lemma 3.2. *Let Assumptions* (A1) *and* (A2) *hold. Then:*

(i) *For any given* $q \in (0,\ 1)$, *a constant* M *can be chosen in such a way that, for all* $s \ge 0$ and $\boldsymbol{x} \in \mathcal{Y}$,

$$P\big(\boldsymbol{x}^1(\widehat{\theta}(s, \boldsymbol{x}, \widehat{\boldsymbol{x}})|s, \boldsymbol{x}) = \widehat{\boldsymbol{x}},\ \boldsymbol{k}(\widehat{\theta}(s, \boldsymbol{x}, \widehat{\boldsymbol{x}})) = \widehat{\boldsymbol{k}}\big) \ge 1 - q > 0.$$

(ii) *There exists a constant* C *such that*

$$\begin{aligned} \frac{M}{z_n} &\le \widehat{\theta}(s, \boldsymbol{x}, \widehat{\boldsymbol{x}}) - s \\ &\le \frac{1}{z_n}\left(\sum_{j=1}^{n}(x_j - \widehat{x}_j) + C[\theta(s, \boldsymbol{x}, \widehat{\boldsymbol{x}}) - s]\right) \end{aligned} \tag{5.23}$$

and

$$\sum_{j=1}^{n} x_j^1(\widehat{\theta}(s, \boldsymbol{x}, \widehat{\boldsymbol{x}})|s, \boldsymbol{x}) \le \sum_{j=1}^{n} x_j + C[\theta(s, \boldsymbol{x}, \widehat{\boldsymbol{x}}) - s]. \tag{5.24}$$

Proof. See Appendix B. □

Step 3. Using the policies $\boldsymbol{u}^0(\boldsymbol{x}, \boldsymbol{k})$ and $\boldsymbol{u}^1(\boldsymbol{x}, \boldsymbol{k})$ given in Steps 1 and 2, now we construct a process $\boldsymbol{x}(t)$ $(t \ge 0)$ and the corresponding control $\boldsymbol{u}(\cdot)$, which satisfies the statement of Theorem 3.2.

Define a sequence of stopping times $\{\widehat{\theta}_\ell\}_{\ell=0}^{\infty}$ and the process $\boldsymbol{x}(t)$ for $\widehat{\theta}_\ell \le t < \widehat{\theta}_{\ell+1}$ $(\ell = 0, 1, 2, \ldots)$ as follows:

$$\widehat{\theta}_0 = 0, \quad \widehat{\theta}_1 = \widehat{\theta}(0, \boldsymbol{x}, \widehat{\boldsymbol{x}}) \quad \text{and} \quad \boldsymbol{x}(t) = \boldsymbol{x}^1(t|0, \boldsymbol{x}) \quad \text{with} \quad 0 \le t < \widehat{\theta}_1.$$

If $\widehat{\theta}_\ell$ is defined for $\ell \ge 1$ and $\boldsymbol{x}(t)$ is defined for $0 \le t < \widehat{\theta}_\ell$, then we let

$$\begin{aligned} \theta_{\ell+1} &= \theta(\widehat{\theta}_\ell, \boldsymbol{x}(\widehat{\theta}_\ell), \widehat{\boldsymbol{x}}), \\ \widehat{\theta}_{\ell+1} &= \widehat{\theta}(\widehat{\theta}_\ell, \boldsymbol{x}(\widehat{\theta}_\ell), \widehat{\boldsymbol{x}}). \end{aligned}$$

Define a control

$$\boldsymbol{u}(t) = \begin{cases} \boldsymbol{u}^0(\boldsymbol{x}(t), \boldsymbol{k}(t)), & \text{if } \widehat{\theta}_{\ell-1} \le t < \theta_\ell, \\ \boldsymbol{u}^1(\boldsymbol{x}(t), \boldsymbol{k}(t)), & \text{if } \theta_\ell \le t < \widehat{\theta}_\ell, \end{cases} \tag{5.25}$$

for $\ell = 1, 2, \ldots$. Let $\boldsymbol{x}(\cdot)$, $t \geq 0$, be the corresponding trajectories with the initial condition $\boldsymbol{x}(0) = \boldsymbol{x}$. According to the first inequality in (5.23), $\widehat{\theta}_\ell \to \infty$ as $\ell \to \infty$. Hence, the process $\boldsymbol{x}(t)$ is defined for all $t \geq 0$. It is clear that this $\boldsymbol{u}(t) \in \mathcal{A}(\boldsymbol{x}, \boldsymbol{k})$. For the process $\boldsymbol{x}(\cdot)$, a Markov time is defined as

$$\tau(\boldsymbol{x}, \boldsymbol{k}, \widehat{\boldsymbol{x}}, \widehat{\boldsymbol{k}}) = \inf\{t \geq 0 : \boldsymbol{x}(t) = \widehat{\boldsymbol{x}},\ \boldsymbol{k}(t) = \widehat{\boldsymbol{k}}\}.$$

Let $\mathcal{B}_\ell = \{\omega : (\boldsymbol{x}(\widehat{\theta}_\ell), \boldsymbol{k}(\widehat{\theta}_\ell)) = (\widehat{\boldsymbol{x}}, \widehat{\boldsymbol{k}})\}$. Using conditional probabilities (see (3.15)), we have, from Lemma 3.2(i), that there exists a positive constant M such that

$$P\left(\bigcap_{\ell_1=1}^{\ell} \mathcal{B}_{\ell_1}^c\right) \leq q^\ell, \quad \ell = 1, 2, \ldots, \tag{5.26}$$

where $\mathcal{B}_{\ell_1}^c$ is the complement set of $\mathcal{B}_{\ell_1}$. Using the definition of $\boldsymbol{x}(t)$ we get that

$$[\tau(\boldsymbol{x}, \boldsymbol{k}, \widehat{\boldsymbol{x}}, \widehat{\boldsymbol{k}})]^r = \sum_{\ell=1}^{\infty} [\widehat{\theta}_\ell]^r I_{\{\cap_{\ell_1=0}^{\ell-1} \mathcal{B}_{\ell_1}^c \cap \mathcal{B}_\ell\}}, \tag{5.27}$$

where $\mathcal{B}_0^c = \Omega$. Using (5.23) and (5.24) we have that, for $\ell = 1, 2, \ldots$,

$$\widehat{\theta}_\ell - \widehat{\theta}_{\ell-1} \leq \frac{1}{z_n}\left[\sum_{j=1}^{n} x_j(\widehat{\theta}_{\ell-1}) - \sum_{j=1}^{n} \widehat{x}_j + C(\theta_\ell - \widehat{\theta}_{\ell-1})\right], \tag{5.28}$$

and

$$\sum_{j=1}^{n} x_j(\widehat{\theta}_\ell) \leq \sum_{j=1}^{n} x_j(\widehat{\theta}_{\ell-1}) + C[\theta_\ell - \widehat{\theta}_{\ell-1}]. \tag{5.29}$$

Using (5.28) and (5.29), there exists a constant $C_1 > 0$ independent of $\boldsymbol{x}$ and $\widehat{\boldsymbol{x}}$ such that, for $\ell = 1, 2, \ldots$,

$$\widehat{\theta}_\ell \leq \frac{C_1(\ell+1)}{z_n}\left(\sum_{j=1}^{n} (x_j - \widehat{x}_j)^+ + \sum_{\ell_1=1}^{\ell} (\theta_{\ell_1} - \widehat{\theta}_{\ell_1-1})\right),$$

this implies

$$(\widehat{\theta}_\ell)^r \leq C_1(\ell+1)^{2r}\left[\sum_{j=1}^{n} \left((x_j - \widehat{x}_j)^+\right)^r + \sum_{\ell_1=1}^{\ell} (\theta_{\ell_1} - \widehat{\theta}_{\ell_1-1})^r\right]. \tag{5.30}$$

Note that $\boldsymbol{x}(\widehat{\theta}_{\ell_1}) \geq \widehat{\boldsymbol{x}}$ for $\ell_1 = 1, 2, \ldots$. Using the Schwarz inequality (Corollary 3 on page 104 of Chow and Teicher [33]), we get, from (5.26) and

Lemma 3.1, that there exists a positive constant $\overline{C}_r$ dependent on r such that

$$E\left(\theta_1^r I_{\{\cap_{\ell_1=1}^{\ell-1}\mathcal{B}_{\ell_1}^c\cap\mathcal{B}_\ell\}}\right) \le q^{(\ell-1)/2}\left(E(\theta_1^{2r})\right)^{1/2} \le \overline{C}_r q^{(\ell-1)/2}\left[1+\sum_{j=1}^n\left((\widehat{x}_j-x_j)^+\right)^r\right], \quad \ell=1,2,\ldots, \tag{5.31}$$

and

$$\begin{aligned} &E\left((\theta_{\ell_1}-\widehat{\theta}_{\ell_1-1})^r I_{\{\cap_{\ell_1=1}^{\ell-1}\mathcal{B}_{\ell_1}^c\cap\mathcal{B}_\ell\}}\right)\\ &\le q^{(\ell-1)/2}\left[E(\theta_{\ell_1}-\widehat{\theta}_{\ell_1-1})^{2r}\right]^{1/2}\\ &\le \overline{C}_r q^{(\ell-1)/2}\left[1+\sum_{j=1}^n\left((\hat{x}_j-x_j)^+\right)^r\right],\\ &\qquad 2\le \ell_1\le \ell=2,3,\ldots. \end{aligned} \tag{5.32}$$

Substituting (5.30) into (5.27), taking expectation, and using (5.31) and (5.32), we get (5.20). The proof of the theorem is completed. □

5.4 Verification Theorem

Our goal is to construct a pair $(\lambda, W(\cdot,\cdot))$ which satisfies (5.16). To get this pair, we use the vanishing discount approach. First, based on Theorem 3.2, we get the following limits:

Theorem 4.1. *Let Assumptions* (A1)–(A3) *hold. There exists a sequence* $\{\rho_\ell : \ell \ge 1\}$ *with* $\rho_\ell \to 0$ *as* $\ell \to \infty$ *such that for* $(\boldsymbol{x}, \boldsymbol{k}) \in \mathcal{Y} \times \mathcal{M}$, *the limits of* $\rho_\ell V^{\rho_\ell}(\boldsymbol{x}, \boldsymbol{k})$ *and* $[V^{\rho_\ell}(\boldsymbol{x}, \boldsymbol{k}) - V^{\rho_\ell}(0, \boldsymbol{k})]$ *exist. Write*

$$\lim_{\ell\to\infty} \rho_\ell V^{\rho_\ell}(\boldsymbol{x}, \boldsymbol{k}) = \widehat{\lambda}, \tag{5.33}$$

$$\lim_{\ell\to\infty} [V^{\rho_\ell}(\boldsymbol{x}, \boldsymbol{k}) - V^{\rho_\ell}(0, \boldsymbol{k})] = V(\boldsymbol{x}, \boldsymbol{k}). \tag{5.34}$$

The limit $V(\boldsymbol{x}, \boldsymbol{k})$ *is convex for any fixed* $\boldsymbol{k} \in \mathcal{M}$.

Proof. For the value function $V^\rho(\boldsymbol{x}, \boldsymbol{k})$ of the discounted cost problem, we define the relative discounted value function

$$\widetilde{V}^\rho(\boldsymbol{x}, \boldsymbol{k}) = V^\rho(\boldsymbol{x}, \boldsymbol{k}) - V^\rho(0, \boldsymbol{k}).$$

By Theorem 3.2 we know that there exists a control policy $\boldsymbol{u}(\cdot) \in \mathcal{A}(\boldsymbol{x}, \boldsymbol{k})$ such that, for each $r \ge 1$,

$$E\left[\tau(\boldsymbol{x}, \boldsymbol{k}, \boldsymbol{x}, \boldsymbol{k})\right]^r \le C_1, \tag{5.35}$$

where $C_1 > 0$ is a constant (which depends on r) and

$$\tau(\boldsymbol{x},\boldsymbol{k},\boldsymbol{x},\boldsymbol{k}) = \inf\{t>0 : (\boldsymbol{x}(t),\boldsymbol{k}(t)) = (\boldsymbol{x},\boldsymbol{k})\},$$

with $\boldsymbol{x}(\cdot)$ being the surplus process corresponding to the control $\boldsymbol{u}(\cdot)$ and the initial condition $(\boldsymbol{x}(0),\boldsymbol{k}(0)) = (\boldsymbol{x},\boldsymbol{k})$. For notation simplicity, in the following we write $\tau(\boldsymbol{x},\boldsymbol{k},\boldsymbol{x},\boldsymbol{k})$ as τ.

By the optimality principle, we have

$$\begin{aligned} V^\rho(\boldsymbol{x},\boldsymbol{k}) &\le E\left\{\int_0^\tau e^{-\rho t} g(\boldsymbol{x}(t),\boldsymbol{u}(t))\,dt + e^{-\rho\tau} V^\rho(\boldsymbol{x}(\tau),\boldsymbol{k}(\tau))\right\} \\ &= E\left\{\int_0^\tau e^{-\rho t} g(\boldsymbol{x}(t),\boldsymbol{u}(t))\,dt + e^{-\rho\tau} V^\rho(\boldsymbol{x},\boldsymbol{k})\right\}. \end{aligned} \tag{5.36}$$

Note that

$$|\boldsymbol{x}(t)| \le \sum_{j=1}^n |x_j| + \left(\sum_{j=1}^n \max_{1\le i\le p} k_j^i + z_n\right) t, \quad \text{for } 0 \le t \le \tau.$$

Thus, by Assumption (A3), we have

$$g(\boldsymbol{x}(t),\boldsymbol{u}(t)) \le C_2\left(1 + t + t^{\beta_{g2}}\right),$$

where C_2 is a positive constant dependent on $\boldsymbol{x}$. It follows from (5.35) and (5.36) that

$$\left[1 - Ee^{-\rho\tau}\right] V^\rho(\boldsymbol{x},\boldsymbol{k}) \le E\int_0^\tau C_2\left(1+t+t^{\beta_{g2}}\right) dt \le C_3, \tag{5.37}$$

for some positive constant C_3 (independent of ρ). Now using the inequality

$$1 - e^{-\rho\tau} \ge \rho\tau - \rho^2\tau^2/2,$$

we can get

$$\left[1 - Ee^{-\rho\tau}\right] V^\rho(\boldsymbol{x},\boldsymbol{k}) \ge \left(E\tau - \rho E[\tau^2]/2\right) \cdot \rho V^\rho(\boldsymbol{x},\boldsymbol{k}). \tag{5.38}$$

From (5.35), we have

$$0 < E\tau < \infty \quad \text{and} \quad 0 < E[\tau^2] < \infty.$$

Take $\rho_0 = E\tau/E[\tau^2]$. By (5.37) and (5.38), we have that, for $0 < \rho \le \rho_0$,

$$\rho V^\rho(\boldsymbol{x},\boldsymbol{k}) \le \frac{C_3}{E\tau - \rho_0 E[\tau^2]/2} = \frac{2C_3}{E\tau} < \infty.$$

Consequently, there exists a sequence $\{\rho_\ell : \ell \ge 1\}$ with $\rho_\ell \to 0$ as $\ell \to \infty$ such that, for $(\boldsymbol{x},\boldsymbol{k}) \in \mathcal{Y}\times\mathcal{M}$,

$$\lim_{\ell\to\infty} \rho_\ell V^{\rho_\ell}(\boldsymbol{x},\boldsymbol{k}) = \widehat{\lambda}. \tag{5.39}$$

Now we prove (5.34). To do this we first show that there is a constant $C_4 > 0$ such that

$$|\widetilde{V}^\rho(\boldsymbol{x},\boldsymbol{k})| \le C_4\left(1+|\boldsymbol{x}|^{\beta_{g2}+1}\right), \tag{5.40}$$

for all $(\boldsymbol{x},\boldsymbol{k}) \in \mathcal{Y}\times\mathcal{M}$ and $\rho > 0$. Without loss of generality we suppose that $V^\rho(\boldsymbol{x},\boldsymbol{k}) \ge V^\rho(0,\boldsymbol{k})$ (the case $V^\rho(\boldsymbol{x},\boldsymbol{k}) \le V^\rho(0,\boldsymbol{k})$ is treated in the same way).

By Theorem 3.2 there exists a control policy $\boldsymbol{u}(\cdot)$ such that, for any $r \ge 1$,

$$E(\tau_1)^r \le C_r\left(1+\sum_{j=1}^n |x_j|^r\right), \tag{5.41}$$

where

$$\tau_1 = \inf\{t > 0 : (\boldsymbol{x}(t),\boldsymbol{k}(t)) = (0,\boldsymbol{k})\},$$

and $\boldsymbol{x}(t)$ is the state process corresponding to $\boldsymbol{u}(t)$ with the initial condition $(\boldsymbol{x}(0),\boldsymbol{k}(0)) = (\boldsymbol{x},\boldsymbol{k})$. Using the dynamic programming principle, similar to (5.36), we have

$$V^\rho(\boldsymbol{x},\boldsymbol{k}) \le E\left\{\int_0^{\tau_1} e^{-\rho t} g(\boldsymbol{x}(t),\boldsymbol{u}(t))\,dt + e^{-\rho\tau_1}V^\rho(0,\boldsymbol{k})\right\}.$$

Therefore,

$$\begin{aligned} |\widetilde{V}^\rho(\boldsymbol{x},\boldsymbol{k})| &= V^\rho(\boldsymbol{x},\boldsymbol{k}) - V^\rho(0,\boldsymbol{k}) \\ &\le E\left[\int_0^{\tau_1} e^{-\rho t} g(\boldsymbol{x}(t),\boldsymbol{u}(t))\,dt\right]. \end{aligned} \tag{5.42}$$

By Assumption (A3), there exists a $C_5 > 0$ such that

$$g(\boldsymbol{x}(t),\boldsymbol{u}(t)) \le C_5\left(1+|\boldsymbol{x}|^{\beta_{g2}}+t+t^{\beta_{g2}}\right), \tag{5.43}$$

where we use the fact that $\boldsymbol{u}(\cdot)$ is bounded. Therefore, (5.41) and (5.43) imply that

$$\begin{aligned} E\int_0^{\tau_1} e^{-\rho t} g(\boldsymbol{x}(t),\boldsymbol{u}(t))\,dt &\le E\int_0^{\tau_1} C_5(1+|\boldsymbol{x}|^{\beta_{g2}}+t+t^{\beta_{g2}})\,dt \\ &\le C_6\left(1+\sum_{j=1}^n |x_j|^{\beta_{g2}+1}\right), \end{aligned}$$

for some $C_6 > 0$. Thus (5.42) gives (5.40).

For $\delta \in (0,\ 1)$, let

$$\mathcal{O}_\delta = \left[\delta,\ \frac{1}{\delta}\right]^{n-1} \times \left[-\frac{1}{\delta},\ \frac{1}{\delta}\right].$$

Based on (5.40) it follows from Lemma C.4 that there is a C_δ such that, for $\boldsymbol{x}, \widehat{\boldsymbol{x}} \in \mathcal{O}_\delta$,

$$|\widetilde{V}^\rho(\boldsymbol{x}, \boldsymbol{k}) - \widetilde{V}^\rho(\widehat{\boldsymbol{x}}, \boldsymbol{k})| \leq C_\delta |\boldsymbol{x} - \widehat{\boldsymbol{x}}|. \tag{5.44}$$

Without loss of generality we assume that C_δ is a decreasing function in δ. For $1 \leq \hat{n} \leq n-1$ and $1 \leq i_1 < \cdots < i_{\hat{n}} \leq n-1$, let

$$\mathcal{O}^{i_1 \dots i_{\hat{n}}} = \{\boldsymbol{x} \in \mathcal{Y} : x_{i_\ell} = 0 \text{ for } \ell = 1, ..., \hat{n}\}$$

and

$$\text{ri}(\mathcal{O}^{i_1 \dots i_{\hat{n}}}) = \{\boldsymbol{x} \in \mathcal{O}^{i_1 \dots i_{\hat{n}}} : x_j > 0,\ j \neq i_\ell \text{ and } 1 \leq j \leq n-1\}.$$

That is, $\text{ri}(\mathcal{O}^{i_1 \dots i_{\hat{n}}})$ is the relative interior of $\mathcal{O}^{i_1 \dots i_{\hat{n}}}$ relative to $[0, \infty)^{n-n_0-1} \times \{0\}^{n_0} \times (-\infty, +\infty)$. Note that the function $V^\rho(\boldsymbol{x}, \boldsymbol{k})$ is still convex on $\mathcal{O}^{i_1 \dots i_{\hat{n}}}$. Let

$$\mathcal{O}_\delta^{i_1 \dots i_{\hat{n}}} = \prod_{\ell=1}^{n-1} \Upsilon_\ell^\delta \times \left[-\frac{1}{\delta}, \frac{1}{\delta}\right]$$

with

$$\Upsilon_\ell^\delta = \begin{cases} \{0\}, & \text{if } \ell \in \{i_1, ..., i_{\hat{n}}\}, \\ [\delta,\ 1/\delta], & \text{if } \ell \notin \{i_1, ..., i_{\hat{n}}\}. \end{cases}$$

Using again Lemma C.4, in view of (5.40), there is a $C_\delta^{i_1 \dots i_{\hat{n}}} > 0$ such that, for $\boldsymbol{x}, \widehat{\boldsymbol{x}} \in \mathcal{O}_\delta^{i_1 \dots i_{\hat{n}}}$,

$$|\widetilde{V}^\rho(\boldsymbol{x}, \boldsymbol{k}) - \widetilde{V}^\rho(\widehat{\boldsymbol{x}}, \boldsymbol{k})| \leq C_\delta^{i_1 \dots i_{\hat{n}}} |\boldsymbol{x} - \widehat{\boldsymbol{x}}|. \tag{5.45}$$

Also we assume that $C_\delta^{i_1 \dots i_{\hat{n}}}$ is a decreasing function in δ. From the arbitrary of δ and (5.44)–(5.45), there exist $V(\boldsymbol{x}, \boldsymbol{k})$ and a sequence of $\{\rho_\ell : \ell \geq 1\}$ with $\rho_\ell \to 0$ as $\ell \to \infty$ such that, for $(\boldsymbol{x}, \boldsymbol{k}) \in \mathcal{Y} \times \mathcal{M}$,

$$\lim_{\ell \to \infty} [V^{\rho_\ell}(\boldsymbol{x}, \boldsymbol{k}) - V^{\rho_\ell}(0, \boldsymbol{k})] = V(\boldsymbol{x}, \boldsymbol{k}). \tag{5.46}$$

Moreover, it follows from the convexity of $V^{\rho_\ell}(\boldsymbol{x}, \boldsymbol{k})$ that the limit function $V(\boldsymbol{x}, \boldsymbol{k})$ is also convex on $\mathcal{Y} \times \mathcal{M}$. Therefore, the proof of the theorem is completed. □

Let $\partial V^{\rho_\ell}(\boldsymbol{x}, \boldsymbol{k})/\partial \boldsymbol{x}$ be the derivative of $V^{\rho_\ell}(\boldsymbol{x}, \boldsymbol{k})$ at the point $\boldsymbol{x}$ when the derivative exists.

Theorem 4.2. *Let Assumptions* (A1)–(A3) *hold. Then:*

(i) $\lambda^*(\boldsymbol{x}, \boldsymbol{k})$ *defined by* (5.15) *does not depend on* $(\boldsymbol{x}, \boldsymbol{k})$.

(ii) *The pair* $(\widehat{\lambda}, V(\boldsymbol{x}, \boldsymbol{k}))$ *defined in Theorem* 4.1 *satisfies* (5.16) *in the interior of* $\mathcal{Y}$.

(iii) *If there exists an open subset* $\widehat{\mathcal{Y}}$ *of* $\mathcal{Y}$ *such that* $b(\mathcal{Y}) \subseteq b(\widehat{\mathcal{Y}})$*, where* $b(\mathcal{Y})$ *and* $b(\widehat{\mathcal{Y}})$ *are the boundary of* $\mathcal{Y}$ *and* $\widehat{\mathcal{Y}}$*, and* $\{\partial V^{\rho_\ell}(\boldsymbol{x}, \boldsymbol{k})/\partial \boldsymbol{x} : \ell \geq 1\}$ *is uniformly equi-Lipschitzian on* $\widehat{\mathcal{Y}}$*, then the pair* $(\widehat{\lambda}, V(\cdot, \cdot))$ *defined in Theorem 4.1 is a solution to (5.16).*

Proof. First we consider (i). Suppose contrariwise that there exist $(\widetilde{\boldsymbol{x}}, \widetilde{\boldsymbol{k}}) \in \mathcal{Y} \times \mathcal{M}$ and $(\widehat{\boldsymbol{x}}, \widehat{\boldsymbol{k}}) \in \mathcal{Y} \times \mathcal{M}$ such that

$$\lambda^*(\widetilde{\boldsymbol{x}}, \widetilde{\boldsymbol{k}}) < \lambda^*(\widehat{\boldsymbol{x}}, \widehat{\boldsymbol{k}}). \tag{5.47}$$

We choose $\delta > 0$ and a control $\widetilde{\boldsymbol{u}}(t) \in \mathcal{A}(\widetilde{\boldsymbol{x}}, \widetilde{\boldsymbol{k}})$ such that

$$\lambda^*(\widetilde{\boldsymbol{x}}, \widetilde{\boldsymbol{k}}) + \delta < \lambda^*(\widehat{\boldsymbol{x}}, \widehat{\boldsymbol{k}}), \tag{5.48}$$

and

$$\limsup_{T\to\infty} \frac{1}{T} E \int_0^T g(\widetilde{\boldsymbol{x}}(t), \widetilde{\boldsymbol{u}}(t))\, dt \leq \lambda^*(\widetilde{\boldsymbol{x}}, \widetilde{\boldsymbol{k}}) + \delta, \tag{5.49}$$

where

$$\frac{d}{dt}\widetilde{\boldsymbol{x}}(t) = \mathcal{L}(\widetilde{\boldsymbol{u}}(t), \boldsymbol{z}), \quad \widetilde{\boldsymbol{x}}(0) = \widetilde{\boldsymbol{x}}.$$

Let $\boldsymbol{u}(t)$, $t \geq 0$, be the one given in Theorem 3.2, and let

$$\tau(\widehat{\boldsymbol{x}}, \widehat{\boldsymbol{k}}, \widetilde{\boldsymbol{x}}, \widetilde{\boldsymbol{k}}) = \inf\{t \geq 0 : (\boldsymbol{x}(t), \boldsymbol{k}(t)) = (\widetilde{\boldsymbol{x}}, \widetilde{\boldsymbol{k}})\},$$

where

$$\frac{d}{dt}\boldsymbol{x}(t) = \mathcal{L}(\boldsymbol{u}(t), \boldsymbol{z}), \quad \boldsymbol{x}(0) = \widehat{\boldsymbol{x}}.$$

We define

$$\widehat{\boldsymbol{u}}(t) = \begin{cases} \boldsymbol{u}(t), & \text{if } t \leq \tau(\widehat{\boldsymbol{x}}, \widehat{\boldsymbol{k}}, \widetilde{\boldsymbol{x}}, \widetilde{\boldsymbol{k}}), \\ \widetilde{\boldsymbol{u}}(t - \tau(\widehat{\boldsymbol{x}}, \widehat{\boldsymbol{k}}, \widetilde{\boldsymbol{x}}, \widetilde{\boldsymbol{k}})), & \text{if } t > \tau(\widehat{\boldsymbol{x}}, \widehat{\boldsymbol{k}}, \widetilde{\boldsymbol{x}}, \widetilde{\boldsymbol{k}}). \end{cases}$$

It directly follows from Theorem 3.2 that

$$\limsup_{T\to\infty} \frac{1}{T} E \int_0^T g(\widehat{\boldsymbol{x}}(t), \widehat{\boldsymbol{u}}(t))\, dt = \limsup_{T\to\infty} \frac{1}{T} E \int_0^T g(\widetilde{\boldsymbol{x}}(t), \widetilde{\boldsymbol{u}}(t))\, dt. \tag{5.50}$$

On the other hand,

$$\limsup_{T\to\infty} \frac{1}{T} E \int_0^T g(\widehat{\boldsymbol{x}}(t), \widehat{\boldsymbol{u}}(t))\, dt \geq \lambda^*(\widehat{\boldsymbol{x}}, \widehat{\boldsymbol{k}}).$$

Thus (5.50) contradicts (5.47). Consequently, (i) is proved.

Now we consider (ii). Let $\mathcal{O}^n$ be the set of all points in the interior part of $\mathcal{Y}$ on which $V(\boldsymbol{x}, \boldsymbol{k})$ is differentiable. From the convexity of $V(\boldsymbol{x}, \boldsymbol{k})$

we know that $\mathcal{O}^n$ is dense in $\mathcal{Y}$. It follows from the properties of convex functions that, for $\boldsymbol{x} \in \mathcal{O}^n$ and any $\boldsymbol{p}$,

$$\lim_{\ell\to\infty} \partial_{\boldsymbol{p}} \widetilde{V}^{\rho_\ell}(\boldsymbol{x},\boldsymbol{k}) = \partial_{\boldsymbol{p}} V(\boldsymbol{x},\boldsymbol{k}). \tag{5.51}$$

From Theorem 3.1,

$$\begin{aligned}\rho_\ell V^{\rho_\ell}(\boldsymbol{x},\boldsymbol{k}) = &\inf_{\boldsymbol{u}\in\mathcal{U}(\boldsymbol{x},\boldsymbol{k})} \left\{\partial_{\mathcal{L}(\boldsymbol{u},\boldsymbol{z})} V^{\rho_\ell}(\boldsymbol{x},\boldsymbol{k}) + g(\boldsymbol{x},\boldsymbol{u})\right\}\\ &+ QV^{\rho_\ell}(\boldsymbol{x},\cdot)(\boldsymbol{k}).\end{aligned}$$

This implies that

$$\begin{aligned}\rho_\ell V^{\rho_\ell}(\boldsymbol{x},\boldsymbol{k}) = &\inf_{\boldsymbol{u}\in\mathcal{U}(\boldsymbol{x},\boldsymbol{k})} \{\partial_{\mathcal{L}(\boldsymbol{u},\boldsymbol{z})} \widetilde{V}^{\rho_\ell}(\boldsymbol{x},\boldsymbol{k}) + g(\boldsymbol{x},\boldsymbol{u})\}\\ &+ Q\widetilde{V}^{\rho_\ell}(\boldsymbol{x},\cdot)(\boldsymbol{k}).\end{aligned} \tag{5.52}$$

Taking the limit on both sides, we have that, for $\boldsymbol{x} \in \mathcal{O}^n$,

$$\widehat{\lambda} = \inf_{\boldsymbol{u}\in\mathcal{U}(\boldsymbol{x},\boldsymbol{k})} \left\{\partial_{\mathcal{L}(\boldsymbol{u},\boldsymbol{z})} V(\boldsymbol{x},\boldsymbol{k}) + g(\boldsymbol{x},\boldsymbol{u})\right\} + QV(\boldsymbol{x},\cdot)(\boldsymbol{k}). \tag{5.53}$$

If $\boldsymbol{x} \notin \mathcal{O}^n$, $\boldsymbol{x}$ is a point of the interior of $\mathcal{Y}$, then for any direction $\boldsymbol{p}$ there exists a sequence $\{\boldsymbol{x}_\ell\}_{\ell=1}^{\infty}$ such that $\boldsymbol{x}_\ell \in \mathcal{O}^n$ and $\partial_{\boldsymbol{p}} V(\boldsymbol{x}_\ell,\boldsymbol{k}) \to \partial_{\boldsymbol{p}} V(\boldsymbol{x},\boldsymbol{k})$. From this fact and from the continuity of $V(\boldsymbol{x},\boldsymbol{k})$, it follows that (5.53) holds for all $\boldsymbol{x}$ in the interior part of $\mathcal{Y}$.

Finally we prove (iii). Consider $b(\mathcal{Y})$ (the boundary of $\mathcal{Y}$). From the uniformly equi-Lipschitzian of $\{\partial V^{\rho_\ell}(\boldsymbol{x},\boldsymbol{k})/\partial \boldsymbol{x} : \ell \geq 1\}$ on $\hat{\mathcal{Y}}$, we know that (5.51) holds for all $\boldsymbol{x} \in b(\mathcal{Y})$. Therefore, we have (5.53) in $b(\mathcal{Y})$. Thus the proof of the theorem is completed. □

Finally, we establish the following verification theorem.

Theorem 4.3. *Under Assumptions* (A1)–(A3), *let* $(\lambda, W(\boldsymbol{x},\boldsymbol{k}))$ *be a solution to the* HJBDD *equation* (5.16). *Then:*

(i) *If there is a control* $\boldsymbol{u}^*(\cdot) \in \mathcal{A}(\boldsymbol{x},\boldsymbol{k})$ *such that*

$$\begin{aligned}&\inf_{\boldsymbol{u}\in\mathcal{U}(\boldsymbol{k}(t))} \left\{\partial_{\mathcal{L}(\boldsymbol{u},\boldsymbol{z})} W(\boldsymbol{x}^*(t),\boldsymbol{k}(t)) + g(\boldsymbol{x}^*(t),\boldsymbol{u})\right\}\\ &\quad = \partial_{\mathcal{L}(\boldsymbol{u}^*(t),\boldsymbol{z})} W(\boldsymbol{x}^*(t),\boldsymbol{k}(t)) + g(\boldsymbol{x}^*(t),\boldsymbol{u}^*(t)),\end{aligned} \tag{5.54}$$

for a.e. $t \geq 0$ *with probability* 1, *where* $\boldsymbol{x}^*(\cdot)$ *is the surplus process corresponding to the control* $\boldsymbol{u}^*(\cdot)$ *with the initial conditions* $\boldsymbol{x}^*(0) = \boldsymbol{x}$ *and* $\boldsymbol{k}(0) = \boldsymbol{k}$, *and*

$$\lim_{T\to\infty} \frac{E[W(\boldsymbol{x}^*(T),\boldsymbol{k}(T))]}{T} = 0, \tag{5.55}$$

then

$$\lambda = J(\boldsymbol{x},\boldsymbol{k},\boldsymbol{u}^*(\cdot)).$$

(ii) *For any* $\boldsymbol{u}(\cdot) \in \mathcal{A}(\boldsymbol{x}, \boldsymbol{k})$, *we have* $\lambda \leq J(\boldsymbol{x}, \boldsymbol{k}, \boldsymbol{u}(\cdot))$, *i.e.,*

$$\limsup_{T\to\infty} \frac{1}{T} E \int_0^T g(\boldsymbol{x}(t), \boldsymbol{u}(t))\, dt \geq \lambda.$$

(iii) *Furthermore, for any (stable) control policy* $\boldsymbol{u}(\cdot) \in \mathcal{S}(\boldsymbol{x}, \boldsymbol{k})$, *we have*

$$\liminf_{T\to\infty} \frac{1}{T} E \int_0^T g(\boldsymbol{x}(t), \boldsymbol{u}(t))\, dt \geq \lambda. \tag{5.56}$$

Note that Theorem 4.3 explains why equation (5.16) is the HJB equation for our problem and why the function $W(\boldsymbol{x}, \boldsymbol{k})$ from Theorem 4.3 is called a potential function.

Proof of Theorem 4.3. Since $(\lambda, W(\cdot, \cdot))$ is a solution to (5.16) and $(\boldsymbol{x}^*(t), \boldsymbol{u}^*(t))$ satisfy condition (5.54), we have

$$\begin{aligned} &\partial_{\mathcal{L}(\boldsymbol{u}^*(t),\boldsymbol{z})} W(\boldsymbol{x}^*(t), \boldsymbol{k}(t)) + QW(\boldsymbol{x}^*(t), \cdot)(\boldsymbol{k}(t)) \\ &\quad = \lambda - g(\boldsymbol{x}^*(t), \boldsymbol{u}^*(t)). \end{aligned} \tag{5.57}$$

Since $W(\boldsymbol{x}, \boldsymbol{k}) \in \mathcal{G}$, we apply Dynkin's formula and use (5.57) to get

$$\begin{aligned} &E\left[W(\boldsymbol{x}^*(T), \boldsymbol{k}(T))\right] \\ &\quad = W(\boldsymbol{x}, \boldsymbol{k}) + E \int_0^T \left[\partial_{\mathcal{L}(\boldsymbol{u}^*(t),\boldsymbol{z})} W(\boldsymbol{x}^*(t), \boldsymbol{k}(t)) \right. \\ &\qquad\qquad\qquad \left. + QW(\boldsymbol{x}^*(t), \cdot)(\boldsymbol{k}(t))\right] dt \\ &\quad = W(\boldsymbol{x}, \boldsymbol{k}) + E \int_0^T \left[\lambda - g(\boldsymbol{x}^*(t), \boldsymbol{u}^*(t)\right] dt \\ &\quad = W(\boldsymbol{x}, \boldsymbol{k}) + \lambda T - E \int_0^T g(\boldsymbol{x}^*(t), \boldsymbol{u}^*(t))\, dt. \end{aligned} \tag{5.58}$$

We can rewrite (5.58) as

$$\begin{aligned} \lambda = &\frac{1}{T}\left[EW(\boldsymbol{x}^*(T), \boldsymbol{k}(T)) - W(\boldsymbol{x}, \boldsymbol{k})\right] \\ &+ \frac{1}{T} E \int_0^T g(\boldsymbol{x}^*(t), \boldsymbol{u}^*(t))\, dt. \end{aligned} \tag{5.59}$$

Using (5.59) and taking the limit as $T \to \infty$, we get

$$\lambda = \limsup_{T\to\infty} \frac{1}{T} E \int_0^T g(\boldsymbol{x}^*(t), \boldsymbol{u}^*(t))\, dt.$$

For the proof of part (iii), if $\boldsymbol{u}(\cdot) \in \mathcal{S}(\boldsymbol{x}, \boldsymbol{k})$, then from $W(\boldsymbol{x}, \boldsymbol{k}) \in \mathcal{G}$, we know that

$$\lim_{T\to\infty} \frac{E[W(\boldsymbol{x}(T), \boldsymbol{k}(T))]}{T} = 0.$$

Moreover, from the HJBDD equation (5.16) we have

$$\partial_{\mathcal{L}(\boldsymbol{u}(t),\boldsymbol{z})} W(\boldsymbol{x}(t), \boldsymbol{k}(t)) + QW(\boldsymbol{x}(t), \cdot)(\boldsymbol{k}(t)) \geq \lambda - g(\boldsymbol{x}(t), \boldsymbol{u}(t)).$$

Now (5.56) can be proved similarly as before.

Finally, we apply Lemma F.3 to show part (ii), i.e., the optimality of the control $\boldsymbol{u}^*(\cdot)$ in the (natural) class of all admissible controls. Let $\boldsymbol{u}(\cdot) \in \mathcal{A}(\boldsymbol{x}, \boldsymbol{k})$ be any policy and let $\boldsymbol{x}(\cdot)$ be the corresponding surplus process. Suppose that

$$J(\boldsymbol{x}, \boldsymbol{k}, \boldsymbol{u}(\cdot)) < \lambda. \tag{5.60}$$

We can apply Lemma F.3 and (5.60) to obtain

$$\limsup_{\rho_\ell \to 0} \rho_\ell J^{\rho_\ell}(\boldsymbol{x}, \boldsymbol{k}, \boldsymbol{u}(\cdot)) < \lambda. \tag{5.61}$$

On the other hand, we know from Theorem 4.1 that

$$\lim_{\rho_\ell \to 0} \rho_\ell V^{\rho_\ell}(\boldsymbol{x}, \boldsymbol{k}) = \lambda.$$

This equation and (5.61) imply the existence of a $\rho > 0$ such that

$$\rho J^{\rho}(\boldsymbol{x}, \boldsymbol{k}, \boldsymbol{u}(\cdot)) < \rho V^{\rho}(\boldsymbol{x}, \boldsymbol{k}),$$

which contradicts the definition of $V^{\rho}(\boldsymbol{x}, \boldsymbol{k})$. Thus (ii) is proved. □

5.5 Notes

This chapter is based on Presman, Sethi and Zhang [101] and Sethi and Zhou [127]. The proof of Lipschitz continuity in Theorem 3.1 is along the same lines as in Section 4.3 or in Presman, Sethi, and Suo [99].

Bai and Gershwin [9, 10] use a decomposition approach to solve the problem of a jobshop. Their approach is based on slicing the machines in the jobshop into partial machines, organizing them into a number of single-product flowshops, and using the heuristic arguments developed for such flowshops in Bai [8] and Bai and Gershwin [11] to obtain hedging points; see Section 4.6. They use these hedging points as data in a linear program in order to construct a feedback policy for the problem. They perform simulations using a program called HIERCSIM developed by Darakananda [40], and conclude that their procedure works well. While Bai and Gershwin [9, 10] do not consider asymptotic optimality, it may be noted nevertheless that their approach does not result in asymptotic optimal controls.

Lasserre [84, 87] considers an integrated deterministic finite horizon jobshop planning and scheduling model. He solves the problem by using a (multipass) decomposition approach which alternates between solving a planning problem with a fixed sequence of products on the machines, and a jobshop scheduling problem for a fixed choice of production plan.

6 Risk-Sensitive Control

6.1 Introduction

This chapter is concerned with long-run average risk-sensitive control. The objective of the problem is to choose the rate of production planning to minimize a risk-sensitive cost over the infinite horizon. This consideration is motivated by the following observations. First, since most manufacturing systems are large complex systems, it is very difficult to establish accurate mathematical models to describe these systems. Modeling errors are inevitable. Second, in practice, an optimal policy for a subdivision of a big corporation is usually not an optimal policy for the whole corporation. Therefore, an optimal solution with regular cost criteria may not be desirable in many real problems. An alternative approach is to consider robust controls. The design of robust controls emphasize system stability rather than optimality. In some manufacturing systems, it is more desirable to consider controls that are robust enough to attenuate uncertain disturbances, which include modeling errors, and therefore to achieve the system stability. Robust control design is particularly important in manufacturing systems with unwanted disturbances, such as those associated with unwanted machine failures. The basic idea of the risk-sensitive control is to consider a risk-sensitive cost function that penalizes heavily on costs associated with values of state and control variables that are extreme. In risk-sensitive control theory, typically an exponential-of-integral cost criterion is considered. Such cost functions penalize heavily state and control trajectories that spend a large amount of time in regions which are far from the origin.

We consider the problem of controlling a manufacturing system with stochastic production capacity and a risk-sensitive criterion. The machine capacity process will be assumed to be an irreducible finite state Markov chain. For simplicity, we consider a single-product manufacturing system facing a constant demand. The control is the production rate, which is subject to a random machine capacity constraint. The goal is to find a control policy which minimizes the long-term growth rate of an expected exponential-of-integral criterion.

In Section 6.2 we formulate the model. Then, in Section 6.3, we show that the minimum growth rate and the associated potential function satisfy the dynamic programming equations specified as a system of partial differential equations. The vanishing discount approach is used for solving the problem. The approach uses a logarithmic transformation introduced by Bensoussan and Nagai [16] to obtain an equivalent problem. In Section 6.4 we give a verification theorem. The chapter is concluded in Section 6.5.

6.2 Problem Formulation

Let us consider a single-product and parallel-machine manufacturing system with stochastic production capacity and constant demand for its production over time. For $t \geq 0$, let $x(t)$, $u(t)$, and z denote the surplus level (the state variable), the production rate (the control variable), and the constant demand rate, respectively. We assume $x(t) \in \Re$, $u(t) \in \Re_+$, $t \geq 0$, and z a positive constant. The dynamics of the system have the following form:

$$\frac{d}{dt}x(t) = -ax(t) + u(t) - z, \quad x(0) = x, \tag{6.1}$$

where $a > 0$ is a constant, representing the deterioration rate (or spoilage rate) of the finished product.

Let $k(t) \in \mathcal{M} = \{0, 1, 2, \ldots, m\}$, $t \geq 0$, denote a Markov chain generated by Q, where $Q = (q_{ij})$, $i, j \in \mathcal{M}$, is an $(m+1) \times (m+1)$ matrix such that $q_{ij} \geq 0$ for $i \neq j$ and $q_{ii} = -\sum_{j \neq i} q_{ij}$. We let $k(t)$ represent the maximum production capacity of the system at time t. The representation for $\mathcal{M}$ usually stands for the case of m identical machines, each with a unit capacity and having two states: up and down.

The production constraint is given by the inequalities:

$$0 \leq u(t) \leq k(t), \quad t \geq 0.$$

Definition 2.1. A production control process $u(\cdot) = \{u(t) : t \geq 0\}$ is *admissible* if: (i) $u(t)$ is $\mathcal{F}_t$ $(= \sigma\{k(s) : 0 \leq s \leq t\})$ progressively measurable;

and (ii) $0 \le u(t) \le k(t)$ for all $t \ge 0$. $\square$

Let $\mathcal{A}(k)$ denote the class of admissible controls with the initial condition $k(0) = k$.

Let $g(x,u)$ denote a cost function of the surplus and the production. For every $u(\cdot) \in \mathcal{A}(k)$, $x(0) = x$, and $k(0) = k$, define

$$J(x,k,u(\cdot)) = \limsup_{T\to\infty} \frac{1}{T}\left[\log E \exp\left(\int_0^T g(x(t),u(t))\,dt\right)\right], \tag{6.2}$$

where $x(\cdot)$ is the surplus process corresponding to the production process $u(\cdot)$. The objective of the problem is to choose $u(\cdot) \in \mathcal{A}(k)$ to minimize $J(x,k,u(\cdot))$. Let

$$\lambda = \inf_{u(\cdot)\in\mathcal{A}(k)} J(x,k,u(\cdot)). \tag{6.3}$$

A motivation for choosing such an exponential cost criterion is that such criteria are sensitive to large values of the exponent which occur with small probability, for example, rare sequences of unusually many machine failures resulting in shortages ($x(t) < 0$).

Remark 2.1. The positive spoilage rate a appears in certain crucial estimates carried out in the next section. It also implies a uniform bound for $x(t)$. Note that the control $u(\cdot)$ is bounded between 0 and m. This implies that a solution $x(\cdot)$ to (6.1) must satisfy

$$|x(t)| \le |x|e^{-at} + (m+z)\int_0^t e^{-a(t-s)}ds \le |x|e^{-at} + \frac{m+z}{a}. \quad \square \tag{6.4}$$

We assume that the cost function $g(x,u)$ and the production capacity process $k(\cdot)$ satisfy the following assumptions:

(A1) $g(x,u) \ge 0$ is continuous, bounded, and uniformly Lipschitz in x.

(A2) Q is irreducible.

Remark 2.2. In manufacturing systems, the running cost function $g(x,u)$ is usually chosen to be of the form $g(x,u) = h(x) + c(u)$ with piecewise linear $h(x)$ and $c(u)$. Note that piecewise linear functions are not bounded as required in (A1). However, this is not important, in view of the uniform bounds on $u(t)$ and on $x(t)$ for initial state $x = x(0)$ in any bounded set.$\square$

In the next section, we discuss the dynamics of the system and the associated Hamilton-Jacobi-Bellman (HJB) equations.

6.3 HJB Equations

First we can write the associated HJB equation as follows:

$$\begin{aligned}\lambda = \inf_{0\le u\le k}&\left\{(-ax+u-z)\frac{\partial\psi(x,k)}{\partial x}+g(x,u)\right\}\\&+\exp(-\psi(x,k))Q\exp(\psi(x,\cdot))(k),\end{aligned} \tag{6.5}$$

where $\psi(\cdot,\cdot)$ is defined on $\Re\times\mathcal{M}$. As in any long-run average-cost problems, an immediate question is if the equation (6.5) has a solution in some sense. In this chapter, we will show that (6.5) indeed has a solution in the viscosity sense. We use a vanishing discount approach. Let $\rho>0$ denote a discount factor and let

$$\widehat{J}^\rho(x,k,u(\cdot)) = \log\left[E\exp\left(\int_0^\infty e^{-\rho t}g(x(t),u(t))\,dt\right)\right].$$

Define

$$\widehat{V}^\rho(x,k) = \inf_{u(\cdot)\in\mathcal{A}(k)}\widehat{J}^\rho(x,k,u(\cdot)).$$

Then, the associated HJB equation has the form:

$$\begin{aligned}\rho\varphi^\rho(x,k) = \inf_{0\le u\le k}&\left\{(-ax+u-z)\frac{\partial\varphi^\rho(x,k)}{\partial x}+g(x,u)\right\}\\&+\exp(-\varphi^\rho(x,k))Q\exp(\varphi^\rho(x,\cdot))(k).\end{aligned} \tag{6.6}$$

Let

$$\psi^\rho(x,k) = \exp\left(\varphi^\rho(x,k)\right).$$

Then, (6.6) becomes

$$\begin{aligned}\rho\psi^\rho(x,k)\log\psi^\rho(x,k) = \inf_{0\le u\le k}&\left\{(-ax+u-z)\frac{\partial\psi^\rho(x,k)}{\partial x}\right.\\&\left.+g(x,u)\psi^\rho(x,k)\right\}+Q\psi^\rho(x,\cdot)(k).\end{aligned} \tag{6.7}$$

We would like to get rid of the term $\psi^\rho(x,k)\log\psi^\rho(x,k)$. One way of doing so is to use the transform device introduced by Bensoussan and Nagai [16] based on the following expression:

$$-r\log r = \inf_y\left\{yr+e^{-(y+1)}\right\}, \quad \text{for any } r>0, \tag{6.8}$$

where the infimum is obtained at $y+1=-\log r$. Letting $r=\psi^\rho(x,k)$, (6.8) is changed into

$$\psi^\rho(x,k)\log\psi^\rho(x,k) = -\inf_y\left\{y\psi^\rho(x,k)+e^{-(y+1)}\right\}.$$

In view of this and (6.7), the discounted HJB equation (6.6) has the form:

$$0 = \inf_{0 \le u \le k,\, y} \left\{ (-ax + u - z) \frac{\partial \psi^{\rho}(x,k)}{\partial x} + Q\psi^{\rho}(x,\cdot)(k) \right. \\ \left. + g(x,u)\psi^{\rho}(x,k) + \rho y \psi^{\rho}(x,k) + \rho e^{-(y+1)} \right\}.$$

By adding $\rho\psi^{\rho}(x,k)$ to both sides of this equation and changing $(y+1)$ to y, we obtain

$$\rho\psi^{\rho}(x,k) = \inf_{0 \le u \le k,\, y} \left\{ (-ax + u - z) \frac{\partial \psi^{\rho}(x,k)}{\partial x} + Q\psi^{\rho}(x,\cdot)(k) \right. \\ \left. + g(x,u)\psi^{\rho}(x,k) + \rho y \psi^{\rho}(x,k) + \rho e^{-y} \right\}. \tag{6.9}$$

Let

$$M_0 = ||g(x,u)|| = \sup_{x,u} |g(x,u)|. \tag{6.10}$$

We consider the following control problem whose value function is a viscosity solution to equation (6.9):

$$\begin{cases} V^{\rho}(x,k) = \inf_{u(\cdot) \in \mathcal{A}(k),\, y(\cdot)} J^{\rho}(x,k,u(\cdot),y(\cdot)), \\ J^{\rho}(x,k,u(\cdot),y(\cdot)) \\ \quad = E \int_0^{\infty} e^{-\rho t} \exp\left(\int_0^t [g(x(s),u(s)) + \rho y(s)]\, ds \right) (\rho e^{-y(t)})\, dt. \\ \text{s.t. } \frac{dx(t)}{dt} = -ax(t) + u(t) - z, \quad x(0) = x, \quad \text{and} \\ y(t) \text{ is } \mathcal{F}_t\text{-measurable and } -M_0/\rho \le y(t) \le 0. \end{cases} \tag{6.11}$$

We next show that $V^{\rho}(x,k)$ is a viscosity solution to (6.9) with some a priori estimates.

Lemma 3.1. *Under Assumptions* (A1) and (A2), *we have the following:*

(i) *For all x and $k \in \mathcal{M}$,*

$$1 \le V^{\rho}(x,k) \le \exp\left(\frac{M_0}{\rho}\right).$$

(ii) *For all x, $\widehat{x}$, and $k \in \mathcal{M}$,*

$$\exp\left(-\frac{M_0|x-\widehat{x}|}{a}\right) \le \frac{V^{\rho}(x,k)}{V^{\rho}(\widehat{x},k)} \le \exp\left(\frac{M_0|x-\widehat{x}|}{a}\right).$$

(iii) *For each $\Delta > 0$, there is a constant $C > 0$ independent of ρ such that, for all k, $\widehat{k} \in \mathcal{M}$, and $|x| \le \Delta$,*

$$e^{-C} \le \frac{V^\rho(x,k)}{V^\rho(x,\widehat{k})} \le e^C.$$

(iv) *$V^\rho(x,k)$ is a viscosity solution to (6.9).*

Proof. We begin with (i). We first show that $V^\rho(x,k) \ge 1$. In view of the nonnegativity of $g(x,u)$, it suffices to show that for all deterministic Borel measurable $y(t)$, $-M_0/\rho \le y(t) \le 0$, $t \ge 0$,

$$\int_0^\infty \rho \exp\left(\int_0^t \rho[y(s)-1]\,ds - y(t)\right) dt \ge 1. \tag{6.12}$$

In fact, note that, for all $T > 0$,

$$\begin{aligned}
&\exp\left(\int_0^T \rho[y(s)-1]\,ds\right) - 1\\
&\quad = \int_0^T \exp\left(\int_0^t \rho[y(s)-1]\,ds\right)[\rho(y(t)-1)]\,dt\\
&\quad \ge \int_0^T \exp\left(\int_0^t \rho[y(s)-1]\,ds\right)[-\rho e^{-y(t)}]\,dt
\end{aligned}$$

because $e^{-x} \ge 1-x$ for all $x \le 0$. Hence,

$$\begin{aligned}
&\int_0^\infty \rho \exp\left(\int_0^t \rho[y(s)-1]\,ds - y(t)\right) dt\\
&\quad = \int_0^\infty \exp\left(\int_0^t \rho[y(s)-1]\,ds\right)(\rho e^{-y(t)})\,dt\\
&\quad = \lim_{T\to\infty} \int_0^T \exp\left(\int_0^t \rho[y(s)-1]\,ds\right)(\rho e^{-y(t)})\,dt\\
&\quad \ge \lim_{T\to\infty} \left[1 - \exp\left(\int_0^T \rho[y(s)-1]\,ds\right)\right]\\
&\quad \ge \lim_{T\to\infty} \left(1 - e^{-\rho T}\right) = 1.
\end{aligned}$$

Thus, the inequality (6.12) follows.

Let $y(t) = -M_0/\rho$ for all $t \ge 0$. Then, for all admissible $u(\cdot)$,

$$\begin{aligned}
V^\rho(x,k) &\le E\int_0^\infty \rho \exp\left(\int_0^t \left(\left[M_0 - \rho\left(\frac{M_0}{\rho}+1\right)\right] ds + \frac{M_0}{\rho}\right) dt\right)\\
&= \exp\left(\frac{M_0}{\rho}\right)\int_0^\infty \rho e^{-\rho t}\,dt\\
&= \exp\left(\frac{M_0}{\rho}\right).
\end{aligned}$$

This proves (i).

We now prove (ii). Let $(u(\cdot), y(\cdot))$ denote a pair of admissible controls and let $x(t)$ and $\widehat{x}(t)$ denote the corresponding trajectories with initial values x and $\widehat{x}$, respectively. Then

$$x(t) - \widehat{x}(t) = (x - \widehat{x})e^{-at}, \quad \text{for all } t \geq 0.$$

In view of this and the Lipschitz property of $g(x, u)$, we have

$$|g(x(t), u(t)) - g(\widehat{x}(t), u(t))| \leq C_1 |x - \widehat{x}| e^{-at},$$

where C_1 is independent of x and $\widehat{x}$. For notational simplification, let

$$\eta(y(\cdot))(t) = \rho \exp\left(\int_0^t \rho[y(s) - 1]\, ds - y(t) \right).$$

Then, we have

$$\begin{aligned}
& E \int_0^\infty \left[\exp\left(\int_0^t g(x(s), u(s))\, ds \right) \right] \eta(y(\cdot))(t)\, dt \\
& \quad \leq E \int_0^\infty \left[\exp\left(\int_0^t g(\widehat{x}(s), u(s))\, ds \right) \right. \\
& \qquad\qquad\qquad \left. \times \exp\left(\int_0^t C_1 |x - \widehat{x}| e^{-as}\, ds \right) \right] \eta(y(\cdot))(t)\, dt \\
& \quad \leq \exp\left(\int_0^\infty C_1 |x - \widehat{x}| e^{-as}\, ds \right) \\
& \qquad\qquad\qquad \times E \int_0^\infty \left[\exp\left(\int_0^t g(\widehat{x}(s), u(s))\, ds \right) \right] \eta(y(\cdot))(t)\, dt \\
& \quad = \exp(C_1 |x - \widehat{x}|) E \int_0^\infty \left[\exp\left(\int_0^t g(\widehat{x}(s), u(s))\, ds \right) \right] \eta(y(\cdot))(t)\, dt.
\end{aligned}$$

Hence,

$$V^\rho(x, k) \leq \left[\exp\left(\frac{C|x - \widehat{x}|}{a} \right) \right] V^\rho(\widehat{x}, k).$$

Similarly we can show the other inequality in (ii).

We now show (iii). Let $k(0) = k$, and let τ denote the first time $k(\cdot)$ jumps to $\widehat{k}$. The optimality principle with the random stopping time τ (see the end of the proof) gives

$$\begin{aligned}
& V^\rho(x, k) \\
& = \inf_{u(\cdot),\, y(\cdot)} E \left\{ \int_0^\tau \left[\exp\left(\int_0^t [g(x(s), u(s)) + \rho(y(s) - 1)]\, ds \right) \right] \rho e^{-y(t)}\, dt \right. \\
& \qquad \left. + \left[\exp\left(\int_0^\tau [g(x(s), u(s)) + \rho(y(s) - 1)]\, ds \right) \right] V^\rho(x(\tau), \widehat{k}) \right\}.
\end{aligned} \tag{6.13}$$

Using $g(x,u) \geq 0$, we have

$$V^\rho(x,k) \geq E\left\{\left[\exp\left(\int_0^\tau \rho(y(s)-1)\,ds\right)\right] V^\rho(x(\tau),\widehat{k})\right\}. \tag{6.14}$$

For x in any bounded interval $[-\Delta, +\Delta]$, there exists a $C_2 > 0$ such that $|x(\tau) - x| \leq C_2(1+|x|) \leq C_2(1+\Delta)$. It follows from (ii) that

$$\frac{V^\rho(x(\tau),\widehat{k})}{V^\rho(x,\widehat{k})} \geq \exp\left(-\frac{M_0 C_2(1+\Delta)}{a}\right). \tag{6.15}$$

We can assume $y(t) \geq -M_0/\rho$, which implies

$$\exp\left(\int_0^\tau \rho[y(s)-1]\,ds\right) \geq \exp(-[M_0+\rho]\,\tau).$$

By Assumption (A3) there is a C_3 such that

$$E\exp(-(M_0+\rho)\tau) \geq e^{-C_3}.$$

Therefore, by (6.14)–(6.15),

$$V^\rho(x,k) \geq V^\rho(x,\widehat{k})e^{-C_4},$$

for some $C_4 > 0$. Interchange k and $\widehat{k}$ to get the opposite inequality.

Finally, it can be shown as in Appendix D that $V^\rho(x,k)$ is a viscosity solution to (6.9) under the constraint $-M_0 \leq y \leq 0$, i.e.,

$$\begin{aligned}\rho V^\rho(x,k) = \inf_{\substack{0\leq u\leq k\\ -M_0\leq y\leq 0}} \bigg\{ &(-ax+u-z)\frac{\partial V^\rho(x,k)}{\partial x} \\ &+ g(x,u)V^\rho(x,k) + QV^\rho(x,\cdot)(k) + \rho y V^\rho(x,k) + \rho e^{-y}\bigg\}.\end{aligned} \tag{6.16}$$

Note that $V^\rho(x,k) \geq 1$ and the infimum in (6.16) is obtained at $y = -\log V^\rho(x,k) \leq 0$, $V^\rho(x,k)$ is also a viscosity solution to (6.9).

In the rest of the proof, we prove (6.13). It suffices to consider τ with finitely many values $0 < t_1 < t_2 < \cdots < t_n$, because one may approximate τ by a step function $\sum t_\ell I_{\{t_\ell \leq \tau < t_{\ell+1}\}}$. Let $\Gamma_\ell = \{\tau = t_\ell\}$. A routine calculation then gives

$$
\begin{aligned}
&E\int_{\tau}^{\infty}\left[\exp\left(\int_0^{\tau}[g(x(s),u(s))+\rho(y(s)-1)]\,ds\right)\right]\rho e^{-y(t)}\,dt\\
&\quad=\sum_{\ell=1}^{n}E\bigg\{I_{\Gamma_\ell}\left[\exp\left(\int_0^{t_\ell}[g(x(s),u(s))+\rho(y(s)-1)]\,ds\right)\right]\\
&\qquad\qquad\times J^{\rho}(x(t_\ell),k(t_\ell),u(\cdot),y(\cdot))\bigg\}\\
&\quad\geq\sum_{\ell=1}^{n}E\bigg\{I_{\Gamma_\ell}\left[\exp\left(\int_0^{t_\ell}[g(x(s),u(s))+\rho(y(s)-1)]\,ds\right)\right]\\
&\qquad\qquad\times V^{\rho}(x(t_\ell),k(t_\ell))\bigg\}\\
&\quad=E\left\{\left[\exp\left(\int_0^{\tau}[g(x(s),u(s))+\rho(y(s)-1)]\,ds\right)\right]V^{\rho}(x(\tau),k(\tau))\right\}.
\end{aligned}
$$

This implies

$$
\begin{aligned}
&V^{\rho}(x,k)\\
&\quad=\inf_{u(\cdot),\,y(\cdot)}E\bigg\{\int_0^{\tau}\left[\exp\left(\int_0^{t}[g(x(s),u(s))+\rho(y(s)-1)]\,ds\right)\right]\rho e^{-y(t)}\,dt\\
&\qquad\qquad+\int_{\tau}^{\infty}\left[\exp\left(\int_0^{t}[g(x(s),u(s))+\rho(y(s)-1)]\,ds\right)\right]\rho e^{-y(t)}\,dt\bigg\}\\
&\quad\geq\inf_{u(\cdot),\,y(\cdot)}E\bigg\{\int_0^{\tau}\left[\exp\left(\int_0^{t}[g(x(s),u(s))+\rho(y(s)-1)]\,ds\right)\right]\rho e^{-y(t)}\,dt\\
&\qquad\qquad+\left[\exp\left(\int_0^{\tau}[g(x(s),u(s))+\rho(y(s)-1)]\,ds\right)\right]V^{\rho}\left(x(\tau),\widehat{k}\right)\bigg\}.
\end{aligned}
\tag{6.17}
$$

It remains to show the opposite inequality. Given an initial $x(0)=x$, formula (6.4) implies that $|x(t)|\leq r$ for some r. Given $\delta_1>0$, partition $\{|x|\leq r\}$ into intervals $B_1,B_2,\ldots,B_\ell$ of length $<\delta_1$ and choose $x_j\in B_j$ for $j=1,\ldots,\ell$. Given $\delta_2>0$, choose admissible $(u_{ij}(\cdot),y_{ij}(\cdot))$ such that

$$J^{\rho}(x_j,i,u_{ij}(\cdot),y_{ij}(\cdot))<V^{\rho}(x_j,i)+\delta_2.$$

Given admissible $(u(\cdot),y(\cdot))$, we define $(\widetilde{u}(\cdot),\widetilde{y}(\cdot))$ by

$$\widetilde{u}(t)=u(t),\quad \widetilde{y}(t)=y(t),\quad \text{for } 0\leq t<\tau,$$

and for $\tau=t_{\ell_1}$, $k(t_{\ell_1})=i$, $x(t_{\ell_1})\in B_j$,

$$\widetilde{u}(t)=u_{ij}(t-t_k\ell_1),\quad \widetilde{y}(t)=y_{ij}(t-t_{\ell_1}),\quad t\geq t_{\ell_1}.$$

Then $\widetilde{u}(\cdot)$ is admissible and $\widetilde{y}(t)$ is $\mathcal{F}_t$-measurable, and a routine calculation by using Lemma 3.1(ii) gives

$$
\begin{aligned}
V^{\rho}(x,k) &\le J^{(\rho)}(x,k,\widetilde{u}(\cdot),\widetilde{y}(\cdot)) \\
&\le E\left\{\int_0^{\tau}\left[\exp\left(\int_0^t [g(x(s),u(s))+\rho(y(s)-1)]\,ds\right)\right]\rho e^{-y(t)}\,dt\right. \\
&\quad \left.+\exp\left(\int_0^{\tau}[g(x(s),u(s))+\rho(y(s)-1)]\,ds\right)V^{\rho}(x(\tau),\widehat{k})\right\} \\
&\quad +F(\delta_1,\delta_2),
\end{aligned}
$$

where $F(\delta_1,\delta_2)\to 0$ as $\delta_1\to 0$ and $\delta_2\to 0$. □

Theorem 3.1. *Let Assumptions* (A1) *and* (A2) *hold. Then the* HJB *equation* (6.5) *has a viscosity solution* $(\lambda, V(x,k))$.

Proof. In view of the logarithmic transformation

$$V^{\rho}(x,k)=\exp(W^{\rho}(x,k)),$$

we have $W^{\rho}(x,k)=\log V^{\rho}(x,k)$. It follows from Lemma 3.1 that:

(i) $0\le \rho W^{\rho}(x,k)\le M_0$, uniformly in ρ;

(ii) $|W^{\rho}(x,k)-W^{\rho}(\widehat{x},k)|\le (M_0/a)|x-\widehat{x}|$ uniformly in ρ; and

(iii) for each $\Delta>0$, $|x|\le\Delta$, $k,\widehat{k}\in\mathcal{M}$, $|W^{\rho}(x,k)-W^{\rho}(x,\widehat{k})|\le C_1$ uniformly in ρ for some $C_1>0$.

Then in view of these and the Arzelà-Ascoli theorem, it is easy to see that for each (x,i), there exists a sequence $\rho_n\to 0$ such that $\rho_n W^{\rho_n}(0,0)\to\lambda$, and for each $k\in\mathcal{M}$,

$$
\begin{aligned}
&W^{\rho_n}(x,k)-W^{\rho_n}(0,0) \\
&\quad =[W^{\rho_n}(x,i)-W^{\rho_n}(0,k)]+[W^{\rho_n}(0,k)-W^{\rho_n}(0,0)] \qquad (6.18)\\
&\quad \to V(x,k)
\end{aligned}
$$

on any compact subset of $\Re\times\mathcal{M}$. Therefore,

$$
\begin{aligned}
\rho_n W^{\rho_n}(x,k) = {}& \rho_n[W^{\rho_n}(x,k)-W^{\rho_n}(0,k)] \\
&+\rho_n[W^{\rho_n}(0,k)-W^{\rho_n}(0,0)]+\rho_n W^{\rho_n}(0,0) \qquad (6.19)\\
\to{}& \lambda.
\end{aligned}
$$

Finally it can be shown, as in Lemma E.2, that the limit $(\lambda, V(x,k))$ defined in (6.18) and (6.19) is a viscosity solution to the HJB equation (6.5). □

Remark 3.1. The proof supplies an approach to construct a viscosity solution of (6.5). We can take the limits in (6.18) and (6.19) to get the solution. □

Corollary 3.1. *Let Assumptions* (A1) *and* (A2) *hold. The pair* $(\lambda, V(x,k))$ *obtained in Theorem* 3.1 *satisfies the following conditions, for some positive constant* C*:*

(i) $0 \le \lambda \le C$; *and*

(ii) $|V(x,k) - V(\widehat{x},k)| \le C|x - \widehat{x}|$.

Proof. It is easy to check from the proof of Theorem 3.1 that (i) holds,

$$|W^{\rho_n}(x,k) - W^{\rho_n}(\widehat{x},k)| \le (M_0/a)|x - \widehat{x}|.$$

Then

$$\begin{aligned}|V(x,k) - V(\widehat{x},k)| &= \lim_{\rho_n \to 0} |W^{\rho_n}(x,k) - W^{\rho_n}(\widehat{x},k)| \\ &\le (M_0/a)|x - \widehat{x}|.\end{aligned}$$

□

Theorem 3.2. *Let Assumptions* (A1) *and* (A2) *hold, and let* $(\widehat{\lambda}, W(x,k))$ *be a viscosity solution to the* HJB *equation* (6.5). *Assume that* $W(x,k)$ *is Lipschitz continuous in* x. *Then,*

$$\widehat{\lambda} = \inf_{u(\cdot)\in\mathcal{A}(k)} J(x,k,u(\cdot)),$$

where $J(x,k,u(\cdot))$ *is defined in* (6.2).

Proof. Since $(\widehat{\lambda}, W(x,i))$ is a viscosity solution to the HJB equation (6.5), we have

$$\begin{aligned}\widehat{\lambda} = \inf_{0\le u\le k} &\left\{(-ax+u-z)\frac{\partial W(x,k)}{\partial x} + g(x,u)\right\} \\ &+ \exp(-W(x,k))Q\exp(W(x,\cdot))(k).\end{aligned} \tag{6.20}$$

We divide the remainder of the proof into two steps.

Step 1. We prove that $\widehat{\lambda} \le \inf_{u(\cdot)\in\mathcal{A}(k)} J(x,k,u(\cdot))$.
Let $\zeta(x,k) = \exp(W(x,k))$. Then equation (6.20) becomes

$$\begin{aligned}\widehat{\lambda}\zeta(x,k) = \inf_{0\le u\le k} &\left\{(-ax+u-z)\frac{\partial \zeta(x,k)}{\partial x} + g(x,u)\zeta(x,k)\right. \\ &\left. + Q\zeta(x,\cdot)(k)\right\}.\end{aligned}$$

It is equivalent to

$$\begin{aligned}0 = \inf_{0\le u\le k} &\left\{(-ax+u-z)\frac{\partial \zeta(x,k)}{\partial x} + [g(x,u) - \widehat{\lambda}]\zeta(x,k)\right. \\ &\left. + Q\zeta(x,\cdot)(k)\right\}.\end{aligned}$$

It is easy to see that $\zeta(x,k)$ is a viscosity solution to the following time-dependent equation for $\xi(T,x,k)$:

$$\begin{cases} \dfrac{\partial \xi(T,x,k)}{\partial T} = \inf_{0\le u\le i}\Big\{(-ax+u-z)\dfrac{\partial \xi(T,x,k)}{\partial x} \\ \qquad\qquad + [g(x,u)-\widehat{\lambda}]\xi(T,x,k) + Q\xi(T,x,\cdot)(k)\Big\}, \\ \xi(0,x,k) = \zeta(x,k). \end{cases} \tag{6.21}$$

As is shown in Appendix E, this HJB equation has a unique viscosity solution. Moreover, if we define

$$\widehat{\xi}(T,x,k) = \inf_{u(\cdot)\in\mathcal{A}(k)} E\left\{\zeta(x(T),k(T))\exp\left(\int_0^T [g(x(t),u(t))-\widehat{\lambda}]\,dt\right)\right\},$$

then, using the optimality principle, it can be shown that $\widehat{\xi}(T,x,k)$ is also a viscosity solution to (6.21). Thus, $\widehat{\xi}(T,x,k)=\zeta(x,k)$ for all $T\ge 0$, namely,

$$\zeta(x,k) = \inf_{u(\cdot)\in\mathcal{A}(k)} E\left\{\zeta(x(T),k(T))\exp\left(\int_0^T [g(x(t),u(t))-\lambda]\,dt\right)\right\}. \tag{6.22}$$

It follows that, for all $u(\cdot)\in\mathcal{A}(k)$,

$$\begin{aligned}\zeta(x,k) &\le E\left\{\zeta(x(T),k(T))\exp\left(\int_0^T [g(x(t),u(t))-\widehat{\lambda}]\,dt\right)\right\}\\ &= E\left\{\zeta(x(T),k(T))\exp\left(\int_0^T g(x(t),u(t))\,dt\right)\right\}\exp(-\widehat{\lambda}T).\end{aligned}$$

Taking the logarithm of both sides, we have

$$\log\zeta(x,k) \le \log E\left\{\zeta(x(T),k(T))\exp\left(\int_0^T g(x(t),u(t))\,dt\right)\right\} - \widehat{\lambda}T. \tag{6.23}$$

Recall the Lipschitz property of $W(x,k)$ in x. It follows that for all x and $\widehat{x}$, there is a positive constant C_1 such that

$$\frac{\zeta(\widehat{x},k)}{\zeta(x,k)} = \exp(W(\widehat{x},k)-W(x,k)) \le \exp(C_1|\widehat{x}-x|).$$

Replacing $\widehat{x}$ by $x(T)$ and k by $k(T)$, we obtain

$$\zeta(X(T),k(T)) \le \zeta(x,k(T))\exp(C_1|x(T)-x|).$$

Note also that $|x(T)-x|\le C_2(1+|x|)$ for some positive constant C_2 (see (6.4)), we have

$$\zeta(X(T),k(T)) \le \left(\max_{i\in\mathcal{M}}\zeta(x,i)\right)\exp(C_1C_2(1+|x|)). \tag{6.24}$$

Combining this inequality with (6.23), we obtain

$$\log \zeta(x,k) \le \log\left(\max_{i\in\mathcal{M}} \zeta(x,i)\right) + C_1C_2(1+|x|) + \log E\exp\int_0^T g(x(t),u(t))\,dt - \widehat{\lambda}T.$$

Dividing both sides by T and letting $T\to\infty$ yield

$$\widehat{\lambda} \le J(x,k,u(\cdot)), \qquad \text{for all } u(\cdot)\in\mathcal{A}(k).$$

Step 2. We show that $\widehat{\lambda} \ge \inf_{u(\cdot)\in\mathcal{A}(k)} J(x,k,u(\cdot))$.
Let

$$\eta(T,x,k) = \inf_{u(\cdot)\in\mathcal{A}(k)} \log E\exp\left(\int_0^T g(x(t),u(t))\,dt\right).$$

We first show that

$$\widehat{\lambda} = \lim_{T\to\infty} \frac{1}{T}\eta(T,x,k), \tag{6.25}$$

uniformly for x in any compact set.

In fact, as in (6.24), we can show that there exists a positive constant C_3 such that, for all $x=x(0)$ and $T>0$,

$$\exp(-C_3(1+|x|)) \le \zeta(x(T),k(T)) \le \exp(C_3(1+|x|)).$$

In view of this and (6.22), we have

$$\begin{aligned}
&\exp(-C_3(1+|x|)) \inf_{u(\cdot)\in\mathcal{A}(k)} E\exp\left(\int_0^T g(x(t),u(t))\,dt\right)\\
&\quad\le \zeta(x,k)\exp\left(\widehat{\lambda}T\right)\\
&\quad\le \exp(C_3(1+|x|)) \inf_{u(\cdot)\in\mathcal{A}(k)} E\exp\left(\int_0^T g(x(t),u(t))\,dt\right).
\end{aligned}$$

Taking logarithm on both sides and noting that

$$\inf_{u(\cdot)\in\mathcal{A}(k)} \log(\cdots) = \log \inf_{u(\cdot)\in\mathcal{A}(k)} (\cdots),$$

we obtain

$$\begin{aligned}
-C_3(1+|x|) + \eta(T,x,k) &\le \log\zeta(x,k) + \widehat{\lambda}T\\
&\le C_3(1+|x|) + \eta(T,x,k).
\end{aligned}$$

Dividing both sides by T and letting $T\to\infty$, we arrive at (6.25).

In view of (6.4), for any fixed $r > 0$, there exists $r_1 > 0$ such that $|x(t)| \le r_1$ for all $t \ge 0$, $k \in \mathcal{M}$, and $|x| \le r$ with $x(0) = x$. Therefore, for each $\delta > 0$, there exists T_0 such that

$$\left| \widehat{\lambda} - \frac{\eta(T_0, x, k)}{T_0} \right| \le \delta,$$

for all $k \in \mathcal{M}$ and $|x| \le r_1$. Hence,

$$\eta(T_0, x, k) \le \widehat{\lambda} T_0 + T_0 \delta, \tag{6.26}$$

for all $k \in \mathcal{M}$ and $|x| \le r_1$.

On $[0,\ T_0)$, choose an admissible $u^{(1)}(t)$ such that

$$\begin{aligned} E \exp\left(\int_0^{T_0} g(x(t), u^{(1)}(t))\, dt \right) &\le \exp(\eta(T_0, x, k) + \delta T_0) \\ &\le \exp(\widehat{\lambda} T_0 + 2\delta T_0). \end{aligned}$$

Let $\widehat{\mathcal{F}}_{T_0} = \sigma\{(x(t), k(t)) : t \le T_0\}$. On $[T_0,\ 2T_0)$, if we choose $u^{(2)}(t)$ to be $\sigma\{k(s) : T_0 \le s \le t\}$-measurable, then

$$E\left\{ \exp\left(\int_{T_0}^{2T_0} g(x(t), u^{(2)}(t))\, dt \right) \middle| \widehat{\mathcal{F}}_{T_0} \right\}$$

is a function of $(T_0, x(T_0), k(T_0))$. More precisely, if we let

$$\Phi(T_0, x, k, u(\cdot)) = E\left\{ \exp\left(\int_{T_0}^{2T_0} g(x(t), u(t))\, dt \right) \middle| x(T_0) = x, k(T_0) = i \right\},$$

then

$$\Phi(T_0, x(T_0), k(T_0), u^{(2)}(\cdot)) = E\left\{ \exp\left(\int_{T_0}^{2T_0} g(x(t), u^{(2)}(t))\, dt \right) \middle| \widehat{\mathcal{F}}_{T_0} \right\}.$$

Moreover, by changing the variable t to $(t - T_0)$, we have

$$\eta(T_0, x, k) = \inf_{u(\cdot) \in \mathcal{A}(k)} \log \Phi(T_0, x, k, u(\cdot)).$$

Similarly as in the proof of Lemma 3.1(ii), we can show that, for some positive constant C_4,

$$|\eta(T, \widehat{x}, k) - \eta(T, x, k)| \le C_4 T |\widehat{x} - x| d,$$

for all T, $\widehat{x}$, x, and $k \in \mathcal{M}$.

Let B_1, B_2, ..., B_ℓ be a partition of $\{x : |x| \le r_1\}$. For any given $\delta > 0$, if the diameter of the B_j's is small enough, then for all $\widehat{x}$ and x in B_j and $u(\cdot) \in \mathcal{A}(k)$,

$$|\eta(T_0, \widehat{x}, k) - \eta(T_0, x, k)| \le \delta T_0$$

and

$$\frac{\Phi(T_0, \widehat{x}, k, u(\cdot))}{\Phi(T_0, x, k, u(\cdot))} \le e^{\delta T_0}.$$

For $j = 1, 2, \ldots, \ell$, pick $x_j \in B_j$. For each (j, i), choose $u_{j,i}^{(2)}(t)$ on $[T_0, 2T_0)$ such that

$$\Phi(T_0, x_j, k) \le \exp(\eta(T_0, x_j, k) + \delta T_0) \le \exp\left(\widehat{\lambda} T_0 + 2\delta T_0\right).$$

On $[0, 2T_0)$, define

$$u(t) = \begin{cases} u^{(1)}(t), & \text{if } 0 \le t < T_0, \\ \displaystyle\sum_{j,i} I_{\{(x(T_0),k(T_0)) \in B_j \times \{i\}\}} u_{j,i}^{(2)}(t), & \text{if } T_0 \le t < 2T_0. \end{cases} \tag{6.27}$$

It follows that

$$\begin{aligned} &E\left\{\exp\left(\int_{T_0}^{2T_0} g(x(t), u(t))\, dt\right) \middle| \widehat{\mathcal{F}}_{T_0}\right\} \\ &\quad = \sum_{j,i} I_{\{x(T_0) \in B_j\}} I_{\{k(T_0) = k\}} \Phi(T_0, x(T_0), k, u_{j,i}^{(2)}(t)) \\ &\quad \le \sum_{j,i} I_{\{x(T_0) \in B_j\}} I_{\{k(T_0) = k\}} \Phi(T_0, x_j, k, u_{j,i}^{(2)}(t)) e^{\delta T_0} \\ &\quad \le \sum_{j,i} I_{\{x(T_0) \in B_j\}} I_{\{k(T_0) = k\}} \exp\left(\widehat{\lambda} T_0 + 3\delta T_0\right). \end{aligned}$$

Note that

$$\begin{aligned} &E \exp\left(\int_0^{2T_0} g(x(t), u(t))\, dt\right) \\ &= E\left\{\exp\left(\int_0^{T_0} g(x(t), u(t))\, dt\right) E\left[\exp\left(\int_{T_0}^{2T_0} g(x(t), u(t))\, dt\right) \middle| \widehat{\mathcal{F}}_{T_0}\right]\right\}. \end{aligned}$$

It follows that

$$E \exp\left(\int_0^{2T_0} g(x(t), u(t))\, dt\right) \le \exp\left(2\widehat{\lambda} T_0 + 5\delta T_0\right).$$

Continuing this procedure on $[(N-1)T_0, NT_0)$ for $N = 3, \ldots$, we can construct an admissible control $u(t)$ as in (6.27) such that

$$E \exp\left(\int_0^{NT_0} g(x(t), u(t))\, dt\right) \leq \exp\left(N\widehat{T}_0 + \delta(3N-1)T_0\right).$$

Hence,

$$\begin{aligned}\frac{1}{NT_0} \log E \exp\left(\int_0^{NT_0} g(x(t), u(t))\, dt\right) &\leq \widehat{\lambda} + \frac{\delta(3N-1)T_0}{NT_0} \\ &\to \widehat{\lambda} + 3\delta.\end{aligned}$$

Since δ is arbitrary, the inequality $\widehat{\lambda} \geq \inf_{u(\cdot)\in\mathcal{A}(k)} J(x, k, u(\cdot))$ follows. □

This theorem implies that λ in a viscosity solution $(\lambda, V(x,k))$ is unique.

6.4 Verification Theorem

Based on the HJB equation (6.5) and its solution discussed in Theorems 3.1 and 3.2, we next give a verification theorem. In order to incorporate nondifferentiability of the value function, we consider superdifferential and subdifferential of the function defined in Section 3.4.

Theorem 4.1. *Let Assumptions* (A1) *and* (A2) *hold, and let* $(\mu, W(x,k))$ *be a viscosity solution to the* HJB *equation* (6.5). *Assume that* $W(x,k)$ *is Lipschitz continuous in* x. *Let* $\zeta(x,k) = \exp(W(x,k))$. *Suppose that there are* $u^*(\cdot)$, $x^*(\cdot)$, *and* $r^*(t)$ *such that*

$$\frac{dx^*(t)}{dt} = -ax^*(t) + u^*(t) - z, \quad x^*(0) = x,$$

$r^*(t) \in D^+W(x^*(t), k(t))$, *and*

$$\begin{aligned}\mu\zeta(x^*(t), k(t)) &= (-ax^*(t) + u^*(t) - z)r^*(t) \\ &\quad + g(x^*(t), u^*(t))\zeta(x^*(t), k(t)) + Q\zeta(x^*(t), \cdot)(k(t)),\end{aligned} \tag{6.28}$$

a.e. in t *and with probability one. Then control* $u^*(\cdot)$ *is optimal, i.e.,* $\lambda = J(x, k, u^*(\cdot))$.

Proof. First of all, note that $(\lambda, W(x,k))$ is a solution of the HJB equation (6.5). In view of the logarithmic transformation (6.8), equation (6.5) is equivalent to

$$\mu\zeta(x,k) = \inf_{0\le u\le k}\left\{(-ax+u-z)\frac{\partial\zeta(x,k)}{\partial x} + g(x,u)\zeta(x,k) + Q\zeta(x,\cdot)(k)\right\}. \tag{6.29}$$

The Lipschitz property of $W(x,k)$ implies that $\zeta(x,k)$ is Lipschitz which, in turn, implies that $\zeta(x(t),k)$ is also Lipschitz in t. For each $t \ge 0$ such that $(d/dt)\zeta(x(t),k)$ exists and

$$\int_t^{t+\delta} [-ax(s)+u(s)-z]\,ds = [-ax(t)+u(t)-z]\delta + o(\delta),$$

we have

$$\begin{aligned}
&\frac{d\zeta(x(t),k)}{dt}\\
&= \lim_{\delta\to 0^+}\frac{1}{\delta}\left[\zeta(x(t+\delta),k)-\zeta(x(t),k)\right]\\
&= \lim_{\delta\to 0^+}\frac{1}{\delta}\left\{\zeta\left(x(t)+\int_t^{t+\delta}[-ax(s)+u(s)-z]\,ds, k\right)-\zeta(x(t),k)\right\}\\
&= \lim_{\delta\to 0^+}\frac{1}{\delta}\left\{\zeta(x(t)+\delta[-ax(t)+u(t)-z]+o(\delta),k)-\zeta(x(t),k)\right\}\\
&= \lim_{\delta\to 0^+}\frac{1}{\delta}\left\{\zeta(x(t)+\delta[-ax(t)+u(t)-z],k)-\zeta(x(t),k)\right\}\\
&\le [-ax(t)+u(t)-z]r,
\end{aligned} \tag{6.30}$$

for $r \in D^+\zeta(x(t),k)$. In view of (6.30) and the proof of the Feynman-Kac formula (see Fleming and Soner [56]), we can show that, for any $T \ge 0$,

$$\begin{aligned}
&E\left\{\zeta(x^*(T),k(T))\exp\left(\int_0^T [g(x^*(t),u^*(t))-\mu]\,dt\right)\right\} - \zeta(x,k)\\
&= E\int_0^T \frac{d}{dt}\left\{\zeta(x^*(t),k(t))\exp\left(\int_0^t [g(x^*(s),u^*(s))-\mu]\,ds\right)dt\right\}\\
&\le E\int_0^T \left[\exp\left(\int_0^t (g(x^*(s),u^*(s))-\mu)\,ds\right)\right]\\
&\quad\times\{[g(x^*(t),u^*(t))-\mu]\,\zeta(x^*(t),k(t))\\
&\quad+[-ax^*(t)+u^*(t)-z]r^*(t)+Q\zeta(x^*(t),\cdot)(k(t))\}\,dt = 0.
\end{aligned} \tag{6.31}$$

Note that for any given initial value x, the corresponding trajectory $x(t)$ is bounded. Thus, for each x, there exist positive constants C_1 and C_2 such that

$$0 < C_1 \le \zeta(x^*(T),k(T)) \le C_2, \qquad \text{for all } T \ge 0.$$

Hence, it follows from (6.29) and (6.31) that

$$C_1 E\left[\exp\left(\int_0^T g(x^*(t), u^*(t))\, dt\right)\right] \exp(-\mu T) \le \zeta(x, i).$$

Taking logarithm on both sides and dividing by T leads to

$$\frac{\log C_1}{T} + \frac{1}{T}\log E\exp\left(\int_0^T g(x^*(t), u^*(t))\, dt\right) - \mu \le \frac{\zeta(x, k)}{T}.$$

Letting $T \to \infty$ yields

$$\mu \ge \limsup_{T\to\infty} \frac{1}{T}\log E\exp\left(\int_0^T g(x^*(t), u^*(t))\, dt\right).$$

Hence, in view of Theorem 3.2, $\lambda = J(x, k, u^*(\cdot))$. □

6.5 Notes

Theorems 3.1, 3.2, and 4.1 on the risk-sensitive control problems are due to Fleming and Zhang [58].

In our model, we assume a positive deterioration rate a for items in storage (formula (6.1)). This corresponds to a stability condition typically imposed for disturbance attenuation problems on an infinite time horizon (see Fleming and McEneaney [52]), and this assumption is used in the proof of technical estimates in Lemma 3.1. Nevertheless, it would be interesting to weaken the assumption that $a > 0$.

The risk-sensitive approach has been applied to the so-called disturbance attenuation problem; see, for example, Whittle [144], Fleming and McEneaney [52], Glover and Doyle [64], and references therein. In Fleming and McEneaney [52], risk-sensitive control problems of controlled diffusions are considered. By using the associated dynamic programming equations, they show that as the system noise goes to zero, the value function of the risk-sensitive control problem converges to the value function of a differential game problem.

Part III:

Near-Optimal Controls

7 Near-Optimal Control of Parallel-Machine Systems

7.1 Introduction

In this chapter we consider a manufacturing system which consists of a number of parallel machines that are subject to breakdown and repair. The problem is to obtain the rate of production over time in order to meet the demand at the minimum long-run average expected cost of production and surplus over the infinite horizon. An exact optimal solution of the problem is difficult to obtain. Our goal is to find near-optimal controls to run these systems. The idea is to derive a limiting control problem (by letting the rate of machine breakdown and repair approach infinity), which is simpler to solve than the original problem. This limiting problem is obtained by replacing the stochastic machine availability process by the average total production capacity of machines and by appropriately modifying the objective function. Using the optimal (or near-optimal) control for the limiting problem, one can construct an approximately optimal control of the original, more complex, problem.

We assume that the production capacity process is a fast-changing process. By a fast-changing process, we mean a process that is changing so rapidly that from any initial condition, it reaches its stationary distribution in a time period during which there are few, if any, fluctuations in the other processes. For example, in the case of a fast-changing Markov process, the state distribution converges rapidly to a distribution close to its stationary distribution. In the case of a fast-changing deterministic process, the time-average of the process quickly reaches a value near its limiting long-run average value. Furthermore, it is possible to associate a *time constant* with each of these processes, namely the reciprocal of the rate of this convergence. The time constant is related to the time it takes the process to cover a specified fraction of the distance between its current value and its equilibrium value, or the time required for the initial distribution to become sufficiently close to the stationary distribution.

We show that the minimum average cost of the original problem converges to that of a limiting problem. We then construct asymptotic optimal controls for the original problem by using near-optimal controls of the limiting problem. We also derive the rate of convergence and error bounds associated with some of the constructed controls.

The plan of this chapter is as follows. In Section 7.2 we provide a precise formulation of the parallel machines, multiproduct model introduced in Section 2.2. The total production capacity of the machines is given by a finite state Markov chain parametrized by a small number $\varepsilon > 0$. The demand is assumed to be constant for convenience in exposition. In Section 7.3 we discuss some elementary properties of the associated minimum average cost. In Section 7.4 we consider the convergence rate of the minimum average cost. In Section 7.5 an asymptotic optimal feedback control is constructed by using an optimal feedback control of the limiting problem. In Section 7.6 we consider the minimum cost problem without attrition and prove the convergence of the minimum average cost as $\varepsilon \to 0$. The chapter is concluded in Section 7.7.

7.2 Problem Formulation

Let us consider a manufacturing system that produces n distinct products using m identical parallel machines. With the production rate $\boldsymbol{u}(t) \in \Re^n_+$, the total surplus $\boldsymbol{x}(t) \in \Re^n$, and a constant demand rate $\boldsymbol{z} \in \Re^n_+$, the system dynamics satisfy the differential equation

$$\frac{d}{dt}\boldsymbol{x}(t) = -\mathrm{diag}(\boldsymbol{a})\boldsymbol{x}(t) + \boldsymbol{u}(t) - \boldsymbol{z}, \quad \boldsymbol{x}(0) = \boldsymbol{x} \in \Re^n, \tag{7.1}$$

where $\boldsymbol{a} = (a_1, ..., a_n)'$ is a constant vector with $a_j > 0$ and $\mathrm{diag}(\boldsymbol{a}) = \mathrm{diag}(a_1, ..., a_n)$ is a diagonal matrix. Recall that surplus $x_j(t)$ of a product type j means inventory when $x_j(t) > 0$ and backlog when $x_j(t) < 0$. The attrition rate a_j represents the deterioration rate of the inventory of the product type j when $x_j(t) > 0$, and it represents a rate of cancellation of backlogged orders when $x_j(t) < 0$. We assume symmetric deterioration and cancellation rates for product j only for convenience in exposition. It is easy to extend our results when a_j is a function of $\boldsymbol{x}$,

$$a_j(\boldsymbol{x}) = \begin{cases} \widehat{a}_j, & \text{if } x_j \geq 0, \\ \overline{a}_j, & \text{if } x_j < 0, \end{cases}$$

where $\widehat{a}_j > 0$ and $\overline{a}_j > 0$. $\widehat{a}_j$ denotes the deterioration rate and $\overline{a}_j$ denotes the order cancellation rate. We rewrite (7.1) as

$$\frac{d}{dt}\boldsymbol{x}(t) = \mathcal{L}(\boldsymbol{x}(t), \boldsymbol{u}(t), \boldsymbol{z}), \quad \boldsymbol{x}(0) = \boldsymbol{x},$$

where

$$\mathcal{L}(\boldsymbol{x}(t), \boldsymbol{u}(t), \boldsymbol{z}) = -\text{diag}(\boldsymbol{a})\boldsymbol{x}(t) + \boldsymbol{u}(t) - \boldsymbol{z}.$$

We now define the random machine capacity process on the probability space $(\Omega, \mathcal{F}, P)$. Let $\mathcal{M} = \{0, 1, ..., m\}$ denote the set of machine capacity states, and let

$$k(\varepsilon, t) \in \mathcal{M}, \quad t \geq 0,$$

denote a Markov chain generated by

$$Q^\varepsilon = Q^{(1)} + \frac{1}{\varepsilon} Q^{(2)},$$

where $\varepsilon > 0$ is a small parameter and $Q^{(\ell)} = (q_{ij}^{(\ell)})$, $i, j \in \mathcal{M}$, is an $(m+1) \times (m+1)$ matrix such that $q_{ij}^{(\ell)} \geq 0$ for $i \neq j$ and $q_{ii}^{(\ell)} = -\sum_{j \neq i} q_{ij}^{(\ell)}$ for $\ell = 1, 2$. We let $k(\varepsilon, t)$ represent the machine capacity state at time t. Since only a finite amount of production capacity is available at any given time t, it imposes an upper bound on the production rate $\boldsymbol{u}(t)$. Specifically, the production rate constraints can be written as

$$u_j(t) \geq 0, \quad j = 1, 2, ..., n, \qquad \sum_{j=1}^{n} u_j(t) \leq k(\varepsilon, t), \quad t \geq 0. \tag{7.2}$$

Definition 2.1. A production process $\boldsymbol{u}(\cdot) = \{\boldsymbol{u}(t) : t \geq 0\}$ is *admissible* if: (i) $\boldsymbol{u}(t)$ is adapted to the filtration $\{\mathcal{F}_t^\varepsilon\}$ with $\mathcal{F}_t^\varepsilon = \sigma(k(\varepsilon, s) : 0 \leq s \leq t)$; and (ii) $u_j(t) \geq 0$, $j = 1, 2, ..., n$, and $\sum_{j=1}^{n} u_j(t)(\omega) \leq k(\varepsilon, t)(\omega)$ for all $t \geq 0$. □

We denote by $\mathcal{A}^\varepsilon(k)$ the set of all admissible controls with the initial condition $k(\varepsilon, 0) = k$.

Remark 2.1. Here we assume that the amount of capacity needed to produce product type j at rate 1 is one. The results established in this chapter can be easily extended to the case in which the amount of capacity needed to produce product type j at rate 1 is any given constant r_j. For this case, the production rate constrained (7.2) can be specified as

$$u_j(t) \geq 0, \quad j = 1, 2, ..., n, \qquad \sum_{j=1}^{n} r_j u_j(t) \leq k(\varepsilon, t), \quad t \geq 0. \quad \square \tag{7.3}$$

Definition 2.2. A function $\boldsymbol{u}(\cdot, \cdot)$ defined on $\Re^n \times \mathcal{M}$ is called an *admissible feedback control* or simply a *feedback control*, if:

(i) for any given initial surplus $\boldsymbol{x}(0) = \boldsymbol{x}$ and production capacity $k(\varepsilon, 0) = k$, the equation

$$\frac{d}{dt}\boldsymbol{x}(t) = \mathcal{L}(\boldsymbol{x}(t), \boldsymbol{u}(\boldsymbol{x}(t), k(\varepsilon, t)), \boldsymbol{z})$$

has a unique solution; and

(ii) the control $\boldsymbol{u}(\cdot) = \{\boldsymbol{u}(t) = \boldsymbol{u}(\boldsymbol{x}(t), k(\varepsilon, t)), t \geq 0\} \in \mathcal{A}^\varepsilon(k)$. □

Let $h(\boldsymbol{x})$ and $c(\boldsymbol{u})$ denote the surplus cost and the production cost, respectively. For any $\boldsymbol{u}(\cdot) \in \mathcal{A}^\varepsilon(k)$, define the expected long-run average cost

$$J^\varepsilon(\boldsymbol{x}, k, \boldsymbol{u}(\cdot)) = \limsup_{T\to\infty} \frac{1}{T} E \int_0^T [h(\boldsymbol{x}(t)) + c(\boldsymbol{u}(t))]\, dt, \tag{7.4}$$

where $\boldsymbol{x}(\cdot)$ is the surplus process corresponding to the production process $\boldsymbol{u}(\cdot)$ with the initial conditions $\boldsymbol{x}(0) = \boldsymbol{x}$ and $k(\varepsilon, 0) = k$. The problem is to obtain $\boldsymbol{u}(\cdot) \in \mathcal{A}^\varepsilon(k)$ that minimizes $J^\varepsilon(\boldsymbol{x}, k, \boldsymbol{u}(\cdot))$. We summarize our control problem as follows:

$$\mathcal{P}^\varepsilon: \begin{cases} \min\ J^\varepsilon(\boldsymbol{x}, k, \boldsymbol{u}(\cdot)), \\ \text{s.t. } \dfrac{d}{dt}\boldsymbol{x}(t) = \mathcal{L}(\boldsymbol{x}(t), \boldsymbol{u}(t), \boldsymbol{z}), \quad \boldsymbol{x}(0) = \boldsymbol{x}, \quad \boldsymbol{u}(\cdot) \in \mathcal{A}^\varepsilon(k), \\ \text{minimum average cost} \quad \lambda^\varepsilon(\boldsymbol{x}, k) = \displaystyle\inf_{\boldsymbol{u}(\cdot)\in\mathcal{A}^\varepsilon(k)} J^\varepsilon(\boldsymbol{x}, k, \boldsymbol{u}(\cdot)). \end{cases} \tag{7.5}$$

We assume that the cost functions $h(\boldsymbol{x})$, $c(\boldsymbol{u})$, and the production capacity process $k(\varepsilon, \cdot)$ satisfy the following assumptions:

(A1) $h(\boldsymbol{x})$ is a nonnegative, convex function with $h(0) = 0$. There are positive constants C_{h1}, C_{h2}, and $\beta_{h1} \geq 1$ such that

$$h(\boldsymbol{x}) \geq C_{h1}|\boldsymbol{x}|^{\beta_{h1}} - C_{h2}, \quad \boldsymbol{x} \in \Re^n.$$

Moreover, there are constants C_{h3} and $\beta_{h2} \geq \beta_{h1}$ such that

$$|h(\boldsymbol{x}) - h(\boldsymbol{y})| \leq C_{h3}(1 + |\boldsymbol{x}|^{\beta_{h2}-1} + |\boldsymbol{y}|^{\beta_{h2}-1})|\boldsymbol{x} - \boldsymbol{y}|, \quad \boldsymbol{x}, \boldsymbol{y} \in \Re^n.$$

(A2) $c(\boldsymbol{u})$ is a nonnegative convex function.

(A3) $Q^{(2)}$ is strongly irreducible.

(A4) The average capacity

$$\bar{k} = \sum_{i=0}^m i\nu_i > \sum_{j=1}^n z_j,$$

where $\boldsymbol{\nu} = (\nu_0, \nu_1, ..., \nu_m)$ is the equilibrium distribution corresponding to generator $Q^{(2)}$.

Remark 2.2. From Theorem 3.7.3 and Lemma B.7, we know that under Assumptions (A3) and (A4), there is an $\varepsilon_0 > 0$ such that for $\varepsilon \in (0, \varepsilon_0]$, the minimum cost $\lambda^\varepsilon(\boldsymbol{x}, k)$ is independent of $\boldsymbol{x}$ and k. Therefore, in the rest of this chapter, we write λ^ε instead of $\lambda^\varepsilon(\boldsymbol{x}, k)$. □

As in Section 6.2, the positive attrition rate $\boldsymbol{a}$ implies a uniform bound on $\boldsymbol{x}(t)$. In view of the fact that the control $\boldsymbol{u}(t)$ is bounded between 0 and m, this implies that any solution $\boldsymbol{x}(\cdot)$ of (7.1) must satisfy

$$\begin{aligned} |x_j(t)| &= \left| x_j e^{-a_j t} + e^{-a_j t} \int_0^t e^{a_j s}[u_j(s) - z_j]\, ds \right| \\ &\leq |x_j| e^{-a_j t} + (m + z_j) \int_0^t e^{-a_j(t-s)}\, ds \\ &\leq |x_j| e^{-a_j t} + \frac{m + z_j}{a_j}, \quad j = 1, ..., n. \end{aligned} \tag{7.6}$$

Thus, under the positive deterioration/cancellation rate, the surplus process $\boldsymbol{x}(\cdot)$ remains bounded.

7.3 The Limiting Control Problem

In this section we derive the limiting control problem as $\varepsilon \to 0$. To do this, let

$$\mathcal{U}(k) = \left\{ (u_1, \cdots, u_n)' \ : \ u_j \geq 0 \text{ and } \sum_{j=1}^{n} u_j \leq k \right\}.$$

The Hamilton-Jacobi-Bellman (HJB) equation for the optimal control problem $\mathcal{P}^\varepsilon$ has the form:

$$\begin{aligned} \lambda^\varepsilon = & \inf_{\boldsymbol{u} \in \mathcal{U}(k)} \left\{ \left\langle \mathcal{L}(\boldsymbol{x}, \boldsymbol{u}, \boldsymbol{z}), \frac{\phi^\varepsilon(\boldsymbol{x}, k)}{\partial \boldsymbol{x}} \right\rangle + c(\boldsymbol{u}) \right\} + h(\boldsymbol{x}) \\ & + \left(Q^{(1)} + \frac{1}{\varepsilon} Q^{(2)} \right) \phi^\varepsilon(\boldsymbol{x}, \cdot)(k), \end{aligned} \tag{7.7}$$

where $\phi^\varepsilon(\boldsymbol{x}, k)$ is a function defined on $\Re^n \times \mathcal{M}$, and

$$Q\phi(\cdot)(k) = \sum_{i \neq k} q_{ki}(\phi(i) - \phi(k))$$

for any function $\phi(\cdot)$ on $\mathcal{M}$.

The HJB equation in terms of directional derivatives (HJBDD) takes the form

$$\begin{aligned} \lambda^\varepsilon = & \inf_{\boldsymbol{u} \in \mathcal{U}(k)} \left\{ \partial_{\mathcal{L}(\boldsymbol{x}, \boldsymbol{u}, \boldsymbol{z})} \phi^\varepsilon(\boldsymbol{x}, k) + c(\boldsymbol{u}) \right\} + h(\boldsymbol{x}) \\ & + \left(Q^{(1)} + \frac{1}{\varepsilon} Q^{(2)} \right) \phi^\varepsilon(\boldsymbol{x}, \cdot)(k). \end{aligned}$$

Our analysis begins with the proof of the boundedness of λ^ε.

Theorem 3.1. *Let Assumptions* (A1)–(A4) *hold. The minimum average expected cost* λ^ε of $\mathcal{P}^\varepsilon$ *is bounded in* ε, *i.e., there exists a constant* $C > 0$ *such that*

$$0 \le \lambda^\varepsilon \le C, \qquad \text{for all } \varepsilon > 0.$$

Proof. According to the definition of λ^ε, it suffices to show that there exist a constant C_1 and a control $\boldsymbol{u}^\varepsilon(\cdot) \in \mathcal{A}^\varepsilon(0)$ such that for the solution $\boldsymbol{x}^\varepsilon(\cdot)$ of

$$\frac{d}{dt}\boldsymbol{x}(t) = \mathcal{L}(\boldsymbol{x}(t), \boldsymbol{u}^\varepsilon(t), \boldsymbol{z}), \quad \boldsymbol{x}(0) = 0, \tag{7.8}$$

we have

$$\limsup_{T\to\infty} \frac{1}{T} E \int_0^T [h(\boldsymbol{x}^\varepsilon(t)) + c(\boldsymbol{u}^\varepsilon(t))]\, dt \le C_1. \tag{7.9}$$

In view of (7.8), we can derive that, for $j = 1, \ldots, n$,

$$x_j^\varepsilon(t) = e^{-a_j t} \int_0^t e^{a_j s}[u_j^\varepsilon(s) - z_j]\, ds,$$

which implies

$$|x_j^\varepsilon(t)| \le \frac{(m + z_j)}{a_j}. \tag{7.10}$$

It is clear under Assumptions (A1) and (A2) that the functions $h(\cdot)$ and $c(\cdot)$ are continuous. Consequently, the control constraints and inequality (7.10) imply (7.9). □

In the remainder of this section, we derive the limiting control problem as $\varepsilon \to 0$. To this purpose, we consider the enlarged control space

$$\begin{aligned}\mathcal{A}^0 = \{U(\cdot) = (\boldsymbol{u}^0(\cdot), \boldsymbol{u}^1(\cdot), ..., \boldsymbol{u}^m(\cdot)) \ : \ \boldsymbol{u}^i(t) \in \mathcal{U}(i) \text{ for all } t \ge 0,\\ \text{and } \boldsymbol{u}^i(\cdot) \text{ is a deterministic process for each } i\}.\end{aligned} \tag{7.11}$$

Then we define the limiting control problem $\mathcal{P}^0$ as follows:

$$\mathcal{P}^0: \begin{cases} \min J(\boldsymbol{x}, U(\cdot)) = \displaystyle\limsup_{T\to\infty} \frac{1}{T} \int_0^T \left[h(\boldsymbol{x}(s)) + \sum_{i=0}^m \nu_i c(\boldsymbol{u}^i(s)) \right] ds, \\ \text{s.t. } \displaystyle\frac{d}{dt}\boldsymbol{x}(t) = \mathcal{L}\left(\boldsymbol{x}(t), \sum_{i=0}^m \nu_i \boldsymbol{u}^i(t), \boldsymbol{z} \right), \quad \boldsymbol{x}(0) = \boldsymbol{x}, \quad U(\cdot) \in \mathcal{A}^0, \\ \text{minimum average cost } \lambda = \displaystyle\inf_{U(\cdot)\in\mathcal{A}^0} J(\boldsymbol{x}, U(\cdot)). \end{cases}$$

The HJB equation for problem $\mathcal{P}^0$ has the form

$$\lambda = \inf_{\boldsymbol{u}^i \in \mathcal{U}(i), i \in \mathcal{M}} \left\{ \left\langle \mathcal{L}\left(\boldsymbol{x}, \sum_{i=0}^{m} \nu_i \boldsymbol{u}^i, \boldsymbol{z}\right), \frac{\partial \phi(\boldsymbol{x})}{\partial \boldsymbol{x}} \right\rangle + \sum_{i=0}^{m} \nu_i c(\boldsymbol{u}^i) \right\} + h(\boldsymbol{x}), \tag{7.12}$$

where $\phi(\boldsymbol{x})$ is a function defined on $\Re^n$. The HJBDD equation takes the form

$$\lambda = \inf_{\boldsymbol{u}^i \in \mathcal{U}(i), i \in \mathcal{M}} \left\{ \partial_{\mathcal{L}(\boldsymbol{x}, \sum_{i=0}^{m} \nu_i \boldsymbol{u}^i, \boldsymbol{z})} \phi(\boldsymbol{x}) + \sum_{i=0}^{m} \nu_i c(\boldsymbol{u}^i) \right\} + h(\boldsymbol{x}).$$

For the sake of notational simplicity, we only deal with the case $n = 1$ (i.e., the single product case) in the remainder of this chapter. In this case, a, z, x, and u are scalars. Using the same method, one can show that all of the results in Sections 7.4 and 7.5 except Theorem 5.2, given here for the single product case, hold also for the multiple product case.

7.4 Convergence of the Minimum Average Expected Cost

In this section we consider the convergence of the minimum average expected cost λ^ε as ε goes to zero, and establish its convergence rate. Armed with Theorem 3.1, we can derive the required convergence result.

Theorem 4.1. *Let Assumptions* (A1)–(A4) *hold. There exists a constant C such that, for all* $\varepsilon > 0$,

$$|\lambda^\varepsilon - \lambda| \le C\varepsilon^{1/2}. \tag{7.13}$$

This implies in particular that $\lim_{\varepsilon \to 0} \lambda^\varepsilon = \lambda$.

Proof. The proof is divided into two parts. First we prove $\lambda^\varepsilon \le \lambda + C\varepsilon^{1/2}$ by constructing an admissible control $u^\varepsilon(t)$ of $\mathcal{P}^\varepsilon$ from the optimal control of the limiting problem $\mathcal{P}^0$ and by estimating the difference between the state trajectories corresponding to these two controls. Then we establish the opposite inequality, namely, $\lambda^\varepsilon \ge \lambda - C\varepsilon^{1/2}$, by constructing a control of the limiting problem $\mathcal{P}^0$ from a near-optimal control of $\mathcal{P}^\varepsilon$ and then using Assumptions (A1)–(A3).

In order to show that

$$\lambda^\varepsilon \le \lambda + C\varepsilon^{1/2}, \tag{7.14}$$

we let $U(\cdot) = (u^0(\cdot), u^1(\cdot), ..., u^m(\cdot)) \in \mathcal{A}^0$, where $\mathcal{A}^0$ is given in (7.11), and we construct the control

$$u^\varepsilon(t) = \sum_{i=0}^{m} I_{\{k(\varepsilon,t)=i\}} u^i(t).$$

Clearly $u^\varepsilon(\cdot) \in \mathcal{A}^\varepsilon(k)$, where $\mathcal{A}^\varepsilon(k)$ is given in Section 7.2. Let $x^\varepsilon(t)$ and $\bar{x}(t)$ denote the corresponding state trajectories of the systems $\mathcal{P}^\varepsilon$ and $\mathcal{P}^0$, respectively. Then

$$\frac{d}{dt}\bar{x}(t) = \mathcal{L}\left(\bar{x}(t), \sum_{i=0}^{m} \nu_i u^i(t), z\right), \quad \bar{x}(0) = x, \tag{7.15}$$

and

$$\frac{d}{dt}x^\varepsilon(t) = \mathcal{L}(x^\varepsilon(t), u^\varepsilon(t), z), \quad x^\varepsilon(0) = x. \tag{7.16}$$

Hence,

$$E|x^\varepsilon(t) - \bar{x}(t)|^2 = E\left[e^{-at}\int_0^t e^{as}\left(u^\varepsilon(s) - \sum_{i=0}^{m} \nu_i u^i(s)\right) ds\right]^2.$$

Writing the right-hand side as a double integral, we have

$$\begin{aligned}
&E\left[e^{-at}\int_0^t e^{as}\left(u^\varepsilon(s) - \sum_{i=0}^{m} \nu_i u^i(s)\right) ds\right]^2 \\
&= e^{-2at}\sum_{i,j=0}^{m} E\int_0^t\int_0^t e^{a(s_1+s_2)}\left[I_{\{k(\varepsilon,s_1)=i\}} - \nu_i\right] \\
&\qquad\qquad \times\left[I_{\{k(\varepsilon,s_2)=j\}} - \nu_j\right]u^i(s_1)u^j(s_2)\,ds_1\,ds_2 \\
&= e^{-2at}\sum_{i,j=0}^{m} E\left(\int_0^t\int_0^t e^{a(s_1+s_2)}\left[I_{\{k(\varepsilon,s_1)=i,k(\varepsilon,s_2)=j\}} - \nu_j I_{\{k(\varepsilon,s_1)=i\}}\right.\right. \\
&\qquad\qquad \left.\left. - \nu_i I_{\{k(\varepsilon,s_2)=j\}} + \nu_i\nu_j\right]u^i(s_1)u^j(s_2)\,ds_1\,ds_2\right) \\
&= e^{-2at}\left(\sum_{i,j=0}^{m} E\int_0^t\int_0^t e^{a(s_1+s_2)}\left[I_{\{k(\varepsilon,s_1)=i,k(\varepsilon,s_2)=j\}} - \nu_i\nu_j\right]\right. \\
&\qquad\qquad \times u^i(s_1)u^j(s_2)\,ds_1\,ds_2 \\
&\qquad - \sum_{i,j=0}^{m} E\int_0^t\int_0^t e^{a(s_1+s_2)}\nu_j\left[I_{\{k(\varepsilon,s_1)=i\}} - \nu_i\right] \\
&\qquad\qquad \times u^i(s_1)u^j(s_2)\,ds_1\,ds_2 \\
&\qquad - \sum_{i,j=0}^{m} E\int_0^t\int_0^t e^{a(s_1+s_2)}\nu_i\left[I_{\{k(\varepsilon,s_2)=j\}} - \nu_j\right] \\
&\qquad\qquad \left. \times u^i(s_1)u^j(s_2)\,ds_1\,ds_2\right)
\end{aligned}$$

$$
\begin{aligned}
&= e^{-2at}\Bigg\{\sum_{i,j=0}^{m} E\int_0^t\left(\int_0^{s_1} e^{a(s_1+s_2)}\big[I_{\{k(\varepsilon,s_2)=j,k(\varepsilon,s_1)=i\}}-\nu_j\nu_i\big]\right. \\
&\qquad\qquad \left.\times u^j(s_2)\,ds_2\right)u^i(s_1)\,ds_1 \\
&\quad+\sum_{i,j=0}^{m} E\int_0^t\left(\int_{s_1}^{t} e^{a(s_1+s_2)}\big[I_{\{k(\varepsilon,s_2)=j,k(\varepsilon,s_1)=i\}}-\nu_j\nu_i\big]\right. \\
&\qquad\qquad \left.\times u^j(s_2)\,ds_2\right)u^i(s_1)\,ds_1 \\
&\quad-\sum_{i,j=0}^{m} E\int_0^t\int_0^t e^{a(s_1+s_2)}\nu_j\big[I_{\{k(\varepsilon,s_1)=i\}}-\nu_i\big] \\
&\qquad\qquad \times u^i(s_1)u^j(s_2)\,ds_1\,ds_2 \\
&\quad-\sum_{i,j=0}^{m}\int_0^t\int_0^t e^{a(s_1+s_2)}\nu_i\big[I_{\{k(\varepsilon,s_2)=j\}}-\nu_j\big] \\
&\qquad\qquad \times u^i(s_1)u^j(s_2)\,ds_1\,ds_2\Bigg\}.
\end{aligned}
\tag{7.17}
$$

Note that for any $s_1 > s_2$, we can write

$$
\begin{aligned}
EI_{\{k(\varepsilon,s_2)=j,k(\varepsilon,s_1)=i\}} &= P(k(\varepsilon,s_2)=j,k(\varepsilon,s_1)=i) \\
&= P(k(\varepsilon,s_1)=i|k(\varepsilon,s_2)=j)P(k(\varepsilon,s_2)=j).
\end{aligned}
$$

In view of this, the first term on the right-hand side of (7.17) can be written as

$$
\begin{aligned}
&E\int_0^t\left(\int_0^{s_1} e^{a(s_1+s_2)}\big[I_{\{k(\varepsilon,s_2)=j,k(\varepsilon,s_1)=i\}}-\nu_i\nu_j\big]u^j(s_2)\,ds_2\right)u^i(s_1)\,ds_1 \\
&= E\left\{\int_0^t e^{as_2}u^j(s_2)\,ds_2\int_{s_2}^t e^{as_1}\big[I_{\{k(\varepsilon,s_2)=j,k(\varepsilon,s_1)=i\}}-\nu_j\nu_i\big]u^i(s_1)\,ds_1\right\} \\
&= \int_0^t e^{as_2}P(k(\varepsilon,s_2)=j)u^j(s_2) \\
&\qquad\qquad \times\int_{s_2}^t e^{as_1}\big[P\big(k(\varepsilon,s_1)=i|k(\varepsilon,s_2)=j\big)-\nu_i\big]u^i(s_1)\,ds_1\,ds_2 \\
&\quad+\int_0^t e^{as_2}\nu_i\big[P(k(\varepsilon,s_2)=j)-\nu_j\big]u^j(s_2)\,ds_2\int_{s_2}^t e^{as_1}u^i(s_1)\,ds_1.
\end{aligned}
\tag{7.18}
$$

In view of Lemma B.3, we have

$$\begin{aligned}
&\int_0^t e^{as_2} P(k(\varepsilon,s_2)=j) u^j(s_2) \\
&\qquad \times \int_{s_2}^t e^{as_1}\big[P(k(\varepsilon,s_1)=i|k(\varepsilon,s_2)=j)-\nu_i\big]u^i(s_1)\,ds_1\,ds_2 \\
&\le \int_0^t e^{as_2} P(k(\varepsilon,s_2)=j)\,ds_2 \int_{s_2}^t e^{as_1} C_1\big[\varepsilon+e^{-\beta_0(s_1-s_2)/\varepsilon}\big]\,ds_1 \qquad (7.19)\\
&\le \int_0^t \frac{C_1\varepsilon}{a} e^{as_2}[e^{at}-e^{as_2}]\,ds_2 + \frac{C_1\varepsilon}{|a\varepsilon-\beta_0|}\int_0^t e^{at}e^{as_2}\,ds_2 \\
&\le \frac{C_1\varepsilon}{a^2}e^{2at} + \frac{C_1\varepsilon}{a|a\varepsilon-\beta_0|}e^{2at},
\end{aligned}$$

for some $C_1>0$, where $\beta_0>0$ is specified in Lemma B.3. Using Lemma B.3 again, we can derive

$$\begin{aligned}
&\int_0^t e^{as_2}\nu_i\big[P(k(\varepsilon,s_2)=j)-\nu_j\big]u^j(s_2)\,ds_2\int_{s_2}^t e^{as_1}u^i(s_1)\,ds_1 \\
&\qquad \le \frac{e^{at}}{a}\int_0^t C_1\big[\varepsilon+e^{-\beta_0 s_2/\varepsilon}\big]e^{as_2}\,ds_2 \qquad (7.20)\\
&\qquad \le \left(\frac{C_1\varepsilon}{a^2}+\frac{C_1\varepsilon}{a|a\varepsilon-\beta_0|}\right)e^{2at}.
\end{aligned}$$

Combining (7.19)–(7.20) yields the following bound on the first term of the right-hand side of (7.17):

$$\begin{aligned}
&E\left\{\int_0^t\left(\int_0^{s_1} e^{a(s_1+s_2)}\big[I_{\{k(\varepsilon,s_1)=i,k(\varepsilon,s_2)=j\}}-\nu_i\nu_j\big]u^j(s_2)\,ds_2\right)u^i(s_1)\,ds_1\right\} \\
&\quad \le \left(\frac{2}{a^2}+\frac{2}{a|a\varepsilon-\beta_0|}\right)C_1\varepsilon e^{2at}.
\end{aligned} \qquad (7.21)$$

Similar to (7.21), we can obtain the following bound on the second, third, and fourth terms on the right-hand side of (7.17), respectively, i.e.,

$$\begin{aligned}
&E\left\{\int_0^t e^{as_1}u^i(s_1)\,ds_1\int_{s_1}^t e^{as_2}\big[I_{\{k(\varepsilon,s_2)=j,k(\varepsilon,s_1)=i\}}-\nu_i\nu_j\big]u^j(s_2)\,ds_2\right\} \\
&\quad \le \left(\frac{2}{a^2}+\frac{2}{a|a\varepsilon-\beta_0|}\right)C_2\varepsilon e^{2at},
\end{aligned} \qquad (7.22)$$

$$E\left\{\int_0^t\int_0^t e^{a(s_1+s_2)}\nu_j\left[I_{\{k(\varepsilon,s_1)=i\}}-\nu_i\right]u^i(s_1)u^j(s_2)\,ds_1\,ds_2\right\}$$
$$\le\left(\frac{2}{a^2}+\frac{2}{a|a\varepsilon-\beta_0|}\right)C_2\varepsilon e^{2at}, \tag{7.23}$$

and

$$E\left\{\int_0^t\int_0^t e^{a(s_1+s_2)}\nu_i\left[I_{\{k(\varepsilon,s_2)=j\}}-\nu_j\right]u^i(s_1)u^j(s_2)\,ds_2\,ds_1\right\}$$
$$\le\left(\frac{2}{a^2}+\frac{2}{a|a\varepsilon-\beta_0|}\right)C_2\varepsilon e^{2at}, \tag{7.24}$$

for some positive constant C_2. By (7.17) and (7.21)–(7.24), there exists a constant $C_3>0$ such that, for all $\varepsilon>0$ and $t\ge 0$,

$$E|x^\varepsilon(t)-\bar{x}(t)|^2\le (C_3)^2\varepsilon.$$

Consequently,

$$E|x^\varepsilon(t)-\bar{x}(t)|\le C_3\varepsilon^{1/2}. \tag{7.25}$$

In view of (7.15) and (7.16), we know that there exists $C_4>0$ such that, for all $t\ge 0$,

$$|x^\varepsilon(t)|\le C_4 \quad\text{and}\quad |\bar{x}(t)|\le C_4. \tag{7.26}$$

Now by Assumption (A1) and (7.25)–(7.26), we have

$$\left|\frac{1}{T}E\int_0^T h(x^\varepsilon(t))\,dt-\frac{1}{T}\int_0^T h(\bar{x}(t))\,dt\right|$$
$$\le\frac{1}{T}E\int_0^T C_{h3}\left(1+|x^\varepsilon(t)|^{\beta_{h2}-1}+|\bar{x}(t)|^{\beta_{h2}-1}\right)|x^\varepsilon(t)-\bar{x}(t)|\,dt$$
$$\le C_5\varepsilon^{1/2}, \tag{7.27}$$

for some $C_5>0$, where C_{h3} and β_{h2} are given by Assumption (A1). In view of $u^i(t)\in[0,\ m]$ and the continuity of $c(\cdot)$, we know that the function $c(u^i(t))$ is bounded on $[0,\ \infty)$. By Lemma B.3, we have

$$\left|\frac{1}{T}E\int_0^T c(u^\varepsilon(t))\,dt-\frac{1}{T}\int_0^T\sum_{i=0}^m \nu_i c(u^i(t))\,dt\right|$$
$$\le\frac{1}{T}\int_0^T\sum_{i=0}^m\left|P(k(\varepsilon,t)=i)-\nu_i\right|\cdot c(u^i(t))\,dt \tag{7.28}$$
$$\le C_6\varepsilon,$$

for some $C_6>0$. By combining (7.27)–(7.28) we get (7.14).

We now show the opposite inequality of (7.14). First we show that for any control $u^\varepsilon(\cdot) \in \mathcal{A}^\varepsilon(k)$, there exists a control $U(\cdot) = (u^0(\cdot), u^1(\cdot), ..., u^m(\cdot)) \in \mathcal{A}^0$ such that $Ex^\varepsilon(t) - \bar{x}(t)$ is small, where $x^\varepsilon(t)$ and $\bar{x}(t)$ are the respective state trajectories under controls $u^\varepsilon(\cdot)$ and $U(\cdot)$ with the same initial condition x. Now we choose $U(\cdot)$ defined by

$$u^i(t) = E[u^\varepsilon(t)|k(\varepsilon, t) = i].$$

Then, we have

$$E[x^\varepsilon(t)] = xe^{-at} + e^{-at} \int_0^t e^{as} \left[\sum_{i=0}^m P(k(\varepsilon, s) = i) u^i(s) - z \right] ds,$$

$$\bar{x}(t) = xe^{-at} + e^{-at} \int_0^t e^{as} \left[\sum_{i=0}^m \nu_i u^i(s) - z \right] ds.$$

Similar to (7.25), by applying Lemma B.3, we obtain

$$|E[x^\varepsilon(t)] - \bar{x}(t)| \le C_7 \varepsilon, \tag{7.29}$$

for some positive constant C_7. In view of the convexity and the local Lipschitz continuity of $h(\cdot)$, inequalities (7.26), (7.29), and Jensen's inequality (cf. Chow and Teicher [33]) yield

$$\begin{aligned} E[h(x^\varepsilon(t))] &\ge h(E[x^\varepsilon(t)]) \\ &= h(\bar{x}(t)) + [h(E[x^\varepsilon(t)]) - h(\bar{x}(t))] \\ &\ge h(\bar{x}(t)) - C_{h3}\big(1 + |E[x^\varepsilon(t)]|^{\beta_{h2}-1} + |\bar{x}(t)|^{\beta_{h2}-1}\big) \\ &\qquad \times \big|E[x^\varepsilon(t)] - \bar{x}(t)\big| \\ &\ge h(\bar{x}(t)) - C_{h3}\big(1 + 2(C_4)^{\beta_{h2}-1}\big) C_7 \varepsilon. \end{aligned} \tag{7.30}$$

In the same way, using Lemma B.3, we can establish

$$\begin{aligned} E[c(u^\varepsilon(t))] &= \sum_{i=0}^m P(k(\varepsilon, t) = i) E\big[c(u^\varepsilon(t))|k(\varepsilon, t) = i\big] \\ &= \sum_{i=0}^m P(k(\varepsilon, t) = i) c(u^i(t)) \\ &\ge \sum_{i=0}^m \nu_i c(u^i(t)) - C_8(\varepsilon + e^{-\beta_0 t/\varepsilon}) \end{aligned} \tag{7.31}$$

for some positive constants C_8, where β_0 is specified in Lemma B.3. By combining (7.30) and (7.31), we obtain

$$\frac{1}{T}E\int_0^T [h(x^\varepsilon(t)) + c(u^\varepsilon(t))]\,dt$$

$$\geq \frac{1}{T}\int_0^T \left[h(\bar{x}(t)) + \sum_{i=0}^{m} \nu_i c(u^i(t))\right] dt - C_9\varepsilon,$$

for some positive constant C_9. The arbitrariness of $u^\varepsilon(t)$ implies

$$\lambda^\varepsilon - \lambda \geq -C\varepsilon,$$

which completes the proof. □

7.5 Asymptotic Optimal Controls

In this section we obtain controls for $\mathcal{P}^\varepsilon$ that are asymptotically optimal. Both open-loop and feedback controls are studied. We first consider open-loop controls.

Theorem 5.1 (Open-loop Control). *Let Assumptions* (A1)–(A4) *hold. Assume that*

$$U(\cdot) = (u^0(\cdot), u^1(\cdot), ..., u^m(\cdot)) \in \mathcal{A}^0$$

is an optimal control for $\mathcal{P}^0$ with the initial condition $\bar{x}(0) = x$. Define

$$u^\varepsilon(t) = \sum_{i=0}^{m} I_{\{k(\varepsilon,t)=i\}} u^i(t).$$

Then $u^\varepsilon(\cdot) \in \mathcal{A}^\varepsilon(k)$, and $u^\varepsilon(\cdot)$ is asymptotically optimal for $\mathcal{P}^\varepsilon$, i.e.,

$$\left|\lambda^\varepsilon - J^\varepsilon(x, k, u^\varepsilon(\cdot))\right| \leq C\varepsilon^{1/2}$$

for some positive constant C independent of ε.

Proof. Observe that

$$0 \leq J^\varepsilon(x, k, u^\varepsilon(\cdot)) - \lambda^\varepsilon = [J^\varepsilon(x, k, u^\varepsilon(\cdot)) - \lambda] + (\lambda - \lambda^\varepsilon).$$

In view of Theorem 4.1, it suffices to show that

$$|J^\varepsilon(x, k, u^\varepsilon(\cdot)) - J(x, U(\cdot))| \leq C_1\varepsilon^{1/2} \tag{7.32}$$

for some positive constant C_1. Using (7.27)–(7.28) in the proof of Theorem 4.1, we know that (7.32) holds, and the proof is completed. □

We next consider feedback controls. We begin with an optimal feedback control for $\mathcal{P}^0$, denoted as a function $U(y) = (u^0(y), u^1(y), ..., u^m(y))$, $y \in$

$\Re$. This is obtained by minimizing the right-hand side of (7.12), i.e., for all $y \in \Re$,

$$\left(-ay + \sum_{i=0}^{m} \nu_i u^i(y) - z\right) \cdot \frac{d}{dx}\phi(y) + \sum_{i=0}^{m} \nu_i c(u^i(y)) + h(y)$$

$$= \inf_{0 \le u^i \le i, i \in \mathcal{M}} \left\{ \left(-ay + \sum_{i=0}^{m} \nu_i u^i - z\right) \cdot \frac{d}{dx}\phi(y) + \sum_{i=0}^{m} \nu_i c(u^i) \right\} + h(y).$$

We then construct the feedback control $u(\cdot,\cdot)$ as follows:

$$u(y, k(\varepsilon,t)) = \sum_{i=0}^{m} I_{\{k(\varepsilon,t)=i\}} u^i(y), \tag{7.33}$$

which is clearly feasible (satisfies the control constraints) for $\mathcal{P}^\varepsilon$. Furthermore, if each $u^i(y)$ is locally Lipschitz in y, then the system

$$\frac{d}{dt}x(t) = -ax(t) + u(x(t), k(\varepsilon,t)) - z \tag{7.34}$$

with the initial conditions $x(0) = \widetilde{x}$ and $k(\varepsilon, 0) = k$ has a unique solution $x^\varepsilon(\cdot)$, and therefore $u(t) = u(x^\varepsilon(t), k(\varepsilon,t))$, $t \ge 0$, is also an admissible control for $\mathcal{P}^\varepsilon$. According to Lemma B.7, there exists an $\varepsilon_0 > 0$ such that $Q^{(1)} + \varepsilon^{-1}Q^{(2)}$ is strongly irreducible for $0 < \varepsilon \le \varepsilon_0$. Let $\boldsymbol{\nu}^\varepsilon = (\nu_0^\varepsilon, \nu_1^\varepsilon, ..., \nu_m^\varepsilon)$ denote the equilibrium distribution of $Q^{(1)} + \varepsilon^{-1}Q^{(2)}$, i.e.,

$$\boldsymbol{\nu}^\varepsilon\left[Q^{(1)} + \varepsilon^{-1}Q^{(2)}\right] = 0 \quad \text{and} \quad \sum_{i=0}^{m} \nu_i^\varepsilon = 1. \tag{7.35}$$

We can now prove the following result, but only in the single product case (see Remark 5.2 below).

Theorem 5.2 (Feedback Control). *Let Assumptions* (A1)–(A4) *hold. Assume $n = 1$, and that the feedback control $U(y)$ of the limiting problem is locally Lipschitz in y. Furthermore, suppose that, for each $\varepsilon \in [0, \varepsilon_0]$, the equation*

$$-ay + \sum_{i=0}^{m} \nu_i^\varepsilon u^i(y) - z = 0 \tag{7.36}$$

has a unique solution θ^ε, called the threshold. Moreover, suppose that, for $y \in (\theta^\varepsilon, \infty)$,

$$-ay + \sum_{i=0}^{m} \nu_i^\varepsilon u^i(y) - z < 0, \tag{7.37}$$

and for $y \in (-\infty, \theta^\varepsilon)$,

$$-ay + \sum_{i=0}^{m} \nu_i^\varepsilon u^i(y) - z > 0, \tag{7.38}$$

where $\nu^0 = \nu$. *Then the feedback control policy given by* (7.33) *is asymptotically optimal, i.e.,*

$$\lim_{\varepsilon \to 0} \left| J^\varepsilon(\widetilde{x}, k, u^\varepsilon(\cdot)) - \lambda \right| = 0,$$

where $u^\varepsilon(t) = u(x^\varepsilon(t), k(\varepsilon, t))$, $t \geq 0$.

Next we give two remarks and one example before proving the theorem.

Remark 5.1. Under the conditions given in Theorem 5.2, we know that the equation

$$\frac{d}{dt}x(t) = -ax(t) + \sum_{i=0}^{m} \nu_i u^i(x(t)) - z, \quad x(0) = \widetilde{x},$$

has a unique solution $\bar{x}(\cdot)$. Furthermore, let $\bar{x}(t) \to \bar{x}$ as $t \to \infty$. □

Remark 5.2. The uniqueness of the solution of (7.36) satisfying (7.37)–(7.38) guarantees that the differential equation

$$\frac{d}{dt}z^\varepsilon(t) = -az^\varepsilon(t) + \sum_{i=0}^{m} \nu_i^\varepsilon u^i(z^\varepsilon(t)) - z, \quad z^\varepsilon(0) = \widetilde{x}, \tag{7.39}$$

has a solution $z^\varepsilon(\cdot)$. Moreover, let the limit of $z^\varepsilon(t)$ as $t \to \infty$ be denoted as x^ε, i.e.,

$$x^\varepsilon = \lim_{t \to \infty} z^\varepsilon(t). \tag{7.40}$$

This fact, which holds only in the single product case, will be used in the proof of Theorem 5.2. □

Since there are several hypotheses, namely (7.36)–(7.38), in Theorem 5.2, it is useful to provide an example.

Example 5.1. We consider the problem

$$\mathcal{P}^\varepsilon: \begin{cases} \min \; J^\varepsilon(\widetilde{x}, k, u^\varepsilon(\cdot)) = \limsup\limits_{T \to \infty} \dfrac{1}{T} E \displaystyle\int_0^T [x(t)]^2 \, dt, \\[2ex] \text{s.t. } \dfrac{d}{dt}x(t) = -\dfrac{x(t)}{10} + u(t) - \dfrac{1}{4}, \quad x(0) = \widetilde{x}, \quad u(\cdot) \in \mathcal{A}^\varepsilon(k), \\[2ex] \text{minimum average cost } \lambda^\varepsilon = \inf\limits_{u(\cdot) \in \mathcal{A}^\varepsilon(k)} J^\varepsilon(\widetilde{x}, k, u(\cdot)), \end{cases}$$

with $\mathcal{M} = \{0, 1\}$ and the generator for $k(\varepsilon, \cdot)$ is

$$Q^\varepsilon = Q^{(1)} + \frac{1}{\varepsilon} Q^{(2)} = \frac{1}{\varepsilon} \begin{pmatrix} -1 & 1 \\ 1 & -1 \end{pmatrix}.$$

This is clearly a special case of the problem formulated in Section 7.2. In particular, Assumptions (A1)–(A4) hold and $\boldsymbol{\nu}^\varepsilon = \boldsymbol{\nu} = (\nu_0, \nu_1) = (1/2, 1/2)$. Also $u^0(t) \equiv 0$. The limiting problem is

$$
\mathcal{P}^0: \begin{cases} \min\ J(\widetilde{x}, U(\cdot)) = \limsup_{T\to\infty} \dfrac{1}{T}\displaystyle\int_0^T [x(t)]^2\,dt, \\[2ex] \text{s.t. } \dfrac{d}{dt}x(t) = -\dfrac{x(t)}{10} + \dfrac{u^1(t)}{2} - \dfrac{1}{4}, \quad x(0) = \widetilde{x}, \quad U(\cdot) \in \mathcal{A}^0, \\[2ex] \text{minimum average cost } \lambda = \inf_{U(\cdot)\in\mathcal{A}^0} J(\widetilde{x}, U(\cdot)). \end{cases}
$$

Let us set the function

$$
U(y) = (u^0(y), u^1(y)) \equiv (0, 1/2) \tag{7.41}
$$

to be a feedback control for $\mathcal{P}^0$. Clearly, the cost associated with (7.41) is zero. Since zero is the lowest possible cost, our solution is optimal and $\lambda = 0$. Furthermore, since $U(y)$ is locally Lipschitz in y and satisfies hypotheses (7.36)–(7.38), Theorem 5.2 implies that

$$
u(y, i) = \begin{cases} 0, & \text{if } i = 0, \\ u^1(y), & \text{if } i = 1, \end{cases}
$$

is an asymptotically optimal feedback control for $\mathcal{P}^\varepsilon$. □

Remark 5.3. Note that the nearly optimal feedback control in Example 5.1 is not unique. In addition, it is possible to come up with examples involving nonzero production cost, for which Lipschitz feedback controls satisfy (7.36)–(7.38). □

Proof of Theorem 5.2. The proof is divided into two steps.

Step 1. We show that

$$
J^\varepsilon(\widetilde{x}, k, u^\varepsilon(\cdot)) = h(x^\varepsilon) + c\left(\sum_{i=0}^m \nu_i^\varepsilon u^i(x^\varepsilon)\right), \tag{7.42}
$$

where x^ε is given by (7.40). In view of (7.6), we know that for each $\omega \in \Omega$, the solution $x^\varepsilon(\cdot)(\omega)$ corresponding to (7.34) satisfies

$$
|x^\varepsilon(\cdot)(\omega)| \le |\widetilde{x}| + \frac{m+z}{a}.
$$

For each $x \in [-|\widetilde{x}| - (m+z)/a, |\widetilde{x}| + (m+z)/a]$, let

$$
z^\varepsilon(x, t) = xe^{-at} + e^{-at}\int_0^t e^{as}\left[\sum_{i=0}^m \nu_i^\varepsilon u^i(z^\varepsilon(x, s)) - z\right] ds. \tag{7.43}
$$

Then,

$$\lim_{t\to\infty} z^\varepsilon(\widetilde{x},t) = \lim_{t\to\infty} z^\varepsilon(x^\varepsilon,t) = x^\varepsilon.$$

Since the optimal feedback control $U(y) = (u^0(y),...,u^m(y))$ for the limiting problem $\mathcal{P}^0$ is assumed to be locally Lipschitz, for any $\delta > 0$, there exists an $N(\varepsilon,\delta)$ such that, for $t \geq N(\varepsilon,\delta)$,

$$|z^\varepsilon(x,t) - x^\varepsilon| < \delta, \quad \text{for } x = \widetilde{x} \text{ and } x^\varepsilon. \tag{7.44}$$

By Assumption (A1) and (7.44), there is a $T(\varepsilon) > N(\varepsilon,\delta)$ such that, for $x = \widetilde{x}$ and x^ε,

$$\begin{aligned}
&\frac{1}{T(\varepsilon)}\left|\int_0^{T(\varepsilon)} h(z^\varepsilon(x,t))\,dt - \int_0^{T(\varepsilon)} h(x^\varepsilon)\,dt\right| \\
&\leq \frac{1}{T(\varepsilon)}\left|\int_0^{N(\varepsilon,\delta)} h(z^\varepsilon(x,t))\,dt - \int_0^{N(\varepsilon,\delta)} h(x^\varepsilon)\,dt\right| \\
&\quad + \frac{1}{T(\varepsilon)}\left|\int_{N(\varepsilon,\delta)}^{T(\varepsilon)} h(z^\varepsilon(x,t))\,dt - \int_{N(\varepsilon,\delta)}^{T(\varepsilon)} h(x^\varepsilon)\,dt\right| \\
&\leq \frac{1}{T(\varepsilon)}\left|\int_0^{N(\varepsilon,\delta)} h(z^\varepsilon(x,t))\,dt - N(\varepsilon,\delta)h(x^\varepsilon)\right| \\
&\quad + \frac{C_{h3}}{T(\varepsilon)}\int_{N(\varepsilon,\delta)}^{T(\varepsilon)} \delta\left(1 + |z^\varepsilon(x,t)|^{\beta_{h2}-1} + |x^\varepsilon|^{\beta_{h2}-1}\right) dt \\
&\leq C_1\left[1 + 2\left(|\widetilde{x}| + \frac{m+z}{a}\right)^{\beta_{h2}-1}\right]\delta,
\end{aligned} \tag{7.45}$$

for some $C_1 > 0$, where C_{h3} and β_{h2} are given by Assumption (A1). Now define $T_\ell(\varepsilon)$, $\ell = 1,2,...$, recursively by

$$T_0(\varepsilon) = 0, \quad T_\ell(\varepsilon) = T(\varepsilon) + T_{\ell-1}(\varepsilon), \quad \ell = 1,2,...,$$

$$\chi^{\varepsilon,\ell}(t) = \left(I_{\{k(\varepsilon,T_{\ell-1}(\varepsilon)+t)=0\}}, I_{\{k(\varepsilon,T_{\ell-1}(\varepsilon)+t)=1\}}, ..., I_{\{k(\varepsilon,T_{\ell-1}(\varepsilon)+t)=m\}}\right),$$
$$\ell = 1,2,..., \quad t \in [0,\ T(\varepsilon)].$$

Note that the process $k(\varepsilon,\cdot)$ is ergodic. Thus we can select $\Omega_0 \subset \Omega$ such that $P(\Omega_0) = 1$, and for each $\omega \in \Omega_0$,

$$\chi^{\varepsilon,\ell}(\omega,\cdot) \to \nu^\varepsilon \ \text{ in } L^2[0,\ T(\varepsilon)]. \tag{7.46}$$

Furthermore, let

$$x^{\varepsilon,\ell}(t) = x^\varepsilon(T_{\ell-1}(\varepsilon) + t), \quad t \in [0,\ T(\varepsilon)], \quad \ell = 1,2,...,$$

where $x^\varepsilon(\cdot)$ is defined by (7.34) with the initial condition $x^\varepsilon(0) = \widetilde{x}$. This implies that

$$\begin{aligned}x^{\varepsilon,\ell}(t) = e^{-at}\Bigg(&\widetilde{x}e^{-aT_{\ell-1}(\varepsilon)} \\ &+ \int_0^{T_{\ell-1}(\varepsilon)} e^{-a(T_{\ell-1}(\varepsilon)-s)}\big[u(x^\varepsilon(s), k(\varepsilon, s)) - z\big]\,ds\Bigg) \\ &+ e^{-at}\int_0^t e^{as}\big[u(x^\varepsilon(T_{\ell-1}(\varepsilon)+s), k(\varepsilon, T_{\ell-1}(\varepsilon)+s)) - z\big]\,ds.\end{aligned} \tag{7.47}$$

Using (7.47), we can see that for any $t_1, t_2 \in [0,\ T(\varepsilon)]$ with $t_1 \le t_2$, we have

$$\begin{aligned}&\big|x^{\varepsilon,\ell}(t_2) - x^{\varepsilon,\ell}(t_1)\big| \\ &\le (e^{-at_1} - e^{-at_2})\left(|\widetilde{x}| + \frac{m+z}{a}\right) \\ &\quad + \left|e^{-at_2}\int_{t_1}^{t_2} e^{as}\left[\sum_{i=0}^m I_{\{k(\varepsilon,T_{\ell-1}(\varepsilon)+s)=i\}}u^i(x^{\varepsilon,\ell}(s)) - z\right] ds\right| \\ &\quad + (e^{-at_1} - e^{-at_2})\left|\int_0^{t_1} e^{as}\left[\sum_{i=0}^m I_{\{k(\varepsilon,T_{\ell-1}(\varepsilon)+s)=i\}}u^i(x^{\varepsilon,\ell}(s)) - z\right] ds\right| \\ &\le C_2|t_2 - t_1|,\end{aligned} \tag{7.48}$$

for some positive constant C_2 independent of ε. By the Arzelà-Ascoli theorem and (7.46), there exists a $K(\varepsilon)$ such that for each $\omega \in \Omega_0$ and each $\ell \ge K(\varepsilon)$,

$$\sup_{0\le t\le T(\varepsilon)} \big|x^{\varepsilon,\ell}(t)(\omega) - z^\varepsilon(\widetilde{x}, t)\big| \le \delta \tag{7.49}$$

where $z^\varepsilon(\widetilde{x}, t)$ and δ are given in (7.43) and (7.44), respectively. Similar to (7.45), it follows from (7.49) that, for $\ell \ge K(\varepsilon)$,

$$\begin{aligned}&\frac{1}{T(\varepsilon)}\left|E\int_0^{T(\varepsilon)} h(x^{\varepsilon,\ell}(t))\,dt - \int_0^{T(\varepsilon)} h(z^\varepsilon(\widetilde{x}, t))\,dt\right| \\ &\qquad \le C_3\left[1 + 2\left(|\widetilde{x}| + \frac{2(m+z)}{a}\right)^{\beta_{h2}-1}\right]\delta,\end{aligned} \tag{7.50}$$

for some $C_3 > 0$. Combining (7.45) and (7.50) yields

$$\frac{1}{T(\varepsilon)}\left|E\int_0^{T(\varepsilon)} h(x^{\varepsilon,\ell}(t))\,dt - \int_0^{T(\varepsilon)} h(x^{\varepsilon})\,dt\right| \le C_4\left[1+\left(|\widetilde{x}|+\frac{2(m+z)}{a}\right)^{\beta_{h2}-1}\right]\delta, \tag{7.51}$$

for some $C_4 > 0$ and $\ell \ge K(\varepsilon)$. On the other hand, for any large T with $T = (K(\varepsilon)+L(\varepsilon))T(\varepsilon)+\widehat{T}(\varepsilon)$, where $L(\varepsilon)$ is a positive integer, and $\widehat{T}(\varepsilon) \in (0,\ T(\varepsilon))$, we have

$$\begin{aligned}\frac{1}{T}\int_0^T h(x^{\varepsilon}(t))\,dt = \frac{1}{T}\Bigg[\int_0^{K(\varepsilon)T(\varepsilon)} h(x^{\varepsilon}(t))\,dt\\ + \sum_{\ell=1}^{L(\varepsilon)}\int_0^{T(\varepsilon)} h(x^{\varepsilon,K(\varepsilon)+\ell}(t))\,dt + \int_{(K(\varepsilon)+L(\varepsilon))T(\varepsilon)}^{T} h(x^{\varepsilon}(t))\,dt\Bigg].\end{aligned} \tag{7.52}$$

Using the boundedness of $x^{\varepsilon}(\cdot)$ we can claim that

$$\lim_{T\to\infty}\frac{1}{T}E\left(\int_0^{K(\varepsilon)T(\varepsilon)} h(x^{\varepsilon}(t))\,dt + \int_{(K(\varepsilon)+L(\varepsilon))T(\varepsilon)}^{T} h(x^{\varepsilon}(t))\,dt\right) = 0. \tag{7.53}$$

The arbitrariness of δ and (7.51)–(7.53) imply

$$\lim_{T\to\infty}\frac{1}{T}E\int_0^T h(x^{\varepsilon}(t))\,dt = h(x^{\varepsilon}), \tag{7.54}$$

where x^{ε} is given by (7.40). Similarly, we have

$$\lim_{T\to\infty}\frac{1}{T}E\int_0^T c(u(x^{\varepsilon}(t),k(\varepsilon,t)))\,dt = c\left(\sum_{i=0}^{m}\nu_i^{\varepsilon}u^i(x^{\varepsilon})\right). \tag{7.55}$$

Finally, (7.54)–(7.55) imply (7.42).

Step 2. We establish

$$\lim_{\varepsilon\to 0}\left[h(x^{\varepsilon})+c\left(\sum_{i=0}^{m}\nu_i^{\varepsilon}u^i(x^{\varepsilon})\right)\right] = \lambda. \tag{7.56}$$

By the same method as that used in Step 1, we can show that

$$\lambda = h(\bar{x})+c\left(\sum_{i=0}^{m}\nu_i u^i(\bar{x})\right),$$

where $\bar{x}$ is defined in Remark 5.1. It follows from $\lim_{\varepsilon\to 0}\nu^{\varepsilon} = \nu$ that $\lim_{\varepsilon\to 0} x^{\varepsilon} = \bar{x}$. Therefore we obtain (7.56).

In view of (7.42) and (7.56) derived in Steps 1 and 2, respectively, the result follows. □

Remark 5.4. Our results are based on the assumption that ε is small. In practice, the structure of $Q^\varepsilon = Q^{(1)} + \varepsilon^{-1}Q^{(2)}$ needs to be identified from physical considerations. When this is not possible, one could employ numerical algorithms for grouping developed in the singular perturbation literature; see Phillips and Kokotovic [98]. □

7.6 Parallel-Machine Systems without Attrition

Consider the manufacturing system given in Section 7.2, but without a positive inventory deterioration/cancellation rate for each product, i.e., $\boldsymbol{a} = 0$. Formally, the system dynamics is

$$\frac{d}{dt}\boldsymbol{x}(t) = \boldsymbol{u}(t) - \boldsymbol{z}, \quad \boldsymbol{x}(0) = \boldsymbol{x}. \tag{7.57}$$

Furthermore, we make the following assumption on the total capacity $k(\varepsilon, t)$ for the manufacturing system at time $t \geq 0$.

(A5) The generator of $k(\varepsilon, \cdot)$ is given by $Q^\varepsilon = Q/\varepsilon$, i.e., $Q^{(1)} = 0$ and $Q^{(2)} = Q$, with Q being strongly irreducible.

The control problem is as follows:

$$\mathcal{P}^\varepsilon: \begin{cases} \min J^\varepsilon(\boldsymbol{x}, k, \boldsymbol{u}(\cdot)) = \limsup\limits_{T\to\infty} \dfrac{1}{T} E \displaystyle\int_0^T [h(\boldsymbol{x}(t)) + c(\boldsymbol{u}(t))]\, dt, \\ \text{s.t. } \dfrac{d}{dt}\boldsymbol{x}(t) = \boldsymbol{u}(t) - \boldsymbol{z}, \quad \boldsymbol{x}(0) = \boldsymbol{x}, \quad \boldsymbol{u}(\cdot) \in \mathcal{A}^\varepsilon(k), \\ \text{minimum average cost } \lambda^\varepsilon = \inf\limits_{\boldsymbol{u}(\cdot)\in\mathcal{A}^\varepsilon(k)} J^\varepsilon(\boldsymbol{x}, k, \boldsymbol{u}(\cdot)). \end{cases} \tag{7.58}$$

To carry out an asymptotic analysis of the minimum long-run average expected cost λ^ε, we introduce a corresponding control problem $\mathcal{P}^{\varepsilon,\rho}$ with the cost discounted at a rate $\rho > 0$:

$$\mathcal{P}^{\varepsilon,\rho}: \begin{cases} \min\ J^{\varepsilon,\rho}(\boldsymbol{x}, k, \boldsymbol{u}^\varepsilon(\cdot)) = E \displaystyle\int_0^\infty e^{-\rho t}[h(\boldsymbol{x}(t)) + c(\boldsymbol{u}(t))]\, dt, \\ \text{s.t. } \dfrac{d}{dt}\boldsymbol{x}(t) = \boldsymbol{u}(t) - \boldsymbol{z}, \quad \boldsymbol{x}(0) = \boldsymbol{x}, \quad \boldsymbol{u}(\cdot) \in \mathcal{A}^\varepsilon(k), \\ V^{\varepsilon,\rho}(\boldsymbol{x}, k) = \inf\limits_{\boldsymbol{u}(\cdot)\in\mathcal{A}^\varepsilon(k)} J^{\varepsilon,\rho}(\boldsymbol{x}, k, \boldsymbol{u}(\cdot)). \end{cases} \tag{7.59}$$

The associated HJB equation is

$$\begin{aligned}\rho\phi^{\varepsilon,\rho}(\boldsymbol{x},k) = &\min_{\boldsymbol{u}\in\mathcal{U}(k)}\left\{\left\langle \boldsymbol{u}-\boldsymbol{z}, \frac{\partial\phi^{\varepsilon,\rho}(\boldsymbol{x},k)}{\partial\boldsymbol{x}}\right\rangle + c(\boldsymbol{u})\right\}\\ &+h(\boldsymbol{x}) + \frac{1}{\varepsilon}Q\phi^{\varepsilon,\rho}(\boldsymbol{x},\cdot)(k),\end{aligned} \tag{7.60}$$

for any $k \in \mathcal{M}$, where $\phi^{\varepsilon,\rho}(\boldsymbol{x},k)$ is defined on $\Re^n \times \mathcal{M}$.

In order to study the long-run average-cost control problem using the vanishing discount approach, just as what we did in Chapter 3, we must first obtain some estimates for the value function $V^{\varepsilon,\rho}(\boldsymbol{x},k)$ in the neighborhood of $\rho = 0$.

Lemma 6.1. *Let Assumptions* (A4) *and* (A5) *hold. For any $r \geq 1$ and any $(\boldsymbol{x},k) \in \Re^n \times \mathcal{M}$, $\widehat{\boldsymbol{x}} \in \Re^n$, there exist a control $\boldsymbol{u}^\varepsilon(\cdot) \in \mathcal{A}^\varepsilon(k)$ and a positive constant C_r, independent of ε, $(\boldsymbol{x},k)$, and $\widehat{\boldsymbol{x}}$, such that*

$$E\big[\tau(\varepsilon,\boldsymbol{x},\widehat{\boldsymbol{x}},k)\big]^r \leq C_r\left(1+\sum_{i=1}^n |x_i - \widehat{x}_i|^r\right),$$

where

$$\tau(\varepsilon,\boldsymbol{x},\widehat{\boldsymbol{x}},k) = \inf\{t \geq 0 : \boldsymbol{x}^\varepsilon(t) = \widehat{\boldsymbol{x}}\}$$

and $\boldsymbol{x}^\varepsilon(t), t \geq 0$, is the surplus process corresponding to the control $\boldsymbol{u}^\varepsilon(\cdot)$ and the initial condition $(\boldsymbol{x}^\varepsilon(0), k(\varepsilon,0)) = (\boldsymbol{x},k)$.

Proof. The proof is same as the proof of Lemma 3.7.1, except that here we need to use the inequality

$$E\exp\left(\frac{1}{\sqrt{t+1}}\left|\int_0^t [k(\varepsilon,s) - \bar{k}]\,ds\right|\right) \leq C_1, \quad \text{for any } t \text{ and } \varepsilon,$$

instead of (3.11). □

Furthermore, we have the following lemma.

Lemma 6.2. *Let Assumptions* (A4) *and* (A5) *hold. For any $(\boldsymbol{x},k), (\widehat{\boldsymbol{x}},\widehat{k}) \in \Re^n \times \mathcal{M}$, and $r \geq 1$, there exist a control policy $\boldsymbol{u}^\varepsilon(\cdot)$ and a positive constant $\widehat{C}_r$ independent of ε, $(\boldsymbol{x},k)$ and $(\widehat{\boldsymbol{x}},\widehat{k})$ such that*

$$E\big[\tau(\varepsilon,\boldsymbol{x},\widehat{\boldsymbol{x}},k,\widehat{k})\big]^r \leq \widehat{C}_r\left(1+\sum_{i=1}^n |\widehat{x}_i - x_i|^r\right),$$

where

$$\tau(\varepsilon,\boldsymbol{x},\widehat{\boldsymbol{x}},k,\widehat{k}) = \inf\big\{t \geq 0 : (\boldsymbol{x}^\varepsilon(t), k(\varepsilon,t)) = (\widehat{\boldsymbol{x}},\widehat{k})\big\},$$

and $\boldsymbol{x}^{\varepsilon}(\cdot)$ *is the surplus process corresponding to the control* $\boldsymbol{u}^{\varepsilon}(\cdot)$ *and initial condition* $(\boldsymbol{x}^{\varepsilon}(0), k(\varepsilon, 0)) = (\boldsymbol{x}, k)$.

Proof. The proof is same as the proof of Theorem 3.7.3, except that here we use the inequality

$$P(k(\varepsilon,t) = j | k(\varepsilon,0) = i) \geq \nu_j/2, \quad \text{for } t \geq t_0 \text{ and } \varepsilon \in (0,\ \varepsilon_0],$$

instead of (3.14). □

With Lemma 6.2 in hand, we prove the following result.

Theorem 6.1. *Let Assumptions* (A1)–(A2) *and* (A4)–(A5) *hold. There exist constants* $\rho_0 > 0$ *and* $\varepsilon_0 > 0$ *such that:*

(i) $\{\rho V^{\varepsilon,\rho}(0,0)\ :\ 0 < \rho \leq \rho_0,\ 0 < \varepsilon \leq \varepsilon_0\}$ *is bounded;*

(ii) *for* $\varepsilon \in (0,\ \varepsilon_0]$ *and* $\rho \in (0,\ \rho_0]$, *the function*

$$\widetilde{V}^{\varepsilon,\rho}(\boldsymbol{x}, k) = V^{\varepsilon,\rho}(\boldsymbol{x}, k) - V^{\varepsilon,\rho}(0,0)$$

is convex in $\boldsymbol{x}$; *and*

(iii) *for* $\varepsilon \in (0,\ \varepsilon_0]$ *and* $\rho \in (0,\ \rho_0]$, $\widetilde{V}^{\varepsilon,\rho}(\boldsymbol{x}, k)$ *is locally Lipschitz continuous in* $\boldsymbol{x}$, *i.e., there exists a constant* C, *independent of* ρ *and* ε, *such that*

$$\left|\widetilde{V}^{\varepsilon,\rho}(\boldsymbol{x}, k) - \widetilde{V}^{\varepsilon,\rho}(\widehat{\boldsymbol{x}}, k)\right| \leq C\big(1 + |\boldsymbol{x}|^{\beta_{h2}-1} + |\widehat{\boldsymbol{x}}|^{\beta_{h2}-1}\big)|\boldsymbol{x} - \widehat{\boldsymbol{x}}|,$$

for $k \in \mathcal{M}$ *and all* $\boldsymbol{x}, \widehat{\boldsymbol{x}} \in \Re^n$, *where* β_{h2} *is given in Assumption* (A1).

Proof. We can prove the theorem by going along the lines of the proofs of Theorems 3.3.2 and 3.3.3. The details are omitted. □

With the help of Theorem 6.1, we are ready to deal with the convergence of λ^{ε}, the minimum average cost of the problem $\mathcal{P}^{\varepsilon}$, with respect to ε. First we can formally write the HJB equation associated with $\mathcal{P}^{\varepsilon}$,

$$\begin{aligned}\lambda^{\varepsilon} = &\inf_{\boldsymbol{u} \in \mathcal{U}(k)} \left\{ \left\langle \boldsymbol{u} - \boldsymbol{z}, \frac{\partial \phi^{\varepsilon}(\boldsymbol{x}, k)}{\partial \boldsymbol{x}} \right\rangle + c(\boldsymbol{u}) \right\} \\ &+ h(\boldsymbol{x}) + \frac{1}{\varepsilon} Q \phi^{\varepsilon}(\boldsymbol{x}, \cdot)(k),\end{aligned} \tag{7.61}$$

for any $k \in \mathcal{M}$ and $\boldsymbol{x} \in \Re^n$, where λ^{ε} is a constant and $\phi^{\varepsilon}(\boldsymbol{x}, k)$ is a real-valued function on $\Re^n \times \mathcal{M}$.

According to (i) and (iii) of Theorem 6.1, for a fixed $\varepsilon \in (0,\ \varepsilon_0]$ and any subsequence of $\{\rho\}$, denoted by $\{\rho_\ell\}$, there is a further subsequence of $\{\rho_\ell\}$, still denoted by $\{\rho_\ell\}$, such that

$$\lambda^{\varepsilon,*} = \lim_{\rho_\ell \to 0} \rho_\ell V^{\varepsilon,\rho_\ell}(0,0) \tag{7.62}$$

and

$$\widetilde{V}^{\varepsilon}(\boldsymbol{x},k) = \lim_{\rho_\ell \to 0} \widetilde{V}^{\varepsilon,\rho_\ell}(\boldsymbol{x},k), \tag{7.63}$$

for all $(\boldsymbol{x},k) \in \Re^n \times \mathcal{M}$. Furthermore, we have the following lemma.

Lemma 6.3. *Let Assumptions* (A1)–(A2) *and* (A4)–(A5) *hold. There exists* $\varepsilon_0 > 0$ *such that for* $\varepsilon \in (0,\ \varepsilon_0]$, *we have the following:*

(i) *the function* $\widetilde{V}^{\varepsilon}(\boldsymbol{x},k)$ *given in* (7.63) *is convex in* $\boldsymbol{x}$. *It is also locally Lipschitz continuous in* $\boldsymbol{x}$, *i.e., there is a constant* C *such that*

$$|\widetilde{V}^{\varepsilon}(\boldsymbol{x},k) - \widetilde{V}^{\varepsilon}(\widehat{\boldsymbol{x}},k)| \le C(1 + |\boldsymbol{x}|^{\beta_{h2}-1} + |\widehat{\boldsymbol{x}}|^{\beta_{h2}-1})|\boldsymbol{x} - \widehat{\boldsymbol{x}}|,$$

for all $\boldsymbol{x}, \widehat{\boldsymbol{x}} \in \Re^n$ *and* $k \in \mathcal{M}$, *where* β_{h2} *is given by Assumption* (A1);

(ii) $(\lambda^{\varepsilon,*}, \widetilde{V}^{\varepsilon}(\boldsymbol{x},k))$ *is a viscosity solution to the* HJB *equation* (7.61);

(iii) *there exists a constant* $\widehat{C}$ *such that, for all* $\boldsymbol{x} \in \Re^n$ *and* $i,k \in \mathcal{M}$,

$$\frac{1}{\varepsilon}\left|\widetilde{V}^{\varepsilon}(\boldsymbol{x},i) - \widetilde{V}^{\varepsilon}(\boldsymbol{x},k)\right| \le \widehat{C}(1 + |\boldsymbol{x}|^{\beta_{h2}+1}).$$

Proof. (i) From the convexity of $V^{\varepsilon,\rho}(\boldsymbol{x},k)$, we get the convexity of $\widetilde{V}^{\varepsilon}(\boldsymbol{x},k)$. According to (7.63), it follows from (iii) of Theorem 6.1 that the function $\widetilde{V}^{\varepsilon}(\boldsymbol{x},k)$ is locally Lipschitz continuous in $\boldsymbol{x}$.

(ii) To justify $(\lambda^{\varepsilon,*}, \widetilde{V}^{\varepsilon}(\boldsymbol{x},k))$ to be a viscosity solution to (7.61), let $\boldsymbol{x}_0 \in \Re^n$ and $\boldsymbol{r} \in D^{-}\widetilde{V}^{\varepsilon}(\boldsymbol{x}_0,k)$ for any fixed $k \in \mathcal{M}$. We choose a continuously differential function $f_k(\cdot)$ and a neighborhood $\mathcal{N}(\boldsymbol{x}_0)$ of $\boldsymbol{x}_0$ such that

$$\left.\frac{\partial f_k(\boldsymbol{x})}{\partial \boldsymbol{x}}\right|_{\boldsymbol{x}=\boldsymbol{x}_0} = \boldsymbol{r},$$
$$0 = \widetilde{V}^{\varepsilon}(\boldsymbol{x}_0,k) - f_k(\boldsymbol{x}_0) = \inf_{\boldsymbol{x}\in\mathcal{N}(\boldsymbol{x}_0)} \{\widetilde{V}^{\varepsilon}(\boldsymbol{x},k) - f_k(\boldsymbol{x})\},$$

and $\boldsymbol{x}_0$ is the only minimum point of $\widetilde{V}^{\varepsilon}(\boldsymbol{x},k) - f_k(\boldsymbol{x})$ on $\mathcal{N}(\boldsymbol{x}_0)$. Now choose $\boldsymbol{x}_0^{\ell} \in \mathcal{N}(\boldsymbol{x}_0)$ such that

$$V^{\varepsilon,\rho_\ell}(\boldsymbol{x}_0^{\ell},k) - f_k(\boldsymbol{x}_0^{\ell}) = \inf_{\boldsymbol{x}\in\mathcal{N}(\boldsymbol{x}_0)} \{V^{\varepsilon,\rho_\ell}(\boldsymbol{x},k) - f_k(\boldsymbol{x})\}.$$

Then, as $\rho_\ell \to 0$,

$$\boldsymbol{x}_0^{\ell} \to \boldsymbol{x}_0. \tag{7.64}$$

Moreover,

$$\left.\frac{\partial f_k(\boldsymbol{x})}{\partial \boldsymbol{x}}\right|_{\boldsymbol{x}=\boldsymbol{x}_0^{\ell}} \in D^{-}V^{\varepsilon,\rho_\ell}(\boldsymbol{x}_0^{\ell},k). \tag{7.65}$$

Note that $V^{\varepsilon,\rho_\ell}(\boldsymbol{x},k)$ is the unique viscosity solution of the HJB equation (7.60). Then (7.65) implies

$$\rho_\ell V^{\varepsilon,\rho_\ell}(\boldsymbol{x}_0,k) \geq \inf_{\boldsymbol{u}\in\mathcal{U}(k)} \left\{ \left\langle \boldsymbol{u}-\boldsymbol{z}, \frac{\partial f_k(\boldsymbol{x})}{\partial \boldsymbol{x}} \right\rangle \bigg|_{\boldsymbol{x}=\boldsymbol{x}_0^\ell} + c(\boldsymbol{u}) \right\}$$
$$+ h(\boldsymbol{x}_0^{\rho_\ell}) + \frac{1}{\varepsilon} QV^{\varepsilon,\rho_\ell}(\boldsymbol{x}_0^\ell,\cdot)(k).$$

Hence, in view of the identity $\sum_{i=0}^m q_{ki} V^{\varepsilon,\rho_\ell}(0,0) = 0$, we have

$$\rho_\ell V^{\varepsilon,\rho_\ell}(0,0) + \rho_\ell \left[V^{\varepsilon,\rho_\ell}(\boldsymbol{x}_0,k) - V^{\varepsilon,\rho_\ell}(0,0) \right]$$
$$\geq \inf_{\boldsymbol{u}\in\mathcal{U}(k)} \left\{ \left\langle \boldsymbol{u}-\boldsymbol{z}, \frac{\partial f_k(\boldsymbol{x})}{\partial \boldsymbol{x}} \right\rangle \bigg|_{\boldsymbol{x}=\boldsymbol{x}_0^\ell} + c(\boldsymbol{u}) \right\}$$
$$+ h(\boldsymbol{x}_0^\ell) + \frac{1}{\varepsilon} \sum_{i=0}^m q_{ki} \left[V^{\varepsilon,\rho_\ell}(\boldsymbol{x}_0,i) - V^{\varepsilon,\rho_\ell}(0,0) \right].$$

Letting $\rho_\ell \to 0$ and using (7.64), we obtain

$$\lambda^{\varepsilon,*} \geq \inf_{\boldsymbol{u}\in\mathcal{U}(k)} \{ \langle \boldsymbol{u}-\boldsymbol{z}, \boldsymbol{r} \rangle + c(\boldsymbol{u}) \} + h(\boldsymbol{x}_0) + \frac{1}{\varepsilon} \sum_{i=0}^m q_{ki} \widetilde{V}^\varepsilon(\boldsymbol{x}_0,i),$$

for all $\boldsymbol{r} \in D^-\widetilde{V}^\varepsilon(\boldsymbol{x}_0,k)$. Similarly, we can show that the opposite inequality for $\boldsymbol{r} \in D^+\widetilde{V}^\varepsilon(\boldsymbol{x}_0,k)$ holds, and thus prove that (ii) holds.

(iii) First we consider the point of differentiability of $\widetilde{V}^\varepsilon(\boldsymbol{x},k)$ for each k. Let $\boldsymbol{x}_0$ be the differentiable point of $\widetilde{V}^\varepsilon(\boldsymbol{x},k)$ for each k. Then by (ii), equation (7.61) holds, and we have $\sum_{i=0}^m q_{ki} \widetilde{V}^\varepsilon(\boldsymbol{x}_0,i) = \varepsilon\zeta(\boldsymbol{x}_0,k)$, where

$$\zeta(\boldsymbol{x}_0,i) = \lambda^{\varepsilon,*} - \inf_{\boldsymbol{u}\in\mathcal{U}(k)} \left\{ \left\langle \boldsymbol{u}-\boldsymbol{z}, \frac{\partial \widetilde{V}^\varepsilon(\boldsymbol{x},k)}{\partial \boldsymbol{x}} \right\rangle \bigg|_{\boldsymbol{x}=\boldsymbol{x}_0} + c(\boldsymbol{u}) \right\} - h(\boldsymbol{x}_0).$$

Observe that, due to (i) and Assumptions (A1)–(A2),

$$|\zeta(\boldsymbol{x}_0,i)| \leq C_1(1+|\boldsymbol{x}_0|^{\beta_{h2}+1}), \quad i \in \mathcal{M}, \tag{7.66}$$

for some positive C_1. Also, the irreducibility of Q implies that the kernel of Q is the one-dimensional subspace spanned by the vector $\mathbf{1} = (1,...,1)' \in \Re^{m+1}$. Hence, for any $i,k \in \mathcal{M}$, we have

$$\frac{1}{\varepsilon} \left| \widetilde{V}^\varepsilon(\boldsymbol{x},i) - \widetilde{V}^\varepsilon(\boldsymbol{x},k) \right| \leq \sup_{\ell\in\mathcal{M}} |\zeta(\boldsymbol{x}_0,\ell)| \leq C_1(1+|\boldsymbol{x}_0|^{\beta_{h2}+1}). \tag{7.67}$$

Recall that $\boldsymbol{x}_0$ is a point of differentiability of $\widetilde{V}^{\varepsilon}(\boldsymbol{x},k)$ for each k. But such points are dense because of the local Lipschitz continuity of $\widetilde{V}^{\varepsilon}(\boldsymbol{x},k)$. Hence, (7.67) holds everywhere. Thus we have (iii). □

From the uniqueness of the solution of the HJB equation (7.61), we know that, under Assumptions (A1)–(A2) and (A4)–(A5),

$$\lim_{\rho\to 0}\rho V^{\varepsilon,\rho}(0,0)=\lambda^{\varepsilon,*}.$$

We now study the limiting behavior as $\varepsilon\to 0$. To do this, we define

$$\mathcal{U}(\bar{k})=\Big\{(u_1,\ldots,u_n)':\ u_i\geq 0 \text{ and } \sum_{i=1}^{n}u_i\leq \bar{k}\Big\},$$

where $\bar{k}$ is the average machine capacity given by Assumption (A4). For $\boldsymbol{u}\in\mathcal{U}(\bar{k})$, define

$$\bar{c}(\boldsymbol{u})=\inf\Big\{\sum_{i=0}^{m}\nu_i c(\boldsymbol{u}^i)\ :\ \boldsymbol{u}^i\in\mathcal{U}(i)\text{ and }\sum_{i=0}^{m}\nu_i\boldsymbol{u}^i=\boldsymbol{u}\Big\}.$$

We introduce the deterministic problem:

$$\mathcal{P}:\begin{cases}\min\ J(\boldsymbol{x},\boldsymbol{u}(\cdot))=\limsup\limits_{T\to\infty}\dfrac{1}{T}\displaystyle\int_0^T[h(\boldsymbol{x}(t))+\bar{c}(\boldsymbol{u}(t))]\,dt,\\ \text{s.t. } \dfrac{d}{dt}\boldsymbol{x}(t)=\boldsymbol{u}(t)-\boldsymbol{z},\quad \boldsymbol{x}(0)=\boldsymbol{x},\quad \boldsymbol{u}(\cdot)\in\mathcal{A},\\ \text{minimum average cost}\lambda=\inf\limits_{\boldsymbol{u}(\cdot)\in\mathcal{A}}J(\boldsymbol{x},\boldsymbol{u}(\cdot)),\end{cases}\tag{7.68}$$

where

$$\mathcal{A}=\big\{\boldsymbol{u}(\cdot):\ \boldsymbol{u}(t)\in\mathcal{U}(\bar{k})\text{ and }\boldsymbol{u}(\cdot)\text{ is a deterministic measurable process}\big\}.$$

The associated HJB equation is given by

$$\widehat{\lambda}=\inf_{\boldsymbol{u}\in\mathcal{U}(\bar{k})}\Big\{\Big\langle\boldsymbol{u}-\boldsymbol{z},\frac{\partial\phi(\boldsymbol{x})}{\partial\boldsymbol{x}}\Big\rangle+\bar{c}(\boldsymbol{u})\Big\}+h(\boldsymbol{x}).\tag{7.69}$$

It follows from the definitions of $\mathcal{U}(\bar{k})$ and $\bar{c}(\cdot)$ that (7.12) is equivalent to (7.69). Here we use (7.69) to prove the following theorem.

Theorem 6.2. *Under Assumptions* (A1)–(A2) *and* (A4)–(A5), *we have* $\lambda^{\varepsilon,*}\to\lambda$ *as* $\varepsilon\to 0$.

Proof. First note that (7.69) has a unique viscosity solution. Now using (i) of Theorem 6.1 and (i) of Lemma 6.3, we have for any subsequence of

$\{\varepsilon\}$ denoted by $\{\varepsilon_\ell\}$, a further subsequence of $\{\varepsilon_\ell\}$, still denoted by $\{\varepsilon_\ell\}$, such that

$$\lambda^* = \lim_{\varepsilon_\ell \to 0} \lambda^{\varepsilon_\ell,*} \tag{7.70}$$

and

$$V(\boldsymbol{x},k) = \lim_{\varepsilon_\ell \to 0} \widetilde{V}^{\varepsilon_\ell}(\boldsymbol{x},k), \tag{7.71}$$

for all $(\boldsymbol{x},k) \in \Re^n \times \mathcal{M}$, where $\lambda^{\varepsilon,*}$ and $\widetilde{V}^\varepsilon(\boldsymbol{x},k)$ are given by (7.62) and (7.63), respectively. From (iii) of Lemma 6.3, we know that $V(\boldsymbol{x},k)$ is independent of k. We therefore write $V(\boldsymbol{x},k)$ as $V(\boldsymbol{x})$. The proof that $(\lambda^*, V(\boldsymbol{x}))$ is a viscosity solution of (7.69) is similar to the proof of (ii) of Lemma 6.3. $\lim_{\varepsilon\to 0}\lambda^\varepsilon = \lambda$ directly follows from the uniqueness of the solution of (7.69). □

Remark 6.1. The results obtained in this section are asymptotic in nature. How good the constructed control is for any given system depends on how small the value of ε associated with the system is. At present, whether ε is sufficiently small or not is a matter of judgment. Computational work on this issue may help sharpen this judgment. Some work along these lines in the discounted case was done by Samaratunga, Sethi, and Zhou [110]. □

Remark 6.2. For parallel machine systems without a positive inventory deterioration/cancellation rate, we only prove that the minimum average cost can be approximated by the minimum average cost of the corresponding limiting problem. The problem of constructing an asymptotic optimal control (open-loop or feedback) from an optimal control of the corresponding limiting problem remains open. This assumption will be used in all subsequent chapters. Additional comments will be given in Chapter 11. □

7.7 Notes

This chapter is based on Sethi, Zhang, and Zhang [119], and Sethi and Zhang [117]. Theorems 4.1, 5.1, and 5.2 are derived in Sethi, Zhang, and Zhang [119]. Theorems 6.1 and 6.2 are obtained in Sethi and Zhang [117].

8

Near-Optimal Control of Dynamic Flowshops

8.1 Introduction

The purpose of this chapter is to obtain near-optimal open-loop controls in dynamic flowshops, defined in Section 2.3 and studied in Chapter 4. A dynamic flowshop consists of $m \geq 2$ machines in tandem and contains internal buffers between any two machines. Since the inventories in any of the internal buffers cannot be allowed to become negative, we must impose nonnegativity constraints on these inventories.

Typically, the presence of these constraints disqualifies the method of constructing asymptotic optimal controls as in the previous chapter. More specifically, while the limiting problem in the flowshop case can be obtained by averaging stochastic machine capacities, the control constructed from the solution of the limiting problem in the parallel machine case may not be feasible for the original flowshop problem in the sense that the corresponding trajectory may not satisfy the state constraints.

Thus, the main difficulty is how to construct an *admissible* control for the original problem from a near-optimal control of the limiting problem in a way which still guarantees the asymptotic optimality. To overcome this difficulty, we introduce a method of "lifting" and "shrinking." The basic idea behind it is as follows. First, we modify a given near-optimal control of the limiting problem by increasing the inventory in the buffer by a small amount. We use this resulting control to construct a "control" for the original problem in the same way as in Chapter 7. The constructed control is not necessarily admissible for the original problem, so we modify it whenever the corresponding state does not satisfy the constraint. The "lifting" procedure in the first step ensures that the average time over which a modification is needed is very small. We also show that the final control constructed in this manner is indeed nearly optimal, although the order of the error bound we obtain is ε^δ for any $0 < \delta < 1/2$, as compared with $\varepsilon^{1/2}$ in the unconstrained case of Chapter 7. The small loss in the sharpness of the error estimate is due to the lifting and modification required to honor the state constraints.

The plan of the rest of the chapter is as follows. In Section 8.2 we recall the formulation of the flowshop problem with m machines under consideration. In Section 8.3 we prove some properties associated with the minimum long-run average cost, and formulate the corresponding limiting problem. Section 8.4 is devoted to proving the asymptotic optimality of the candidate control constructed from the limiting problem. In Section 8.5 we illustrate the procedure of constructing asymptotic optimal controls from the optimal control of the limiting problem. The chapter is concluded in Section 8.6.

8.2 Problem Formulation

Let us consider a manufacturing system producing a single finished product using m machines in tandem that are subject to breakdown and repair; see Figure 8.1.

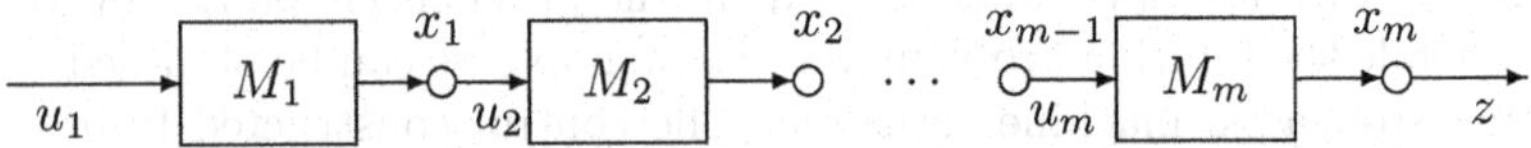

Figure 8.1. A Single Product Dynamic Flowshop with an m-Machine.

We are given a stochastic process $\boldsymbol{k}(\varepsilon, \cdot) = (k_1(\varepsilon, \cdot), \ldots, k_m(\varepsilon, \cdot))$ on the standard probability space $(\Omega, \mathcal{F}, P)$, where $k_j(\varepsilon, t), j = 1, \ldots, m$, is the capacity of the jth machine at time t, and ε is a small scale parameter to be specified later. We use $u_j^\varepsilon(t)$ to denote the input rate to the jth machine, $j = 1, \ldots, m$, and $x_j^\varepsilon(t)$ to denote the number of parts in the buffer between the jth and $(j+1)$th machines, $j = 1, \ldots, m-1$. We assume a constant demand rate z. The difference between cumulative production and cumulative demand, called surplus, is denoted by $x_m^\varepsilon(t)$. If $x_m^\varepsilon(t) > 0$, we have finished goods inventories, and if $x_m^\varepsilon(t) < 0$, we have a backlog.

The dynamics of the system can then be written as follows:

$$\begin{cases} \dfrac{d}{dt}x_j^\varepsilon(t) = -a_j x_j^\varepsilon(t) + u_j^\varepsilon(t) - u_{j+1}^\varepsilon(t), \quad x_j^\varepsilon(0) = x_j, \\ \qquad\qquad\qquad\qquad\qquad\qquad\qquad j = 1, \ldots, m-1, \\ \dfrac{d}{dt}x_m^\varepsilon(t) = -a_m x_m^\varepsilon(t) + u_m^\varepsilon(t) - z, \quad x_m^\varepsilon(0) = x_m, \end{cases} \tag{8.1}$$

where $a_j > 0$, $j = 1, 2, \ldots, m-1$, and a_m are constants. The attrition rate a_j represents the deterioration rate of the inventory of the part type j when $x_j^\varepsilon(t) > 0$ ($j = 1, \ldots, m-1, m$), and a_m represents the rate of cancellation of backlogged orders for finished goods when $x_m^\varepsilon(t) < 0$. We assume symmetric deterioration and cancellation rates for finished goods only for convenience in exposition. It would be easy to extend our results if a_m is a function of $\boldsymbol{x}$,

$$a_m(\boldsymbol{x}) = \begin{cases} \widehat{a}_m, & \text{if } x_m \geq 0, \\ \overline{a}_m, & \text{if } x_m < 0, \end{cases}$$

where $\widehat{a}_m > 0$ denotes the deterioration rate and $\overline{a}_m > 0$ denotes the order cancellation rate. Define $\boldsymbol{a} = (a_1, \ldots, a_m)'$,

$$A = \begin{pmatrix} 1 & -1 & 0 & \cdots & 0 \\ 0 & 1 & -1 & \cdots & 0 \\ \vdots & \vdots & \vdots & \cdots & \vdots \\ 0 & 0 & 0 & \cdots & 1 \end{pmatrix} \quad \text{and} \quad B = \begin{pmatrix} 0 \\ 0 \\ \vdots \\ 0 \\ -1 \end{pmatrix}.$$

Equation (8.1) can be written in the following vector form:

$$\frac{d}{dt}\boldsymbol{x}^\varepsilon(t) = -\text{diag}(\boldsymbol{a})\boldsymbol{x}^\varepsilon(t) + A\boldsymbol{u}^\varepsilon(t) + Bz, \quad \boldsymbol{x}^\varepsilon(0) = \boldsymbol{x}. \tag{8.2}$$

Let

$$\mathcal{L}(\boldsymbol{x}^\varepsilon(t), \boldsymbol{u}^\varepsilon(t), z) = -\text{diag}(\boldsymbol{a})\boldsymbol{x}^\varepsilon(t) + A\boldsymbol{u}^\varepsilon(t) + Bz.$$

Then (8.2) can be changed into

$$\frac{d}{dt}\boldsymbol{x}^\varepsilon(t) = \mathcal{L}(\boldsymbol{x}^\varepsilon(t), \boldsymbol{u}^\varepsilon(t), z), \quad \boldsymbol{x}^\varepsilon(0) = \boldsymbol{x}.$$

Since the number of parts in the internal buffers cannot be negative, we impose the state constraints $x_j^\varepsilon(t) \geq 0$, $j = 1, \ldots, m-1$. To formulate the problem precisely, let $\mathcal{X} = [0, +\infty)^{m-1} \times (-\infty, +\infty) \subseteq \Re^m$ denote the state constraint domain. For $\boldsymbol{k} = (k_1, \ldots, k_m), k_j \geq 0, j = 1, \ldots, m$, let

$$\mathcal{U}(\boldsymbol{k}) = \{\boldsymbol{u} = (u_1, \ldots, u_m)' : 0 \leq u_j \leq k_j, \ j = 1, \ldots, m\}, \tag{8.3}$$

and for $\boldsymbol{x} \in \mathcal{X}$ let

$$\mathcal{U}(\boldsymbol{x}, \boldsymbol{k}) = \{\boldsymbol{u} : \boldsymbol{u} \in \mathcal{U}(\boldsymbol{k}) \text{ and } x_j = 0 \Rightarrow u_j - u_{j+1} \geq 0, \quad j = 1, \ldots, m-1\}. \tag{8.4}$$

Let the σ-algebra $\mathcal{F}_t^\varepsilon = \sigma\{\boldsymbol{k}(\varepsilon, s) : 0 \leq s \leq t\}$. We now define the concept of admissible controls.

Definition 2.1. We say that a control $\boldsymbol{u}(\cdot) = (u_1(\cdot), \ldots, u_m(\cdot))'$ is *admissible* with respect to the initial state vector $\boldsymbol{x} = (x_1, \ldots, x_m)' \in \mathcal{X}$, if each of the following conditions hold:

(i) $\boldsymbol{u}(\cdot)$ is an $\mathcal{F}_t^\varepsilon$-adapted measurable process;

(ii) $\boldsymbol{u}(t) \in \mathcal{U}(\boldsymbol{k}(\varepsilon, t))$ for all $t \geq 0$; and

(iii) the solution $\boldsymbol{x}(t)$ of

$$\frac{d}{dt}\boldsymbol{x}(t) = \mathcal{L}(\boldsymbol{x}(t), \boldsymbol{u}(t), z), \tag{8.5}$$

with $\boldsymbol{x}(0) = \boldsymbol{x}$ satisfies $\boldsymbol{x}(t) = (x_1(t), \ldots, x_m(t))' \in \mathcal{X}$ for all $t \geq 0$. □

Remark 2.1. Condition (iii) is equivalent to $\boldsymbol{u}(t) \in \mathcal{U}(\boldsymbol{x}(t), \boldsymbol{k}(\varepsilon, t))$, $t \geq 0$. □

We use $\mathcal{A}^\varepsilon(\boldsymbol{x}, \boldsymbol{k})$ to denote the set of all admissible controls with respect to $\boldsymbol{x} \in \mathcal{X}$ and $\boldsymbol{k}(\varepsilon, 0) = \boldsymbol{k}$. The problem is to find an admissible control $\boldsymbol{u}(\cdot) \in \mathcal{A}^\varepsilon(\boldsymbol{x}, \boldsymbol{k})$ that minimizes the cost function

$$J^\varepsilon(\boldsymbol{x}, \boldsymbol{k}, \boldsymbol{u}(\cdot)) = \limsup_{T \to \infty} \frac{1}{T} E \int_0^T [h(\boldsymbol{x}(t)) + c(\boldsymbol{u}(t))]\, dt, \tag{8.6}$$

where $h(\cdot)$ defines the cost of surplus and $c(\cdot)$ is the production cost.

We impose the following assumptions on the random process $\boldsymbol{k}(\varepsilon, \cdot) = (k_1(\varepsilon, \cdot), \ldots, k_m(\varepsilon, \cdot))$ and the cost functions $h(\cdot)$ and $c(\cdot)$ throughout this chapter.

(A1) Let $\mathcal{M} = \{\boldsymbol{k}^1, \ldots, \boldsymbol{k}^p\}$ for some given integer $p \geq 1$, where the ith state $\boldsymbol{k}^i = (k_1^i, \ldots, k_m^i)$, with $k_j^i, j = 1, \ldots, m$, denoting the capacity of the jth machine in state i, $i = 1, \ldots, p$. The capacity process $\boldsymbol{k}(\varepsilon, t) \in \mathcal{M}, t \geq 0$, is a Markov chain with the infinitesimal generator

$$Q^\varepsilon = Q^{(1)} + \frac{1}{\varepsilon} Q^{(2)},$$

where $Q^{(1)} = (q^{(1)}_{\boldsymbol{k}^i\boldsymbol{k}^j})$ and $Q^{(2)} = (q^{(2)}_{\boldsymbol{k}^i\boldsymbol{k}^j})$ are matrices such that $q^{(\ell)}_{\boldsymbol{k}^i\boldsymbol{k}^j} \geq 0$ if $j \neq i$, $q^{(\ell)}_{\boldsymbol{k}^i\boldsymbol{k}^i} = -\sum_{j \neq i} q^{(\ell)}_{\boldsymbol{k}^i\boldsymbol{k}^j}$ for $\ell = 1, 2$, and $\varepsilon > 0$ is the given scale parameter assumed to be small.

(A2) $Q^{(2)}$ is strongly irreducible. Let $\boldsymbol{\nu} = (\nu_{\boldsymbol{k}^1}, \ldots, \nu_{\boldsymbol{k}^p})$ denote the equilibrium distribution of $Q^{(2)}$. Furthermore, we assume that the average capacity of the least capable machine exceeds demand, i.e.,

$$\min_{1 \le j \le m} \left\{ \sum_{i=1}^{p} \nu_{\boldsymbol{k}^i} k_j^i \right\} > z. \tag{8.7}$$

(A3) $h(\boldsymbol{x})$ is a nonnegative, convex function with $h(0) = 0$. There are positive constants C_{h1}, C_{h2}, and $\beta_{h1} \ge 1$ such that

$$h(\boldsymbol{x}) \ge C_{h1}|\boldsymbol{x}|^{\beta_{h1}} - C_{h2}, \quad \boldsymbol{x} \in \mathcal{X}.$$

Moreover, there are constants C_{h3} and $\beta_{h2} \ge \beta_{h1}$ such that

$$|h(\boldsymbol{x}) - h(\boldsymbol{y})| \le C_{h3}(1 + |\boldsymbol{x}|^{\beta_{h2}-1} + |\boldsymbol{y}|^{\beta_{h2}-1})|\boldsymbol{x} - \boldsymbol{y}|, \quad \boldsymbol{x}, \boldsymbol{y} \in \mathcal{X}.$$

(A4) $c(\boldsymbol{u})$ is a nonnegative convex function.

Let $\lambda^\varepsilon(\boldsymbol{x}, \boldsymbol{k})$ denote the minimal expected cost, i.e.,

$$\lambda^\varepsilon(\boldsymbol{x}, \boldsymbol{k}) = \inf_{\boldsymbol{u}(\cdot) \in \mathcal{A}^\varepsilon(\boldsymbol{x}, \boldsymbol{k})} J^\varepsilon(\boldsymbol{x}, \boldsymbol{k}, \boldsymbol{u}(\cdot)). \tag{8.8}$$

Here the long-run average-cost criterion is used. Under Assumptions (A1) and (A2), there exists a $\varepsilon_0 > 0$ such that for $\varepsilon \in (0, \ \varepsilon_0]$, the Markov chain $\boldsymbol{k}(\varepsilon, \cdot)$ is strongly irreducible, i.e.,

$$\boldsymbol{\nu}^\varepsilon \left[Q^{(1)} + \frac{1}{\varepsilon} Q^{(2)} \right] = 0 \quad \text{and} \quad \sum_{i=1}^{p} \nu_{\boldsymbol{k}^i}^\varepsilon = 1$$

have a unique positive solution $\boldsymbol{\nu}^\varepsilon$, and

$$\min_{1 \le j \le m} \left\{ \sum_{i=1}^{p} \nu_{\boldsymbol{k}^i}^\varepsilon k_j^i \right\} > z.$$

It follows from Chapter 4 that $\lambda^\varepsilon(\boldsymbol{x}, \boldsymbol{k})$ is independent of the initial condition $(\boldsymbol{x}, \boldsymbol{k})$. Thus we will use λ^ε instead of $\lambda^\varepsilon(\boldsymbol{x}, \boldsymbol{k})$. We use $\mathcal{P}^\varepsilon$ to denote our control problem, i.e.,

$$\mathcal{P}^\varepsilon: \begin{cases} \min \ J^\varepsilon(\boldsymbol{x}, \boldsymbol{k}, \boldsymbol{u}(\cdot)) = \limsup_{T \to \infty} \dfrac{1}{T} E \displaystyle\int_0^T [h(\boldsymbol{x}(t)) + c(\boldsymbol{u}(t))] \, dt, \\ \text{s.t. } \dfrac{d}{dt}\boldsymbol{x}(t) = \mathcal{L}(\boldsymbol{x}(t), \boldsymbol{u}(t), z), \quad \boldsymbol{x}(0) = \boldsymbol{x}, \quad \boldsymbol{u}(\cdot) \in \mathcal{A}^\varepsilon(\boldsymbol{x}, \boldsymbol{k}), \\ \text{minimum average cost } \lambda^\varepsilon = \displaystyle\inf_{\boldsymbol{u}(\cdot) \in \mathcal{A}^\varepsilon(\boldsymbol{x}, \boldsymbol{k})} J^\varepsilon(\boldsymbol{x}, \boldsymbol{k}, \boldsymbol{u}(\cdot)). \end{cases}$$

As in Chapter 7, the positive attrition rate $\boldsymbol{a}$ implies a uniform bound for $\boldsymbol{x}^\varepsilon(t)$. In view of the fact that the control $\boldsymbol{u}^\varepsilon(t)$ is bounded between 0 and $\max\{|\boldsymbol{k}^i|, 1 \le i \le p\}$, this implies that any solution $\boldsymbol{x}^\varepsilon(\cdot)$ of equation (8.1) must satisfy

$$\begin{aligned} |x_j^\varepsilon(t)| &= \left| x_j e^{-a_j t} + e^{-a_j t} \int_0^t e^{a_j s} \left[u_j^\varepsilon(s) - u_{j+1}^\varepsilon(s) \right] ds \right| \\ &\le |x_j| e^{-a_j t} + \left(\max_{1 \le i \le p} \{|\boldsymbol{k}^i|\} + z \right) \int_0^t e^{-a_j(t-s)}\, ds \qquad (8.9) \\ &\le |x_j| e^{-a_j t} + \frac{\max_{1 \le i \le p}\{|\boldsymbol{k}^i|\} + z}{a_j}, \qquad j = 1, \ldots, m, \end{aligned}$$

where $u_{m+1}(s) = z$. Thus under the positive deterioration/cancellation rate, the surplus process $\boldsymbol{x}^\varepsilon(\cdot)$ remains bounded.

8.3 The Limiting Control Problem

In this section we examine elementary properties of the potential function and obtain the limiting control problem as $\varepsilon \to 0$.

The average-cost Hamilton-Jacobi-Bellman (HJB) equation in the directional derivative sense (HJBDD) for the optimal control problem in $\mathcal{P}^\varepsilon$, as shown in Chapter 4, takes the form

$$\begin{aligned} \lambda^\varepsilon = & \inf_{\boldsymbol{u} \in \mathcal{U}(\boldsymbol{x}, \boldsymbol{k})} \left\{ \partial_{\mathcal{L}(\boldsymbol{x}, \boldsymbol{u}, z)} \phi^\varepsilon(\boldsymbol{x}, \boldsymbol{k}) + c(\boldsymbol{u}) \right\} + h(\boldsymbol{x}) \\ & + \left(Q^{(1)} + \frac{1}{\varepsilon} Q^{(2)} \right) \phi^\varepsilon(\boldsymbol{x}, \cdot)(\boldsymbol{k}), \end{aligned} \qquad (8.10)$$

where $\phi^\varepsilon(\cdot,\cdot)$ is the function defined on $\mathcal{X} \times \mathcal{M}$. Our analysis begins with the proof of the boundedness of λ^ε.

Theorem 3.1. *Let Assumptions* (A1)–(A4) *hold. The minimum average expected cost* λ^ε *of* $\mathcal{P}^\varepsilon$ *is bounded in* ε, *i.e., there exists a constant* $C > 0$ *such that*

$$0 \le \lambda^\varepsilon \le C, \quad \text{for all } \varepsilon > 0.$$

Proof. According to the definition of λ^ε, it suffices to show that there exist a constant C_1 and a control $\boldsymbol{u}(\cdot) \in \mathcal{A}^\varepsilon(0, \boldsymbol{k})$ such that the solution $\boldsymbol{x}(t)$ of

$$\frac{d}{dt}\boldsymbol{x}(t) = \mathcal{L}(\boldsymbol{x}(t), \boldsymbol{u}(t), z), \quad \boldsymbol{x}(0) = 0, \qquad (8.11)$$

along with $\boldsymbol{u}(\cdot)$ satisfy

$$\limsup_{T\to\infty} \frac{1}{T} E \int_0^T [h(\boldsymbol{x}(t)) + c(\boldsymbol{u}(t))]\, dt \le C_1. \tag{8.12}$$

In view of (8.11), we can derive

$$x_j(t) = e^{-a_j t} \int_0^t e^{a_j s}[u_j(s) - u_{j+1}(s)]\, ds,$$

for $j = 1, \ldots, m-1$, and

$$x_m(t) = e^{-a_m t} \int_0^t e^{a_m s}[u_m(s) - z]\, ds.$$

Then from (8.3), we conclude that, for $j = 1, \ldots, m$,

$$|x_j(t)| \le \frac{z + \max\{|\boldsymbol{k}^i|, 1 \le i \le p\}}{a_j}. \tag{8.13}$$

It is clear under Assumptions (A3) and (A4) that the functions $h(\boldsymbol{x})$ and $c(\boldsymbol{u})$ are continuous. Recall that $\boldsymbol{u}(t) \ge 0$, $|\boldsymbol{u}(t)| \le \max\{|\boldsymbol{k}^i|, 1 \le i \le p\}$. These facts together with inequality (8.13) imply (8.12). □

In the remainder of this section, we derive the limiting control problem as $\varepsilon \to 0$. Intuitively, as the rates of the machine breakdown and repair approach infinity, the problem $\mathcal{P}^\varepsilon$, which is termed the *original control problem*, can be approximated by a simpler problem called the *limiting control problem*, where the stochastic machine capacity process $\boldsymbol{k}(\varepsilon, \cdot)$ is replaced by a weighted form. The limiting control problem is precisely formulated as follows.

We consider the augmented control

$$U(\cdot) = (\boldsymbol{u}^1(\cdot), \ldots, \boldsymbol{u}^p(\cdot)),$$

where $\boldsymbol{u}^i(t) = (u_1^i(t), \ldots, u_m^i(t))' \in \mathcal{U}(\boldsymbol{k}^i)$ for all t and $\boldsymbol{u}^i(\cdot)$ is a deterministic process for each i. For $\boldsymbol{x} \in \mathcal{X}$, let

$$\mathcal{A}^0(\boldsymbol{x}) = \Bigg\{ U(\cdot) \;:\; \text{the solution } \frac{d}{dt}\boldsymbol{x}(t) = \mathcal{L}\left(\boldsymbol{x}(t), \sum_{i=1}^p \nu_{\boldsymbol{k}^i} \boldsymbol{u}^i(t), z\right), \\ \boldsymbol{x}(0) = \boldsymbol{x} \text{ satisfies that } \boldsymbol{x}(t) \in \mathcal{X} \text{ for all } t \ge 0 \Bigg\}.$$

The objective of the limiting control problem is to choose a control $U(\cdot) \in \mathcal{A}^0(\boldsymbol{x})$ that minimizes

$$J(\boldsymbol{x}, U(\cdot)) = \limsup_{T\to\infty} \frac{1}{T} \int_0^T \left[h(\boldsymbol{x}(s)) + \sum_{i=1}^p \nu_{\boldsymbol{k}^i} c(\boldsymbol{u}^i(s)) \right] ds.$$

We use $\mathcal{P}^0$ to denote the above problem, and will regard this as our limiting control problem. For ease of reference, we rewrite $\mathcal{P}^0$ as follows:

$$\mathcal{P}^0 : \begin{cases} \min \; J(\boldsymbol{x}, U(\cdot)) \\ \quad = \limsup_{T\to\infty} \frac{1}{T} \int_0^T \left[h(\boldsymbol{x}(s)) + \sum_{i=1}^{p} \nu_{\boldsymbol{k}^i} c(\boldsymbol{u}^i(s)) \right] ds, \\ \text{s.t. } \frac{d}{dt}\boldsymbol{x}(t) = \mathcal{L}\left(\boldsymbol{x}(t), \sum_{i=1}^{p} \nu_{\boldsymbol{k}^i} \boldsymbol{u}^i(t), z \right), \\ \quad \boldsymbol{x}(0) = \boldsymbol{x}, \quad U(\cdot) \in \mathcal{A}^0(\boldsymbol{x}), \\ \text{minimum average cost } \lambda = \inf_{U(\cdot)\in\mathcal{A}^0(\boldsymbol{x})} J(\boldsymbol{x}, U(\cdot)). \end{cases}$$

The HJBDD equation associated with $\mathcal{P}^0$ is

$$\lambda = \inf_{\boldsymbol{u}^i \in \mathcal{U}(\boldsymbol{k}^i), \boldsymbol{k}^i \in \mathcal{M}} \left\{ \partial_{\mathcal{L}(\boldsymbol{x}, \sum_{i=1}^{p} \nu_{\boldsymbol{k}^i} \boldsymbol{u}^i, z)} \phi(\boldsymbol{x}) + \sum_{i=1}^{p} \nu_{\boldsymbol{k}^i} c(\boldsymbol{u}^i) \right\} + h(\boldsymbol{x}), \quad (8.14)$$

where $\phi(\cdot)$ is the function defined on $\mathcal{X}$.

8.4 Convergence of the Minimum Average Expected Cost

In this section we consider the convergence of the minimum average expected cost λ^ε as ε goes to zero, and establish its convergence rate. In order to get the required convergence result, we need the following lemma, which is the key to obtaining our main result.

Lemma 4.1. *Let Assumptions* (A3) *and* (A4) *hold for* $\delta \in (0,\ 1/2)$. *Then there exist* $\varepsilon_0 > 0$, $C > 0$, *and* $\boldsymbol{x} = (x_1, \ldots, x_m)' \in \mathcal{X}$, *such that for each given* $\varepsilon \in (0,\ \varepsilon_0]$, *we can choose a control*

$$(\overline{\boldsymbol{u}}^1(\cdot), \ldots, \overline{\boldsymbol{u}}^p(\cdot)) = ((\bar{u}_1^1(\cdot), \ldots, \bar{u}_m^1(\cdot))', \ldots, (\bar{u}_1^p(\cdot), \ldots, \bar{u}_m^p(\cdot))') \in \mathcal{A}^0(\boldsymbol{x}),$$

satisfying, for each $j = 1, \ldots, m$, *and all* $t \geq 0$,

$$\varepsilon^\delta \leq \sum_{i=1}^{p} \nu_{\boldsymbol{k}^i} \bar{u}_j^i(t) \leq \sum_{i=1}^{p} \nu_{\boldsymbol{k}^i} k_j^i - \varepsilon^\delta \quad (8.15)$$

and

$$\lambda + C\varepsilon^\delta \geq \limsup_{T\to\infty} \frac{1}{T} \int_0^T \left[h(\overline{\boldsymbol{x}}(t)) + \sum_{i=1}^{p} \nu_{\boldsymbol{k}^i} c(\overline{\boldsymbol{u}}^i(t)) \right] dt, \quad (8.16)$$

where $\overline{\boldsymbol{x}}(\cdot)$ *is a solution of*

$$\frac{d}{dt}\boldsymbol{x}(t) = \mathcal{L}\left(\boldsymbol{x}(t), \sum_{i=1}^{p} \nu_{\boldsymbol{k}^i}\overline{\boldsymbol{u}}^i(t), z\right), \quad \boldsymbol{x}(0) = \boldsymbol{x}.$$

Proof. First for each fixed $\varepsilon > 0$, we select an $\widetilde{\boldsymbol{x}}$ and

$$\begin{aligned}&(\widetilde{\boldsymbol{u}}^1(\cdot), \ldots, \widetilde{\boldsymbol{u}}^p(\cdot))\\ &\quad = ((\widetilde{u}_1^1(\cdot), \ldots, \widetilde{u}_m^1(\cdot))', \ldots, (\widetilde{u}_1^p(\cdot), \ldots, \widetilde{u}_m^p(\cdot))') \in \mathcal{A}^0(\widetilde{\boldsymbol{x}}),\end{aligned} \tag{8.17}$$

such that

$$\lambda + \varepsilon > \limsup_{T\to\infty} \frac{1}{T}\int_0^T \left[h(\widetilde{\boldsymbol{x}}(t)) + \sum_{i=1}^{p} \nu_{\boldsymbol{k}^i} c(\widetilde{\boldsymbol{u}}^i(t))\right] dt, \tag{8.18}$$

where $\widetilde{\boldsymbol{x}}(t) = (\widetilde{x}_1(t), \ldots, \widetilde{x}_m(t))'$ satisfies

$$\frac{d}{dt}\widetilde{\boldsymbol{x}}(t) = \mathcal{L}\left(\widetilde{\boldsymbol{x}}(t), \sum_{i=1}^{p} \nu_{\boldsymbol{k}^i}\widetilde{\boldsymbol{u}}^i(t), z\right), \quad \widetilde{\boldsymbol{x}}(0) = \widetilde{\boldsymbol{x}}.$$

Based on (8.17) and (8.18), we construct $(\boldsymbol{u}^1(\cdot), \ldots, \boldsymbol{u}^p(\cdot))$ such that (8.15) and (8.16) hold. Let

$$\mathcal{C}(j) = \{i : k_j^i \neq 0\}, \quad j = 1, \ldots, m, \tag{8.19}$$

and

$$C_1 = \min_{1\le j\le m}\left\{\sum_{i=1}^{p} \nu_{\boldsymbol{k}^i} k_j^i\right\} \quad \text{and} \quad C_2 = \frac{1}{\sum_{i\in\mathcal{C}(1)} \nu_{\boldsymbol{k}^i}} \cdot \frac{C_1}{C_1 - 2\varepsilon^\delta}. \tag{8.20}$$

We choose $\varepsilon_1 > 0$ such that, for $\varepsilon \in (0,\ \varepsilon_1]$,

$$C_2\varepsilon^\delta \le \min_{i\in\mathcal{C}(1)} \{k_1^i\}. \tag{8.21}$$

Let

$$\begin{aligned} u_1^i(t) &= \widetilde{u}_1^i(t) \vee (C_2\varepsilon^\delta), \quad \text{for } i \in \mathcal{C}(1),\\ u_1^i(t) &= 0, \quad \text{for } i \notin \mathcal{C}(1),\end{aligned}$$

and

$$u_j^i(t) = \widetilde{u}_j^i(t), \quad \text{for } i = 1, \ldots, p, \quad j = 2, \ldots, m.$$

Clearly, for $i = 1, \ldots, p$,

$$\left|\boldsymbol{u}^i(t) - \widetilde{\boldsymbol{u}}^i(t)\right| \le C_2\varepsilon^\delta. \tag{8.22}$$

Let $\boldsymbol{x}(\cdot)$ be a solution of

$$\frac{d}{dt}\boldsymbol{x}(t) = \mathcal{L}\left(\boldsymbol{x}(t), \sum_{i=1}^{p} \nu_{\boldsymbol{k}^i}\boldsymbol{u}^i(t), z\right), \quad \boldsymbol{x}(0) = \widetilde{\boldsymbol{x}}.$$

Since, for $i = 1, \ldots, p$ and $t \geq 0$,

$$u_1^i(t) \geq \widetilde{u}_1^i(t) \quad \text{and} \quad u_j^i(t) = \widetilde{u}_j^i(t), \quad j = 2, \ldots, m,$$

we have

$$\begin{aligned} x_1(t) &= \widetilde{x}_1 e^{-a_1 t} + \int_0^t e^{-a_1(t-s)} \left[\sum_{i=1}^{p} \nu_{\boldsymbol{k}^i} u_1^i(s) - \sum_{i=1}^{p} \nu_{\boldsymbol{k}^i} u_2^i(s)\right] ds \\ &\geq \widetilde{x}_1 e^{-a_1 t} + \int_0^t e^{-a_1(t-s)} \left[\sum_{i=1}^{p} \nu_{\boldsymbol{k}^i} \widetilde{u}_1^i(s) - \sum_{i=1}^{p} \nu_{\boldsymbol{k}^i} \widetilde{u}_2^i(s)\right] ds \\ &= \widetilde{x}_1(t) \geq 0, \end{aligned} \tag{8.23}$$

$$\begin{aligned} x_j(t) &= \widetilde{x}_j e^{-a_j t} + \int_0^t e^{-a_j(t-s)} \left[\sum_{i=1}^{p} \nu_{\boldsymbol{k}^i} u_j^i(s) - \sum_{i=1}^{p} \nu_{\boldsymbol{k}^i} u_{j+1}^i(s)\right] ds \\ &= \widetilde{x}_j e^{-a_j t} + \int_0^t e^{-a_j(t-s)} \left[\sum_{i=1}^{p} \nu_{\boldsymbol{k}^i} \widetilde{u}_j^i(s) - \sum_{i=1}^{p} \nu_{\boldsymbol{k}^i} \widetilde{u}_{j+1}^i(s)\right] ds \\ &= \widetilde{x}_j(t), \end{aligned} \tag{8.24}$$

for $j = 2, \ldots, m-1$, and

$$\begin{aligned} x_m(t) &= \widetilde{x}_m e^{-a_m t} + \int_0^t e^{-a_m(t-s)} \left[\sum_{i=1}^{p} \nu_{\boldsymbol{k}^i} u_m^i(s) - z\right] ds \\ &= \widetilde{x}_m e^{-a_m t} + \int_0^t e^{-a_m(t-s)} \left[\sum_{i=1}^{p} \nu_{\boldsymbol{k}^i} \widetilde{u}_m^i(s) - z\right] ds \\ &= \widetilde{x}_m(t). \end{aligned} \tag{8.25}$$

Therefore,

$$(\boldsymbol{u}^1(\cdot), \ldots, \boldsymbol{u}^p(\cdot)) = ((u_1^1(\cdot), \ldots, u_m^1(\cdot))', \ldots, (u_1^p(\varepsilon, \cdot), \ldots, u_m^p(\cdot))') \in \mathcal{A}^0(\widetilde{\boldsymbol{x}}).$$

Furthermore, from the definition of $(\boldsymbol{u}^1(\cdot), \ldots, \boldsymbol{u}^p(\cdot))$, we have

$$0 \leq \sum_{i=1}^{p} \nu_{\boldsymbol{k}^i} u_1^i(t) - \sum_{i=1}^{p} \nu_{\boldsymbol{k}^i} \widetilde{u}_1^i(t) \leq C_2 \varepsilon^{\delta}, \tag{8.26}$$

where $\sum_{i=1}^{p} \nu_{\boldsymbol{k}^i} = 1$ is applied. Hence,

$$\begin{aligned} 0 \le x_1(t) - \widetilde{x}_1(t) &\le e^{-a_1 t} \int_0^t e^{a_1 s} C_2 \varepsilon^\delta \, ds \\ &\le C_2 \varepsilon^\delta / a_1. \end{aligned} \tag{8.27}$$

Thus (8.24), (8.25), and (8.27) lead to

$$|\boldsymbol{x}(t) - \widetilde{\boldsymbol{x}}(t)| \le C_2 \varepsilon^\delta / a_1. \tag{8.28}$$

In view of (8.22) and (8.28), Assumptions (A3) and (A4), and the boundedness of $\widetilde{\boldsymbol{x}}(t)$ and $\boldsymbol{x}(t)$, there is a constant $C_3 > 0$ such that

$$\begin{aligned} &\left| \limsup_{T\to\infty} \frac{1}{T} \int_0^T \left[h(\boldsymbol{x}(t)) + \sum_{i=1}^{p} \nu_{\boldsymbol{k}^i} c(\boldsymbol{u}^i(t)) \right] dt \right. \\ &\qquad \left. - \limsup_{T\to\infty} \frac{1}{T} \int_0^T \left[h(\widetilde{\boldsymbol{x}}(t)) + \sum_{i=1}^{p} \nu_{\boldsymbol{k}^i} c(\widetilde{\boldsymbol{u}}^i(t)) \right] dt \right| \\ &\le \limsup_{T\to\infty} \frac{1}{T} \int_0^T \left| \left[h(\boldsymbol{x}(t)) + \sum_{i=1}^{p} \nu_{\boldsymbol{k}^i} c(\boldsymbol{u}^i(t)) \right] \right. \\ &\qquad\qquad \left. - \left[h(\widetilde{\boldsymbol{x}}(t)) + \sum_{i=1}^{p} \nu_{\boldsymbol{k}^i} c(\widetilde{\boldsymbol{u}}^i(t)) \right] \right| dt \\ &\le \limsup_{T\to\infty} \frac{1}{T} \int_0^T C_3 \left[\frac{C_2 \varepsilon^\delta}{a_1} + C_2 \varepsilon^\delta \right] dt \\ &= C_3 C_2 \left(1 + \frac{1}{a_1} \right) \varepsilon^\delta. \end{aligned}$$

Thus, (8.18) implies that

$$\lambda + C_4 \varepsilon^\delta > \limsup_{T\to\infty} \frac{1}{T} \int_0^T \left[h(\boldsymbol{x}(t)) + \sum_{i=1}^{p} \nu_{\boldsymbol{k}^i} c(\boldsymbol{u}^i(t)) \right] dt, \tag{8.29}$$

for some $C_4 > 0$. For $i = 1, \ldots, p$, define

$$\widehat{u}_1^i(t) = \frac{u_1^i(t)}{1 + 2\varepsilon^\delta/(C_1 - 2\varepsilon^\delta)}$$

and

$$\widehat{u}_j^i(t) = \frac{u_j^i(t)}{1 + 2\varepsilon^\delta/(C_1 - 2\varepsilon^\delta)}, \quad j = 2, \ldots, m.$$

Then solve

$$\frac{d}{dt}\widehat{\boldsymbol{x}}(t) = \mathcal{L}\left(\widehat{\boldsymbol{x}}(t), \sum_{i=1}^{p} \nu_{\boldsymbol{k}^i} \widehat{\boldsymbol{u}}^i(t), z \right), \quad \widehat{\boldsymbol{x}}(0) = \frac{\widetilde{\boldsymbol{x}}}{1 + 2\varepsilon^\delta/(C_1 - 2\varepsilon^\delta)},$$

to obtain

$$\begin{aligned}\widehat{x}_j(t) &= \frac{\widetilde{x}_j}{1+2\varepsilon^\delta/(C_1-2\varepsilon^\delta)}e^{-a_jt} \\ &\quad + e^{-a_jt}\int_0^t e^{a_js}\left(\sum_{i=1}^p \nu_{\boldsymbol{k}^i}\widehat{u}_j^i(s) - \sum_{i=1}^p \nu_{\boldsymbol{k}^i}\widehat{u}_{j+1}^i(s)\right)ds \\ &= \frac{x_j(t)}{1+2\varepsilon^\delta/(C_1-2\varepsilon^\delta)} \geq 0, \quad \text{for } j=1,\ldots,m-1,\end{aligned}$$

and

$$\begin{aligned}\widehat{x}_m(t) &= \frac{\widetilde{x}_m}{1+2\varepsilon^\delta/(C_1-2\varepsilon^\delta)}e^{-a_mt} \\ &\quad + e^{-a_mt}\int_0^t e^{a_ms}\left(\sum_{i=1}^p \nu_{\boldsymbol{k}^i}\widehat{u}_m^i(s) - z\right)ds \\ &= \frac{1}{1+2\varepsilon^\delta/(C_1-2\varepsilon^\delta)}x_m(t) - \frac{2z\varepsilon^\delta}{a_mC_1}(1-e^{-a_mt}).\end{aligned}$$

Therefore,

$$\begin{aligned}&(\widehat{\boldsymbol{u}}^1(\cdot),\ldots,\widehat{\boldsymbol{u}}^p(\cdot)) \\ &= ((\widehat{u}_1^1(\cdot),\ldots,\widehat{u}_m^1(\cdot))',\ldots,(\widehat{u}_1^p(\cdot),\ldots,\widehat{u}_m^p(\cdot))') \in \mathcal{A}^0\left(\frac{\widetilde{\boldsymbol{x}}}{1+2\varepsilon^\delta/(C_1-2\varepsilon^\delta)}\right).\end{aligned}$$

Furthermore, from the boundedness of $\boldsymbol{x}(t)$ and $(\boldsymbol{u}^1(\cdot),\ldots,\boldsymbol{u}^p(\cdot))$, we have

$$\begin{aligned}&|\widehat{\boldsymbol{x}}(t)-\boldsymbol{x}(t)| \\ &= \left(1-\frac{1}{1+2\varepsilon^\delta/(C_1-2\varepsilon^\delta)}\right)\sum_{j=1}^{m-1}x_j(t) \\ &\quad + \left|\frac{2z\varepsilon^\delta}{a_mC_1}(1-e^{-a_mt}) + \left(1-\frac{1}{1+2\varepsilon^\delta/(C_1-2\varepsilon^\delta)}\right)x_m(t)\right| \\ &\leq C_5\varepsilon^\delta,\end{aligned} \tag{8.30}$$

and

$$\left|\widehat{\boldsymbol{u}}^i(t)-\boldsymbol{u}^i(t)\right| \leq C_5\varepsilon^\delta, \quad i=1,\ldots,p, \tag{8.31}$$

for some $C_5 > 0$. Similar to (8.29), we can show by (8.30) and (8.31) that there exists a constant $C_6 > 0$ such that

$$\lambda + C_6\varepsilon^\delta > \limsup_{T\to\infty}\frac{1}{T}\int_0^T\left[h(\widehat{\boldsymbol{x}}(t)) + \sum_{i=1}^p \nu_{\boldsymbol{k}^i}c(\widehat{\boldsymbol{u}}^i(t))\right]dt. \tag{8.32}$$

By the definitions of C_1 and C_2 we know that

$$\begin{aligned}\sum_{i=1}^{p} \nu_{\boldsymbol{k}^i} \widehat{u}_1^i(t) &= \sum_{i=1}^{p} \nu_{\boldsymbol{k}^i} \frac{u_1^i(t)}{1+2\varepsilon^\delta/(C_1-2\varepsilon^\delta)} \\ &\geq \frac{C_2\varepsilon^\delta}{1+2\varepsilon^\delta/(C_1-2\varepsilon^\delta)} \sum_{i\in\mathcal{C}(1)} \nu_{\boldsymbol{k}^i} \\ &\geq \varepsilon^\delta,\end{aligned} \tag{8.33}$$

$$\begin{aligned}\sum_{i=1}^{p} \nu_{\boldsymbol{k}^i} \widehat{u}_1^i(t) &= \sum_{i=1}^{p} \nu_{\boldsymbol{k}^i} \frac{u_1^i(t)}{1+2\varepsilon^\delta/(C_1-2\varepsilon^\delta)} \\ &\leq \sum_{i=1}^{p} \nu_{\boldsymbol{k}^i} \frac{k_1^i}{1+2\varepsilon^\delta/(C_1-2\varepsilon^\delta)} \\ &\leq \sum_{i=1}^{p} \nu_{\boldsymbol{k}^i} k_1^i - 2\varepsilon^\delta,\end{aligned} \tag{8.34}$$

and, for $j = 2, \ldots, m$,

$$\begin{aligned}\sum_{i=1}^{p} \nu_{\boldsymbol{k}^i} \widehat{u}_j^i(t) &= \sum_{i=1}^{p} \nu_{\boldsymbol{k}^i} \frac{u_j^i(t)}{1+2\varepsilon^\delta/(C_1-2\varepsilon^\delta)} \\ &\leq \sum_{i=1}^{p} \nu_{\boldsymbol{k}^i} \frac{k_j^i}{1+2\varepsilon^\delta/(C_1-2\varepsilon^\delta)} \\ &\leq \sum_{i=1}^{p} \nu_{\boldsymbol{k}^i} k_j^i - 2\varepsilon^\delta.\end{aligned} \tag{8.35}$$

In order to obtain the left inequality of (8.15) for $j = 2, \ldots, m$, we choose $(v_1^i(t), \ldots, v_m^i(t))'(\geq 0)$, $i = 1, \ldots, p$. First, for $j = 1, \ldots, m$ and $i \notin \mathcal{C}(j)$, set

$$v_j^i(t) = 0.$$

For $i \in \mathcal{C}(j)$, we consider two cases.

$$\textbf{Case 1:} \quad \min_{1\leq j\leq m} \left\{ \sum_{\ell=1}^{p} \nu_{\boldsymbol{k}^\ell} \widehat{u}_j^\ell(t) \right\} \geq \varepsilon^\delta,$$

$$\textbf{Case 2:} \quad \min_{1\leq j\leq m} \left\{ \sum_{\ell=1}^{p} \nu_{\boldsymbol{k}^\ell} \widehat{u}_j^\ell(t) \right\} < \varepsilon^\delta.$$

For Case 1, set $v_j^i(t) = 0$. To deal with Case 2, note that for each $j \in \{1, \ldots, m\}$, there is an $l(j)$ $(\in \{1, \ldots, p\})$ such that $k_j^{l(j)} > 0$. Now choose

$\varepsilon_2 > 0$ ($< \varepsilon_1$) such that for each $\varepsilon \in (0,\ \varepsilon_2]$ and $j = 1, \ldots, m$, $\nu_{\boldsymbol{k}^{l(j)}} k_j^{l(j)} > 2\varepsilon^\delta$. For Case 2, choose

$$\nu_{\boldsymbol{k}^i} v_j^i(t) = \begin{cases} 0, & \text{if } i \neq l(j), \\ \varepsilon^\delta, & \text{if } i = l(j). \end{cases}$$

Define

$$\bar{u}_j^i(t) = \widehat{u}_j^i(t) + v_j^i(t), \quad i = 1, \ldots, p \text{ and } \ j = 1, \ldots, m, \tag{8.36}$$

and

$$(\overline{\boldsymbol{u}}^1(t), \ldots, \overline{\boldsymbol{u}}^p(t)) = ((\bar{u}_1^1(t), \ldots, \bar{u}_m^1(t))', \ldots, (\bar{u}_1^p(t), \ldots, \bar{u}_m^p(t))'),$$

and let $\overline{\boldsymbol{x}}(t)$ be defined by

$$\frac{d}{dt}\overline{\boldsymbol{x}}(t) = \mathcal{L}\left(\overline{\boldsymbol{x}}(t), \sum_{i=1}^p \nu_{\boldsymbol{k}^i} \overline{\boldsymbol{u}}^i(t), z\right), \quad \overline{\boldsymbol{x}}(0) = \frac{\widetilde{\boldsymbol{x}}}{1 + 2\varepsilon^\delta/(C_1 - 2\varepsilon^\delta)}.$$

Set

$$\boldsymbol{x} = \frac{\widetilde{\boldsymbol{x}}}{1 + 2\varepsilon^\delta/(C_1 - 2\varepsilon^\delta)}.$$

We know, in view of (8.33)–(8.35), that $(\overline{\boldsymbol{u}}^1(\cdot), \ldots, \overline{\boldsymbol{u}}^p(\cdot)) \in \mathcal{A}^0(\boldsymbol{x})$ satisfies (8.15) with $\varepsilon_0 = \varepsilon_1 \wedge \varepsilon_2$. Furthermore, from the definition of $(v_1^i(t), \ldots, v_m^i(t))$ and (8.36),

$$\begin{aligned} \bar{x}_j(t) &= \widehat{x}_j(t), \quad j = 1, \ldots, m-1, \quad t \geq 0, \\ \bar{x}_m(t) &= \widehat{x}_m(t) + \frac{1 - e^{-a_m t}}{a_m} \varepsilon^\delta, \qquad t \geq 0. \end{aligned}$$

Therefore, (8.32) gives (8.16). □

From Lemma 4.1, we can get the following result, which states that there is an asymptotic optimal control that keeps the work-in-process in each buffer to be bounded below by a possible quantity.

Lemma 4.2. *Let Assumptions* (A3) *and* (A4) *hold for* $\delta \in (0,\ 1/2)$. *Then there exist an* $\varepsilon_0 > 0$, $C > 0$, $\widehat{C} > 0$, *and* $\boldsymbol{x} = (x_1, \ldots, x_m)' \in \mathcal{X}$ *such that for each given* $\varepsilon \in (0,\ \varepsilon_0]$, *we can choose a control*

$$(\overline{\boldsymbol{u}}^1(\cdot), \ldots, \overline{\boldsymbol{u}}^p(\cdot)) = ((\bar{u}_1^1(\cdot), \ldots, \bar{u}_m^1(\cdot))', \ldots, (\bar{u}_1^p(\cdot), \ldots, \bar{u}_m^p(\cdot))') \in \mathcal{A}^0(\boldsymbol{x})$$

satisfying

$$\min_{1 \leq j \leq m-1} \ \inf_{0 \leq t < \infty} \bar{x}_j(t) \geq C\varepsilon^\delta \tag{8.37}$$

and

$$\lambda + \widehat{C}\varepsilon^\delta \geq \limsup_{T\to\infty} \frac{1}{T}\int_0^T \left[h(\overline{\boldsymbol{x}}(t)) + \sum_{i=1}^p \nu_{\boldsymbol{k}^i} c(\overline{\boldsymbol{u}}^i(t)) \right] dt, \tag{8.38}$$

where $\overline{\boldsymbol{x}}(\cdot)$ *is a solution of*

$$\frac{d}{dt}\boldsymbol{x}(t) = \mathcal{L}\left(\boldsymbol{x}(t), \sum_{i=1}^p \nu_{\boldsymbol{k}^i}\overline{\boldsymbol{u}}^i(t), z \right), \quad \boldsymbol{x}(0) = \boldsymbol{x}.$$

Proof. Let

$$(\boldsymbol{u}^1(\cdot), \ldots, \boldsymbol{u}^p(\cdot)) = ((u_1^1(\cdot), \ldots, u_m^1(\cdot))', \ldots, (u_1^p(\cdot), \ldots, u_m^p(\cdot))') \in \mathcal{A}^0(\boldsymbol{x})$$

satisfy (8.15) and (8.16) in Lemma 4.1, and let $\boldsymbol{x}(\cdot)$ be a solution of

$$\frac{d}{dt}\boldsymbol{x}(t) = \mathcal{L}\left(\boldsymbol{x}(t), \sum_{i=1}^p \nu_{\boldsymbol{k}^i}\boldsymbol{u}^i(t), z \right), \quad \boldsymbol{x}(0) = \boldsymbol{x}.$$

Therefore, there exists

$$(\overline{\boldsymbol{u}}^1(\cdot), \ldots, \overline{\boldsymbol{u}}^p(\cdot)) = ((\bar{u}_1^1(\cdot), \ldots, \bar{u}_m^1(\cdot))', \ldots, (\bar{u}_1^p(\cdot), \ldots, \bar{u}_m^p(\cdot))') \in \mathcal{A}^0(\boldsymbol{x}),$$

such that

$$\sum_{i=1}^p \nu_{\boldsymbol{k}^i}\overline{u}_j^i(t) = \sum_{i=1}^p \nu_{\boldsymbol{k}^i} u_j^i(t) + \frac{\varepsilon^\delta}{j}, \quad j = 1, \ldots, m,$$

and

$$\left|\overline{\boldsymbol{u}}^i(t) - \boldsymbol{u}^i(t)\right| \leq \varepsilon^\delta, \quad i = 1, \ldots, p. \tag{8.39}$$

Let $\overline{\boldsymbol{x}}(\cdot)$ be a solution of

$$\frac{d}{dt}\boldsymbol{x}(t) = \mathcal{L}\left(\boldsymbol{x}(t), \sum_{i=1}^p \nu_{\boldsymbol{k}^i}\overline{\boldsymbol{u}}^i(t), z \right), \quad \boldsymbol{x}(0) = \boldsymbol{x} + \mathbf{1}\varepsilon^\delta.$$

Then, for $j = 1, \ldots, m-1$,

$$\begin{aligned} \overline{x}_j(t) &= x_j(t) + e^{-a_j t}\varepsilon^\delta + \frac{\varepsilon^\delta}{j(j+1)}\int_0^t e^{-a_j(t-s)}\, ds \\ &\geq C_1\varepsilon^\delta, \end{aligned} \tag{8.40}$$

for some $C_1 > 0$, which implies (8.37). Furthermore,

$$\overline{x}_m(t) = x_m(t) + e^{-a_m t}\varepsilon^\delta + \frac{\varepsilon^\delta}{m}\int_0^t e^{-a_m(t-s)}\, ds. \tag{8.41}$$

It follows from the first equality of (8.40) and (8.41) that, for $t \geq 0$,

$$|\overline{\boldsymbol{x}}(t) - \boldsymbol{x}(t)| \leq C_2 \varepsilon^\delta. \tag{8.42}$$

By (8.39) and (8.42), similar to (8.29), we have (8.38). □

With Lemmas 4.1 and 4.2 we can prove our main result.

Theorem 4.1. *Let Assumptions* (A1)–(A4) *hold. Then, for any given* $\delta \in (0,\ 1/2)$, *there exist an* $\varepsilon_0 > 0$ *and a constant* $C > 0$ *such that, for all* $\varepsilon \in (0,\ \varepsilon_0]$,

$$|\lambda^\varepsilon - \lambda| \leq C\varepsilon^\delta.$$

This implies in particular that $\lim_{\varepsilon \to 0} \lambda^\varepsilon = \lambda$.

Proof. We begin with an outline of the major steps in the proof. First we prove $\lambda^\varepsilon < \lambda + C\varepsilon^\delta$ by constructing an admissible control $\boldsymbol{u}^\varepsilon(t)$ of $\mathcal{P}^\varepsilon$ from a near-optimal control of the limiting control problem $\mathcal{P}^0$ and by estimating the difference between the state trajectories corresponding to these two controls. Then we establish the opposite inequality, namely, $\lambda^\varepsilon > \lambda - C\varepsilon^\delta$, by constructing a control of the limiting control problem $\mathcal{P}^0$ from a near-optimal control of $\mathcal{P}^\varepsilon$, and then using Assumptions (A3) and (A4).

In order to show that

$$\lambda^\varepsilon \leq \lambda + C\varepsilon^\delta, \tag{8.43}$$

we can choose, in view of Lemma 4.2,

$$(\overline{\boldsymbol{u}}^1(\cdot), \ldots, \overline{\boldsymbol{u}}^p(\cdot)) = ((\bar{u}_1^1(\cdot), \ldots, \bar{u}_m^1(\cdot))', \ldots, (\bar{u}_1^p(\cdot), \ldots, \bar{u}_m^p(\cdot))') \in \mathcal{A}^0(\boldsymbol{x}),$$

such that

$$\min_{1 \leq j \leq m-1} \inf_{0 \leq t < \infty} \bar{x}_j(t) \geq C_1 \varepsilon^\delta \tag{8.44}$$

and

$$\lambda + C_2 \varepsilon^\delta \geq \limsup_{T \to \infty} \frac{1}{T} \int_0^T \left[h(\overline{\boldsymbol{x}}(t)) + \sum_{i=1}^p \nu_{\boldsymbol{k}^i} c(\overline{\boldsymbol{u}}^i(t)) \right] dt, \tag{8.45}$$

for some $C_1 > 0$ and $C_2 > 0$, where $\overline{\boldsymbol{x}}(\cdot)$ is a solution of

$$\frac{d}{dt}\boldsymbol{x}(t) = \mathcal{L}\left(\boldsymbol{x}(t), \sum_{i=1}^p \nu_{\boldsymbol{k}^i} \overline{\boldsymbol{u}}^i(t), z \right), \quad \boldsymbol{x}(0) = \boldsymbol{x}.$$

We construct the control

$$\overline{\boldsymbol{u}}^\varepsilon(t) = \sum_{i=1}^p I_{\{\boldsymbol{k}(\varepsilon,t) = \boldsymbol{k}^i\}} \overline{\boldsymbol{u}}^i(t),$$

and let $\overline{\boldsymbol{x}}^{\varepsilon}(\cdot)$ be the solution of

$$\frac{d}{dt}\boldsymbol{x}(t) = \mathcal{L}\left(\boldsymbol{x}(t), \overline{\boldsymbol{u}}^{\varepsilon}(t), z\right), \quad \boldsymbol{x}(0) = \boldsymbol{x}.$$

Then, for $j = 1, \ldots, m-1$,

$$\begin{aligned}
&E\left|\bar{x}_j^{\varepsilon}(t) - \bar{x}_j(t)\right|^2 \\
&\quad = E\left|e^{-a_j t}\int_0^t e^{a_j s}\left[\left(\bar{u}_j^{\varepsilon}(s) - \sum_{i=1}^{p}\nu_{\boldsymbol{k}^i}\bar{u}_j^i(s)\right)\right.\right. \\
&\qquad\qquad \left.\left. - \left(\bar{u}_{j+1}^{\varepsilon}(s) - \sum_{i=1}^{p}\nu_{\boldsymbol{k}^i}\bar{u}_{j+1}^i(s)\right)\right] ds\right|^2 \\
&\quad \le 2E\left[e^{-a_j t}\int_0^t e^{a_j s}\left(\bar{u}_j^{\varepsilon}(s) - \sum_{i=1}^{p}\nu_{\boldsymbol{k}^i}\bar{u}_j^i(s)\right) ds\right]^2 \\
&\qquad + 2E\left[e^{-a_j t}\int_0^t e^{a_j s}\left(\bar{u}_{j+1}^{\varepsilon}(s) - \sum_{i=1}^{p}\nu_{\boldsymbol{k}^i}\bar{u}_{j+1}^i(s)\right) ds\right]^2,
\end{aligned}$$

and

$$E\left|\bar{x}_m^{\varepsilon}(t) - \bar{x}_m(t)\right|^2 = E\left[e^{-a_m t}\int_0^t e^{a_m s}\left(\bar{u}_m^{\varepsilon}(s) - \sum_{i=1}^{p}\nu_{\boldsymbol{k}^i}\bar{u}_m^i(s)\right) ds\right]^2.$$

Similar to the proof of (7.21) in Chapter 7, we have

$$E\left|\bar{x}_j^{\varepsilon}(t) - \bar{x}_j(t)\right| \le C_3\varepsilon^{1/2}, \quad j = 1, \ldots, m, \tag{8.46}$$

for some $C_3 > 0$. Consequently, by the boundedness of $\overline{\boldsymbol{x}}^{\varepsilon}(t)$ and $\overline{\boldsymbol{x}}(t)$ and Assumption (A3), we get

$$\begin{aligned}
&\left|\limsup_{T\to\infty}\frac{1}{T}E\int_0^T h(\overline{\boldsymbol{x}}^{\varepsilon}(t))\,dt - \limsup_{T\to\infty}\frac{1}{T}\int_0^T h(\overline{\boldsymbol{x}}(t))\,dt\right| \\
&\quad \le \limsup_{T\to\infty}\frac{C_{h3}}{T}\int_0^T E\big[\left(1 + |\overline{\boldsymbol{x}}^{\varepsilon}(t)|^{\beta_{h2}-1} + |\overline{\boldsymbol{x}}(t)|^{\beta_{h2}-1}\right) \\
&\qquad\qquad \times |\overline{\boldsymbol{x}}^{\varepsilon}(t) - \overline{\boldsymbol{x}}(t)|\big]\,dt \\
&\quad \le C_4\varepsilon^{1/2},
\end{aligned} \tag{8.47}$$

for some $C_4 > 0$. It follows from Lemma B.3 and Assumption (A4) that

$$\begin{aligned}
&\left| \limsup_{T\to\infty} \frac{1}{T} E \int_0^T c(\overline{\boldsymbol{u}}^\varepsilon(t))\, dt - \limsup_{T\to\infty} \frac{1}{T} \int_0^T \sum_{i=1}^p \nu_{\boldsymbol{k}^i} c(\overline{\boldsymbol{u}}^i(t))\, dt \right| \\
&= \left| \limsup_{T\to\infty} \frac{1}{T} E \int_0^T \sum_{i=1}^p I_{\{\boldsymbol{k}(\varepsilon,t)=\boldsymbol{k}^i\}} c(\overline{\boldsymbol{u}}^i(t))\, dt \right. \\
&\qquad \left. - \limsup_{T\to\infty} \frac{1}{T} \int_0^T \sum_{i=1}^p \nu_{\boldsymbol{k}^i} c(\overline{\boldsymbol{u}}^i(t))\, dt \right| \qquad (8.48) \\
&\le \limsup_{T\to\infty} \frac{1}{T} \int_0^T \sum_{i=1}^p \left| P(\boldsymbol{k}(\varepsilon,t) = \boldsymbol{k}^i) - \nu_{\boldsymbol{k}^i} \right| \cdot c(\overline{\boldsymbol{u}}^i(t))\, dt \\
&\le C_5 \varepsilon,
\end{aligned}$$

for some $C_5 > 0$. Thus, combining (8.45) and (8.47)–(8.48), we see that there is a constant $C_6 > 0$ such that

$$\limsup_{T\to\infty} \frac{1}{T} E \int_0^T [h(\overline{\boldsymbol{x}}^\varepsilon(t)) + c(\overline{\boldsymbol{u}}^\varepsilon(t))]\, dt \le \lambda + C_6(\varepsilon + \varepsilon^\delta + \varepsilon^{1/2}). \qquad (8.49)$$

Note that in general, $\overline{\boldsymbol{u}}^\varepsilon(\cdot) \notin \mathcal{A}^\varepsilon(\boldsymbol{x}, \boldsymbol{k}(\varepsilon,0))$. So starting from $\overline{\boldsymbol{u}}^\varepsilon(\cdot)$, we must construct $\boldsymbol{u}^\varepsilon(\cdot) \in \mathcal{A}^\varepsilon(\boldsymbol{x}, \boldsymbol{k}(\varepsilon,0))$ such that it and its corresponding state trajectory satisfy (8.49). Consequently, we get (8.43). To do this, let

$$M_1 = \max_{\substack{1\le i\le p \\ 1\le j\le m}} \{k_j^i\} \quad \text{and} \quad M_2 = \max_{1\le j\le m} \left\{ \frac{1}{a_j} \ln \left(\frac{C_1 a_j \varepsilon^\delta}{5M_1} \right)^{-1} \right\},$$

where C_1 is given in (8.44). There is an $\varepsilon_0 > 0$ such that, for $\varepsilon \in (0,\ \varepsilon_0]$, $M_2 > 0$. We show that there exists a control $\boldsymbol{u}^\varepsilon(\cdot) \in \mathcal{A}^\varepsilon(\boldsymbol{x}, \boldsymbol{k}(\varepsilon,0))$ such that for $t > 2M_2$ and $j = 1, \ldots, m$,

$$E\left| u_j^\varepsilon(t) - \overline{u}_j^\varepsilon(t) \right| \le C_7 \exp\left\{ -M_3 \varepsilon^{-(1-2\delta)/2} (1+M_2)^{-3/2} \right\}, \qquad (8.50)$$

for some $C_7 > 0$ and $M_3 > 0$. Let $\boldsymbol{x}^\varepsilon(\cdot)$ be the solution of

$$\frac{d}{dt}\boldsymbol{x}(t) = \mathcal{L}(\boldsymbol{x}(t), \boldsymbol{u}^\varepsilon(t), z), \quad \boldsymbol{x}(0) = \boldsymbol{x}.$$

Hence, Assumptions (A3) and (A4) imply that, for $T > 2M_2$,

$$
\begin{aligned}
E\int_0^T & |c(\boldsymbol{u}^\varepsilon(t)) - c(\overline{\boldsymbol{u}}^\varepsilon(t))|\,dt \\
& \le C_8 \int_0^T E\,|\boldsymbol{u}^\varepsilon(t) - \overline{\boldsymbol{u}}^\varepsilon(t)|\,dt \\
& \le C_8 \int_0^{2M_2} E\,|\boldsymbol{u}^\varepsilon(t) - \overline{\boldsymbol{u}}^\varepsilon(t)|\,dt \\
& \quad + C_8 C_7 \int_{2M_2}^T \exp\big\{ - M_3 \varepsilon^{-(1-2\delta)/2} (1+M_2)^{-3/2} \big\} dt
\end{aligned}
\tag{8.51}
$$

and

$$
\begin{aligned}
E\int_0^T & |h(\boldsymbol{x}^\varepsilon(t)) - h(\overline{\boldsymbol{x}}^\varepsilon(t))|\,dt \\
& \le C_9 \int_0^T E\,|\boldsymbol{x}^\varepsilon(t) - \overline{\boldsymbol{x}}^\varepsilon(t)|\,dt \\
& \le C_9 \sum_{j=1}^m \int_0^T e^{-a_j t} \left(\int_0^t e^{a_j s} E|u_j^\varepsilon(s) - \bar{u}_j^\varepsilon(s)|\,ds \right) dt \\
& \le \frac{C_9}{a_j} \sum_{j=1}^m \int_0^T E|u_j^\varepsilon(s) - \bar{u}_j^\varepsilon(s)|\,ds,
\end{aligned}
\tag{8.52}
$$

for some $C_8 > 0$ and $C_9 > 0$. Therefore, we know, in view of (8.49)–(8.50) and (8.51)–(8.52), that (8.43) holds. Thus, it suffices to show that there is $\boldsymbol{u}^\varepsilon(\cdot) \in \mathcal{A}^\varepsilon(\boldsymbol{x}, \boldsymbol{k}(\varepsilon, 0))$ which satisfies (8.50). We will modify $\overline{\boldsymbol{u}}^\varepsilon(t)$ to $\boldsymbol{u}^\varepsilon(t)$ in such a way that (8.50) holds and $\boldsymbol{u}^\varepsilon(\cdot) \in \mathcal{A}^\varepsilon(\boldsymbol{x}, \boldsymbol{k}(\varepsilon, 0))$. This modification is based on the estimation of

$$
P\left(\bar{x}_j^\varepsilon(t) < 0\right), \quad t > 2M_2, \quad j = 1, \ldots, m-1.
$$

Thus we use (8.44) first to establish the following inequality. For $j =$

$1, \ldots, m-1$,

$$
\begin{aligned}
&P\left(\bar{x}_j^\varepsilon(t) \le \frac{C_1\varepsilon^\delta}{2}\right) \\
&\quad \le P\left(\bar{x}_j^\varepsilon(t) \le \bar{x}_j(t) - \frac{C_1\varepsilon^\delta}{2}\right) \quad \text{(by (8.44))} \\
&\quad = P\left(\bar{x}_j(t) - \bar{x}_j^\varepsilon(t) \ge \frac{C_1\varepsilon^\delta}{2}\right) \\
&\quad \le P\left(\left|\bar{x}_j(t) - \bar{x}_j^\varepsilon(t)\right| \ge \frac{C_1\varepsilon^\delta}{2}\right) \\
&\quad \le P\left(\left|\int_0^t e^{-a_j(t-s)}\left[\sum_{i=1}^p \left(I_{\{\boldsymbol{k}(\varepsilon,s)=\boldsymbol{k}^i\}} - \nu_{\boldsymbol{k}^i}\right)\bar{u}_j^i(s)\right] ds\right| \ge \frac{C_1\varepsilon^\delta}{4}\right) \\
&\qquad + P\left(\left|\int_0^t e^{-a_j(t-s)}\left[\sum_{i=1}^p (I_{\{\boldsymbol{k}(\varepsilon,s)=\boldsymbol{k}^i\}} - \nu_{\boldsymbol{k}^i})\bar{u}_{j+1}^i(s)\right] ds\right| \ge \frac{C_1\varepsilon^\delta}{4}\right).
\end{aligned}
\tag{8.53}
$$

Note that, for $t > 2M_2$ and $j = 1, \ldots, m$,

$$
\begin{aligned}
&\left|\int_0^t e^{-a_j(t-v)}\left[\sum_{i=1}^p (I_{\{\boldsymbol{k}(\varepsilon,v)=\boldsymbol{k}^i\}} - \nu_{\boldsymbol{k}^i})\bar{u}_j^i(v)\right] dv\right| \\
&\quad \le \left|\int_0^{t-M_2} e^{-a_j(t-v)}\left[\sum_{i=1}^p (I_{\{\boldsymbol{k}(\varepsilon,v)=\boldsymbol{k}^i\}} - \nu_{\boldsymbol{k}^i})\bar{u}_j^i(v)\right] dv\right| \\
&\qquad + \left|\int_{t-M_2}^t e^{-a_j(t-v)}\left[\sum_{i=1}^p (I_{\{\boldsymbol{k}(\varepsilon,v)=\boldsymbol{k}^i\}} - \nu_{\boldsymbol{k}^i})\bar{u}_j^i(v)\right] dv\right| \\
&\quad \le M_1 \int_0^{t-M_2} e^{-a_j(t-v)}\, dv \\
&\qquad + \left|\int_{t-M_2}^t e^{-a_j(t-v)}\left[\sum_{i=1}^p (I_{\{\boldsymbol{k}(\varepsilon,v)=\boldsymbol{k}^i\}} - \nu_{\boldsymbol{k}^i})\bar{u}_j^i(v)\right] dv\right| \\
&\quad \le \left|\int_{t-M_2}^t e^{-a_j(t-v)}\left[\sum_{i=1}^p (I_{\{\boldsymbol{k}(\varepsilon,v)=\boldsymbol{k}^i\}} - \nu_{\boldsymbol{k}^i})\bar{u}_j^i(v)\right] dv\right| + \frac{C_1\varepsilon^\delta}{5}.
\end{aligned}
\tag{8.54}
$$

Therefore, it follows from Corollary B.3 and (8.53)–(8.54) that, for $t > 2M_2$ and $j = 1, \ldots, m-1$,

$$
P\left(\bar{x}_j^\varepsilon(t) \le \frac{C_1\varepsilon^\delta}{2}\right) \le \widehat{C}_1 \exp\left\{-M_3\varepsilon^{-(1-2\delta)/2}(1+M_2)^{-3/2}\right\}, \tag{8.55}
$$

for some $\widehat{C}_1 > 0$ and $M_3 > 0$. This implies that

$$P\left(\bar{x}_j^\varepsilon(t) \le 0\right) \le \widehat{C}_1 \exp\left\{ - M_3 \varepsilon^{-(1-2\delta)/2}(1+M_2)^{-3/2}\right\}. \tag{8.56}$$

Based on (8.56), we use the induction argument to construct the desired $\boldsymbol{u}^\varepsilon(t)$ from $\overline{\boldsymbol{u}}^\varepsilon(t)$. First, for $m=2$, let

$$\mathcal{B}_1^\varepsilon = \left\{t : \bar{x}_1^\varepsilon(t) - \inf_{0\le s\le t} \bar{x}_1^\varepsilon(s) = 0 \text{ and } \bar{x}_1^\varepsilon(t) < 0\right\}.$$

Define

$$(u_1^\varepsilon(t), u_2^\varepsilon(t)) = \begin{cases} (\bar{u}_1^\varepsilon(t), \bar{u}_2^\varepsilon(t)), & t \notin \mathcal{B}_1^\varepsilon, \\ (\bar{u}_1^\varepsilon(t) \wedge \bar{u}_2^\varepsilon(t), \bar{u}_1^\varepsilon(t) \wedge \bar{u}_2^\varepsilon(t)), & t \in \mathcal{B}_1^\varepsilon, \end{cases} \tag{8.57}$$

and let

$$x_1^\varepsilon(t) = x_1 e^{-a_1 t} + e^{-a_1 t}\int_0^t e^{a_1 s}\left[u_1^\varepsilon(s) - u_2^\varepsilon(s)\right] ds,$$

$$x_2^\varepsilon(t) = x_2 e^{-a_2 t} + e^{-a_2 t}\int_0^t e^{a_2 s}\left[u_2^\varepsilon(s) - z\right] ds.$$

We know that $(x_1^\varepsilon(t), x_2^\varepsilon(t))' \in \mathcal{X}$, and

$$\begin{aligned} E|\boldsymbol{u}^\varepsilon(t) - \overline{\boldsymbol{u}}^\varepsilon(t)| &= E\left[|\boldsymbol{u}^\varepsilon(t) - \overline{\boldsymbol{u}}^\varepsilon(t)| I_{\{\bar{x}_1^\varepsilon(t)<0\}}\right] \\ &\le \widehat{C}_2 P\left(\bar{x}_1^\varepsilon(t) < 0\right), \end{aligned}$$

for some $\widehat{C}_2 > 0$. Thus, (8.56) implies that there is a control $\boldsymbol{u}^\varepsilon(\cdot)$ such that (8.50) is true and $\boldsymbol{u}^\varepsilon(\cdot) \in \mathcal{A}^\varepsilon(\boldsymbol{x}, \boldsymbol{k}(\varepsilon, 0))$ for $m=2$. At the same time, it follows from (8.57) that

$$u_1^\varepsilon(t) \le \bar{u}_1^\varepsilon(t), \quad u_2^\varepsilon(t) \le \bar{u}_2^\varepsilon(t). \tag{8.58}$$

In order to apply induction, we modify the first buffer. From the case $m=2$, there exist $\widehat{u}_1^\varepsilon(t)$ and $\widehat{u}_2^\varepsilon(t)$ with

$$\widehat{u}_1^\varepsilon(t) \le \bar{u}_1^\varepsilon(t) \quad \text{and} \quad \widehat{u}_2^\varepsilon(t) \le \bar{u}_2^\varepsilon(t), \tag{8.59}$$

such that, for $i = 1,2$,

$$E\left|\widehat{u}_i^\varepsilon(t) - \bar{u}_i^\varepsilon(t)\right| \le \widehat{C}_2 P\left(\bar{x}_1^\varepsilon(t) < 0\right), \tag{8.60}$$

and

$$\widehat{x}_1^\varepsilon(t) = x_1 e^{-a_1 t} + e^{-a_1 t}\int_0^t e^{a_1 s}\left[\widehat{u}_1^\varepsilon(s) - \widehat{u}_2^\varepsilon(s)\right] ds \ge 0. \tag{8.61}$$

Now consider the system

$$\begin{cases} \widehat{u}_j^\varepsilon(t) = \bar{u}_j^\varepsilon(t), \quad j = 3, \ldots, m, \\ \widehat{x}_j^\varepsilon(t) = x_j e^{-a_j t} + \displaystyle\int_0^t e^{-a_j(t-s)} \left[\widehat{u}_j^\varepsilon(s) - \widehat{u}_{j+1}^\varepsilon(s)\right] ds, \\ \qquad\qquad\qquad\qquad\qquad\qquad\qquad\qquad j = 2, \ldots, m-1, \\ \widehat{x}_m^\varepsilon(t) = x_m e^{-a_m t} + \displaystyle\int_0^t e^{-a_m(t-s)} \left[\widehat{u}_m^\varepsilon(s) - z\right] ds. \end{cases} \tag{8.62}$$

We want to use an induction on this system. For this, we need to show that the inequality (8.56) for $\widehat{\boldsymbol{x}}^\varepsilon(t)$ holds. By (8.62), we only need to show that (8.55) holds for $j = 2$. Note that, for $t > 2M_2$, by (8.59),

$$\begin{aligned} \widehat{x}_2^\varepsilon(t) &= \bar{x}_2^\varepsilon(t) + \int_0^t e^{-a_2(t-s)} \left[\widehat{u}_2^\varepsilon(s) - \bar{u}_2^\varepsilon(s)\right] ds \\ &= \bar{x}_2^\varepsilon(t) + \int_0^{t-M_2} e^{-a_2(t-s)} \left[\widehat{u}_2^\varepsilon(s) - \bar{u}_2^\varepsilon(s)\right] ds \\ &\quad + \int_{t-M_2}^t e^{-a_2(t-s)} \left[\widehat{u}_2^\varepsilon(s) - \bar{u}_2^\varepsilon(s)\right] ds \\ &\geq \bar{x}_2^\varepsilon(t) - M_1 \int_0^{t-M_2} e^{-a_2(t-s)} ds \\ &\quad + \int_{t-M_2}^t e^{-a_2(t-s)} \left[\widehat{u}_2^\varepsilon(s) - \bar{u}_2^\varepsilon(s)\right] I_{\{\bar{x}_1^\varepsilon(s)<0\}} ds \\ &\geq \bar{x}_2^\varepsilon(t) - \frac{C_1\varepsilon^\delta}{5} - M_1 \int_{t-M_2}^t e^{-a_2(t-s)} I_{\{\bar{x}_1^\varepsilon(s)<0\}} ds. \end{aligned} \tag{8.63}$$

Thus,

$$\begin{aligned} &P\left(\widehat{x}_2^\varepsilon(t) < \frac{C_1\varepsilon^\delta}{5}\right) \\ &\leq P\left(\bar{x}_2^\varepsilon(t) - M_1 \int_{t-M_2}^t e^{-a_2(t-s)} I_{\{\bar{x}_1^\varepsilon(s)<0\}} ds \leq \frac{2C_1\varepsilon^\delta}{5}\right) \\ &\leq P\left(\bar{x}_2^\varepsilon(t) \leq \frac{C_1\varepsilon^\delta}{2}\right) \\ &\quad + P\left(\bar{x}_2^\varepsilon(t) > \frac{C_1\varepsilon^\delta}{2},\ \bar{x}_2^\varepsilon(t) - M_1 \int_{t-M_2}^t e^{-a_2(t-s)} I_{\{\bar{x}_1^\varepsilon(s)<0\}} ds \leq \frac{2C_1\varepsilon^\delta}{5}\right) \\ &\leq P\left(\bar{x}_2^\varepsilon(t) \leq \frac{C_1\varepsilon^\delta}{2}\right) \\ &\quad + P\left(M_1 \int_{t-M_2}^t e^{-a_2(t-s)} I_{\{\bar{x}_1^\varepsilon(s)<0\}} ds \geq \frac{C_1\varepsilon^\delta}{10}\right). \end{aligned} \tag{8.64}$$

Note that

$$\begin{aligned} &P\left(M_1 \int_{t-M_2}^{t} e^{-a_2(t-s)} I_{\{\bar{x}_1^\varepsilon(s)<0\}}\, ds \geq \frac{C_1\varepsilon^\delta}{10}\right) \\ &\quad \leq \frac{10M_1}{C_1\varepsilon^\delta} E \int_{t-M_2}^{t} e^{-a_2(t-s)} I_{\{\bar{x}_1^\varepsilon(s)<0\}}\, ds \\ &\quad \leq \frac{10M_1}{a_2 C_1 \varepsilon^\delta} \widehat{C}_1 \exp\left\{ - M_3 \varepsilon^{-(1-2\delta)/2} (1+M_2)^{-3/2}\right\}. \end{aligned} \tag{8.65}$$

Combining (8.55), (8.64), and (8.65), we have

$$P\left(\widehat{x}_2^\varepsilon(t) < \frac{C_1\varepsilon^\delta}{5}\right) \leq \widehat{C}_3 \exp\left\{ - M_3 \varepsilon^{-(1-2\overline{\delta})/2} (1+M_2)^{-3/2}\right\} d, \tag{8.66}$$

for some $\widehat{C}_3 > 0$ and $\overline{\delta} \in (0,\ 1/2)$. By induction on $(m-1)$, there exist

$$(u_2^\varepsilon(t), \ldots, u_m^\varepsilon(t))'$$

such that (8.50) holds, and for $j = 2, \ldots, m-1$,

$$x_j^\varepsilon(t) = x_j e^{-a_j t} + \int_0^t e^{-a_j(t-s)} \left[u_j^\varepsilon(s) - u_{j+1}^\varepsilon(s)\right] ds \geq 0. \tag{8.67}$$

Furthermore, for $j = 2, \ldots, m$,

$$u_j^\varepsilon(t) \leq \widehat{u}_j^\varepsilon(t). \tag{8.68}$$

If we let $u_1^\varepsilon(t) = \widehat{u}_1^\varepsilon(t)$, then from (8.61) and (8.68), we have

$$\begin{aligned} x_1^\varepsilon(t) &= x_1 e^{-a_1 t} + \textstyle\int_0^t e^{-a_1(t-s)} \left[u_1^\varepsilon(s) - u_2^\varepsilon(s)\right] ds \\ &\geq x_1 e^{-a_1 t} + \int_0^t e^{-a_1(t-s)} \left[\widehat{u}_1^\varepsilon(s) - \widehat{u}_2^\varepsilon(s)\right] ds \geq 0. \end{aligned} \tag{8.69}$$

Consequently, by combining (8.67) and (8.69), we see that there is a control $\boldsymbol{u}^\varepsilon(\cdot)$ such that (8.50) is true and $\boldsymbol{u}^\varepsilon(\cdot) \in \mathcal{A}^\varepsilon(\boldsymbol{x}, \boldsymbol{k}(\varepsilon, 0))$ holds for m.

We now show that

$$\lambda^\varepsilon \geq \lambda - C\varepsilon^\delta. \tag{8.70}$$

Similar to Lemma 4.2, we can prove that there exists a control $\boldsymbol{u}^\varepsilon(\cdot) \in \mathcal{A}^\varepsilon(\boldsymbol{x}, \boldsymbol{k})$ such that

$$\min_{1\leq j\leq m-1} \inf_{0\leq t<\infty} E[x_j^\varepsilon(t)] \geq \widehat{C}_4 \varepsilon^\delta \tag{8.71}$$

and

$$\limsup_{T\to\infty} \frac{1}{T} E \int_0^T \left[h(\boldsymbol{x}^\varepsilon(t)) + c(\boldsymbol{u}^\varepsilon(t))\right] dt \leq \lambda^\varepsilon + \widehat{C}_5 \varepsilon^{1/2}, \tag{8.72}$$

for some $\widehat{C}_4 > 0$ and $\widehat{C}_5 > 0$, where $\boldsymbol{x}^\varepsilon(t)$ is the state trajectory under the control $\boldsymbol{u}^\varepsilon(t)$. Define

$$\boldsymbol{u}^i(t) = E\left[\boldsymbol{u}^\varepsilon(t)|\boldsymbol{k}(\varepsilon,t) = \boldsymbol{k}^i\right], \quad i = 1,\ldots,p.$$

Then, for $j = 1,\ldots,m-1$,

$$E[x_j^\varepsilon(t)] = x_j e^{-a_j t} + \int_0^t e^{-a_j(t-s)} \left[\sum_{i=1}^p P(\boldsymbol{k}(\varepsilon,s) = \boldsymbol{k}^i) u_j^i(s) - \sum_{i=1}^p P(\boldsymbol{k}(\varepsilon,s) = \boldsymbol{k}^i) u_{j+1}^i(s)\right] ds,$$

and

$$E[x_m^\varepsilon(t)] = x_m e^{-a_m t} + \int_0^t e^{-a_m(t-s)} \left[\sum_{i=1}^p P(\boldsymbol{k}(\varepsilon,s) = \boldsymbol{k}^i) u_m^i(s) - z\right] ds.$$

Let $\boldsymbol{x}(\cdot)$ be the solution of

$$\frac{d}{dt}\boldsymbol{x}(t) = \mathcal{L}\left(\boldsymbol{x}(t), \sum_{i=1}^p \nu_{\boldsymbol{k}^i} \boldsymbol{u}^i(t), z\right), \quad \boldsymbol{x}(0) = \boldsymbol{x}.$$

Similar to (8.46), there is a constant $\widehat{C}_6 > 0$ such that

$$\left|E[x_j^\varepsilon(t)] - x_j(t)\right| \le \widehat{C}_6 \varepsilon^{1/2}, \quad j = 1,\ldots,m.$$

Consequently, it follows from (8.71) that, for sufficiently small ε,

$$(\boldsymbol{u}^1(\cdot),\ldots,\boldsymbol{u}^p(\cdot)) \in \mathcal{A}^0(\boldsymbol{x}).$$

In view of the convexity and the local Lipschitz continuity of $h(\cdot)$, Jensen's inequality and Assumption (A3) yield

$$\begin{aligned} E[h(\boldsymbol{x}^\varepsilon(t))] &\ge h(E[\boldsymbol{x}^\varepsilon(t)]) \\ &= h(\boldsymbol{x}(t)) + [h(E[\boldsymbol{x}^\varepsilon(t)]) - h(\boldsymbol{x}(t))] \\ &\ge h(\boldsymbol{x}(t)) - C_{h3}\left(1 + |E[\boldsymbol{x}^\varepsilon(t))]|^{\beta_{h2}-1} + |\boldsymbol{x}(t)|^{\beta_{h2}-1}\right) \\ &\qquad \times |E[\boldsymbol{x}^\varepsilon(t)] - \boldsymbol{x}(t)| \\ &\ge h(\boldsymbol{x}(t)) - \widehat{C}_7 \varepsilon^{1/2}, \end{aligned} \tag{8.73}$$

for some $\widehat{C}_7 > 0$. In the same way, using Lemma B.3, we can establish

$$\begin{aligned} E[c(\boldsymbol{u}^\varepsilon(t))] &= \sum_{i=1}^{p} P(\boldsymbol{k}(\varepsilon,t)=\boldsymbol{k}^i)E[c(\boldsymbol{u}^\varepsilon(t))|\boldsymbol{k}(\varepsilon,t)=\boldsymbol{k}^i] \\ &\geq \sum_{i=1}^{p} P(\boldsymbol{k}(\varepsilon,t)=\boldsymbol{k}^i)c(\boldsymbol{u}^i(t)) \\ &\geq \sum_{i=1}^{p} \nu_{\boldsymbol{k}^i} c(\boldsymbol{u}^i(t)) - \widehat{C}_8(\varepsilon + e^{-\beta_0 t/\varepsilon}), \end{aligned} \tag{8.74}$$

for some positive constant $\widehat{C}_8$, where β_0 is specified in Lemma B.3. By combining (8.73) and (8.74), we obtain

$$\begin{aligned} &\limsup_{T\to\infty} \frac{1}{T} E \int_0^T [h(\boldsymbol{x}^\varepsilon(t)) + c(\boldsymbol{u}^\varepsilon(t))]\,dt \\ &\geq \limsup_{T\to\infty} \frac{1}{T}\int_0^T \left[h(\overline{\boldsymbol{x}}(t)) + \sum_{i=1}^{p} \nu_{\boldsymbol{k}^i} c(\overline{\boldsymbol{u}}^i(t))\right] dt - \widehat{C}_9\varepsilon, \end{aligned}$$

for some positive constant $\widehat{C}_9$. The inequality (8.72) implies that, for a sufficiently small ε,

$$\lambda^\varepsilon - \lambda \geq -C\varepsilon^{1/2},$$

which completes the proof. □

8.5 Construction of a Near-Optimal Control

In this section, based on the proof of Lemmas 4.1, 4.2, and Theorem 4.1, we give a procedure to construct an asymptotic optimal control in four steps.

Step 1. Pick an ε-optimal control $(\widetilde{\boldsymbol{u}}^1(\cdot),\ldots,\widetilde{\boldsymbol{u}}^p(\cdot)) \in \mathcal{A}^0(\boldsymbol{x})$ for $\mathcal{P}^0$, i.e.,

$$\limsup_{T\to\infty} \frac{1}{T}\int_0^T \left[h(\widetilde{\boldsymbol{x}}(t)) + \sum_{i=1}^{p} \nu_{\boldsymbol{k}^i} c(\widetilde{\boldsymbol{u}}^i(t))\right] dt < \lambda + \varepsilon,$$

where $\widetilde{\boldsymbol{x}}(\cdot)$ is the solution of

$$\frac{d}{dt}\boldsymbol{x}(t) = \mathcal{L}\left(\boldsymbol{x}(t), \sum_{i=1}^{p} \nu_{\boldsymbol{k}^i}\widetilde{\boldsymbol{u}}^i(t), z\right), \quad \boldsymbol{x}(0) = \boldsymbol{x}.$$

For $j = 1,\ldots,m$, let $\mathcal{C}(j) = \{i : k_j^i \neq 0\}$. Furthermore, let

$$M = \min_{1\leq j\leq m}\left\{\sum_{i=1}^{p} \nu_{\boldsymbol{k}^i} k_j^i\right\} \quad \text{and} \quad \widehat{M} = \frac{M}{M - 2\varepsilon^\delta}\cdot\frac{1}{\sum_{i\in\mathcal{C}(1)} \nu_{\boldsymbol{k}^i}}.$$

For a sufficiently small ε, define

$$u_1^i(t) = \begin{cases} \widetilde{u}_1^i(t) \vee (\widehat{M}\varepsilon^\delta), & \text{if } i \in \mathcal{C}(1), \\ 0, & \text{otherwise}, \end{cases}$$

$$u_j^i(t) = \widetilde{u}_j^i(t), \quad i = 1, \ldots, p, \quad j = 2, \ldots, m-1.$$

Then we see that the control

$$(\boldsymbol{u}^1(\cdot), \ldots, \boldsymbol{u}^p(\cdot)) \in \mathcal{A}^0(\boldsymbol{x}).$$

This step will be called *partial pathwise lifting*.

Step 2. Define

$$\widehat{u}_j^i(t) = \frac{u_j^i(t)}{1 + 2\varepsilon^\delta/(M - 2\varepsilon^\delta)}, \quad i = 1, \ldots, p, \quad j = 1, \ldots, m.$$

Then we can see that the control

$$(\widehat{\boldsymbol{u}}^1(\cdot), \ldots, \widehat{\boldsymbol{u}}^p(\cdot)) \in \mathcal{A}^0\left(\frac{\boldsymbol{x}}{1 + 2\varepsilon^\delta/(M - 2\varepsilon^\delta)}\right).$$

This step will be called *pathwise shrinking*.

Step 3. Choose $(v_1^i(t), \ldots, v_m^i(t))'(\geq 0)$, $i = 1, \ldots, p$. For $j = 1, \ldots, m$ and $i \notin \mathcal{C}(j)$, set $v_j^i(t) = 0$. For $j = 1, \ldots, m$ and $i \in \mathcal{C}(j)$, if

$$\min_{1 \leq j \leq m} \left\{ \sum_{i=1}^p \nu_{\boldsymbol{k}^i} \widehat{u}_j^i(t) \right\} \geq \varepsilon^\delta,$$

then set $v_j^i(t) = 0$. If, on the other hand,

$$\min_{1 \leq j \leq m} \left\{ \sum_{i=1}^p \nu_{\boldsymbol{k}^i} \widehat{u}_j^i(t) \right\} < \varepsilon^\delta,$$

then set

$$\nu_{\boldsymbol{k}^i} v_j^i(t) = \begin{cases} 0, & \text{if } i \neq l(j), \\ \varepsilon, & \text{if } i = l(j), \end{cases}$$

where $l(j)$ is an index such that $k_j^{l(j)} > 0$ $(j = 1, \ldots, m)$.

Define

$$\overline{u}_j^i(t) = \widehat{u}_j^i(t) + v_j^i(t), \quad i = 1, \ldots, p \text{ and } j = 1, \ldots, m,$$

and

$$(\overline{\boldsymbol{u}}^1(t), \ldots, \overline{\boldsymbol{u}}^p(t)) = ((\overline{u}_1^1(t), \ldots, \overline{u}_m^1(t))', \ldots, (\overline{u}_1^p(t), \ldots, \overline{u}_m^p(t))').$$

We choose $(\check{\boldsymbol{u}}^1(t), \ldots, \check{\boldsymbol{u}}^p(t))$ such that

$$\sum_{i=1}^{p} \nu_{\boldsymbol{k}^i} \check{u}_j^i(t) = \sum_{i=1}^{p} \nu_{\boldsymbol{k}^i} \widetilde{u}_j^i(t) + \frac{\varepsilon^\delta}{j}, \qquad j = 1, \ldots, m.$$

This step will be called *total pathwise lifting.*

Step 4. Set

$$\check{\boldsymbol{u}}^\varepsilon(t) = (\check{u}_1^\varepsilon(t), \ldots, \check{u}_m^\varepsilon) = \sum_{i=1}^{p} I_{\{\boldsymbol{k}(\varepsilon,t)=\boldsymbol{k}^i\}} \check{\boldsymbol{u}}^i(t),$$

and

$$\check{x}_1^{\varepsilon,1}(t) = (x_1 + \varepsilon^\delta) e^{-a_1 t} + \int_0^t e^{-a_1(t-s)} [\check{u}_1^\varepsilon(s) - \check{u}_2^\varepsilon(s)]\, ds.$$

Define

$$\mathcal{B}_1^\varepsilon = \left\{ t \geq 0 : \check{x}_1^{\varepsilon,1}(t) - \inf_{0 \leq s \leq t} \check{x}_1^{\varepsilon,1}(s) = 0 \text{ and } \check{x}_1^{\varepsilon,1}(t) < 0 \right\}.$$

Let

$$(u_1^{\varepsilon,1}(t), u_2^{\varepsilon,1}(t)) = \begin{cases} (\check{u}_1^\varepsilon(t), \check{u}_2^\varepsilon(t)), & \text{if } t \notin \mathcal{B}_1^\varepsilon, \\ (\check{u}_1^\varepsilon(t) \wedge \check{u}_2^\varepsilon(t), \check{u}_1^\varepsilon(t) \wedge \check{u}_2^\varepsilon(t)), & \text{if } t \in \mathcal{B}_1^\varepsilon. \end{cases} \tag{8.75}$$

Set

$$\check{x}_2^{\varepsilon,2}(t) = (x_2 + \varepsilon^\delta) e^{-a_2 t} + \int_0^t e^{-a_2(t-s)} [u_2^{\varepsilon,1}(s) - \check{u}_3^\varepsilon(s)]\, ds.$$

Define

$$\mathcal{B}_2^\varepsilon = \left\{ t \geq 0 : \check{x}_2^{\varepsilon,2}(t) - \inf_{0 \leq s \leq t} \check{x}_2^{\varepsilon,2}(s) = 0 \text{ and } \check{x}_2^{\varepsilon,2}(t) < 0 \right\}.$$

Let

$$(u_2^{\varepsilon,2}(t), u_3^{\varepsilon,2}(t)) = \begin{cases} (u_2^{\varepsilon,2}(t), \check{u}_3^\varepsilon(t)), & \text{if } t \notin \mathcal{B}_2^\varepsilon, \\ (u_2^{\varepsilon,2}(t) \wedge \check{u}_3^\varepsilon(t), u_2^{\varepsilon,2}(t) \wedge \check{u}_3^\varepsilon(t)), & \text{if } t \in \mathcal{B}_1^\varepsilon. \end{cases} \tag{8.76}$$

Sub-step ℓ ($\ell = 2, \ldots, m-1$): Set

$$\check{x}_\ell^{\varepsilon,\ell}(t) = (x_\ell + \varepsilon^\delta) e^{-a_\ell t} + \int_0^t e^{-a_\ell(t-s)} [u_\ell^{\varepsilon,\ell-1}(s) - \check{u}_{\ell+1}^\varepsilon(s)]\, ds.$$

Define

$$\mathcal{B}_\ell^\varepsilon = \left\{ t \geq 0 : \check{x}_\ell^{\varepsilon,\ell}(t) - \inf_{0 \leq s \leq t} \check{x}_\ell^{\varepsilon,\ell}(s) = 0 \text{ and } \check{x}_\ell^{\varepsilon,\ell}(t) < 0 \right\}.$$

Let

$$(u_\ell^{\varepsilon,\ell}(t), u_{\ell+1}^{\varepsilon,\ell}(t)) = \begin{cases} (u_\ell^{\varepsilon,\ell}(t), \check{u}_{\ell+1}^{\varepsilon}(t)), & \text{if } t \notin \mathcal{B}_\ell^\varepsilon, \\ (u_\ell^{\varepsilon,\ell}(t) \wedge \check{u}_{\ell+1}^{\varepsilon}(t), u_\ell^{\varepsilon,\ell}(t) \wedge \check{u}_{\ell+1}^{\varepsilon}(t)), & \text{if } t \in \mathcal{B}_\ell^\varepsilon. \end{cases} \tag{8.77}$$

Then we get $\boldsymbol{u}^\varepsilon(t) = (u_1^{\varepsilon,1}(t), \ldots, u_{m-1}^{\varepsilon,m-1}(t), u_m^{\varepsilon,m-1}(t))$.

8.6 Notes

This chapter is based on Sethi, Zhang, and Zhang [122]. The idea of partial path lifting and path shrinking used in Section 7.4 is first used in Sethi, Zhang, and Zhang [120]; see also Sethi, Zhang, and Zhou [126].

Sometimes, when buffers are of small sizes, we may need to impose upper bounds on inventory levels. Sethi, Zhang, and Zhang [124] and Fong and Zhou [59] treat this situation with a finite internal buffer. The extension involves an additional procedure termed *squeezing*. This is achieved by rescaling both the time and the space.

Perkins and Kumar [97] consider the problem of a deterministic dynamic m-machine flowshop with reliable machines. They use undiscounted linear inventory and backlog costs over the infinite horizon. They reduce the optimization problem to a set of quadratic programming problems under the assumption that the inventory costs are nondecreasing along the route of production, and obtain explicitly the form of the optimal input rate at each of the machines. Their problem is a special case of the limiting flowshop problem of this chapter.

9

Near-Optimal Controls of Dynamic Jobshops

9.1 Introduction

In this chapter we revisit the jobshops discussed in Chapter 5. The problem is to determine rates of production to meet demand for finished products at the minimum long-run average cost of production, inventories, and backlogs.

As in Chapter 8, the main difficulty is how to construct an *admissible* control for the original problem from a near-optimal control of the limiting problem in a way which still guarantees asymptotic optimality. To overcome the difficulty, we still use the method of "lifting" and "shrinking." We show that the resulting control constructed in this manner is nearly optimal with an error bound of order ε^δ for $0 < \delta < 1/2$.

The chapter is organized as follows. In Section 9.2 we give a mathematical description of the problem. In Section 9.3 we discuss some elementary properties of the associated minimum long-run average cost. In Section 9.4 we consider the convergence rate of the minimum average cost. In Section 9.5 we construct asymptotic optimal open-loop controls for the original problem. Finally, Section 9.6 concludes the chapter.

9.2 The Optimization Problem

Consider a jobshop $(V, A, \mathcal{K})$ as defined in Chapter 5. Let $u^\varepsilon_{i,j}(t)$ be a control at time t associated with arc (i, j), $(i, j) \in A$ and $j \neq n+1$. Suppose we are given a stochastic process $\boldsymbol{k}(\varepsilon, t) = (k_1(\varepsilon, t), ..., k_{m_c}(\varepsilon, t))$ on a probability space $(\Omega, \mathcal{F}, P)$ with $k_j(\varepsilon, t)$ representing the capacity of the jth machine at time t, $j = 1, ..., m_c$, where ε is a small parameter to be precisely specified later. The state space of $\boldsymbol{k}(\varepsilon, t)$ is

$$\mathcal{M} = \{\boldsymbol{k}^1, \dots, \boldsymbol{k}^p\}, \quad \boldsymbol{k}^i = (k_1^i, \dots, k_{m_c}^i), \quad i = 1, \dots, p.$$

The controls $u_{i,j}^{\varepsilon}(t)$ with $(i,j) \in K_\ell$ and $j \neq n+1$ must satisfy the following constraints:

$$0 \leq \sum_{(i,j) \in K_\ell} u_{i,j}^{\varepsilon}(t) \leq k_\ell(\varepsilon, t) \quad \text{for all} \quad t \geq 0, \quad \ell = 1, ..., m_c, \tag{9.1}$$

where we have assumed that the required machine capacity r_{ij} (for unit production rate of type j from part type i) equals 1, for convenience in exposition. The analysis in this chapter can be readily extended to the case when the required machine capacity for the unit production rate of part j from part i is any given positive constant. In this chapter, we still use notation introduced in Section 5.2. In what follows, we will set

$$u_{i,j}(t) = 0, \quad \text{for } (i,j) \notin A, \quad -n_0 + 1 \leq i \leq m, \quad 1 \leq j \leq n.$$

Let

$$\mathcal{U} = \left\{(u_{i,j})_{(n_0+m) \times n} : u_{i,j} \geq 0, \ u_{i,j} = 0 \text{ and } (i,j) \notin A\right\}.$$

In order to simplify expressions, we introduce the following notation. For $\ell = -n_0 + 1, \dots, 0$ and $j = 1, \dots, m$, let

$$\boldsymbol{u}_\ell = \begin{pmatrix} u_{\ell,1} \\ \vdots \\ u_{\ell,n} \end{pmatrix}, \quad \boldsymbol{u}_j = \begin{pmatrix} u_{j,j+1} \\ \vdots \\ u_{j,n} \end{pmatrix}, \tag{9.2}$$

$$\boldsymbol{u} = \begin{pmatrix} \boldsymbol{u}_{-n_0+1} \\ \vdots \\ \boldsymbol{u}_m \end{pmatrix}, \quad \text{and} \quad \boldsymbol{z} = \begin{pmatrix} z_{m+1} \\ \vdots \\ z_n \end{pmatrix}. \tag{9.3}$$

Note that $\boldsymbol{u}$ is a $(n_0 n + \sum_{\ell=1}^{m}(n - \ell))$-dimensional vector. Let $\widehat{\mathcal{U}}$ be the set of all nonnegative vectors given by (9.2) and (9.3). It follows from the definition of $\mathcal{U}$ that there is a one-to-one mapping between $\mathcal{U}$ and $\widehat{\mathcal{U}}$. In the following, to represent the control variables, it is convenient to use vectors in $\widehat{\mathcal{U}}$ rather than matrices in $\mathcal{U}$. Furthermore, for $\boldsymbol{k} = (k_1, ..., k_{m_c})$, let

$$\mathcal{U}(\boldsymbol{k}) = \left\{ \boldsymbol{u} : \boldsymbol{u} \in \widehat{\mathcal{U}}, \ 0 \leq \sum_{(i,j) \in K_\ell} u_{i,j} \leq k_\ell, \ \ell = 1, \dots, m_c \right\},$$

and for $\boldsymbol{x} \in \mathcal{Y} = \Re_+^m \times \Re^{n-m}$ and $\boldsymbol{k} = (k_1, ..., k_{m_c})$,

$$\mathcal{U}(\boldsymbol{x}, \boldsymbol{k}) = \Bigg\{ \boldsymbol{u} : \boldsymbol{u} \in \mathcal{U}(\boldsymbol{k}) \text{ and } x_\ell = 0 \\ \Rightarrow \sum_{i=-n_0+1}^{\ell-1} u_{i,\ell} - \sum_{i=\ell+1}^{n} u_{\ell,i} \geq 0, \ \ell = 1, ..., m \Bigg\}.$$

We denote the surplus at time t in buffer j by $x_j^\varepsilon(t)$, $j = 1, \ldots, n$. Note that if $x_j^\varepsilon(t) > 0$, $j = 1, ..., n$, we have an inventory in buffer j, and if $x_j^\varepsilon(t) < 0$, $j = m+1, ..., n$, we have a shortage of the finished product j. The dynamics of the system are

$$\begin{cases} \dfrac{d}{dt} x_j^\varepsilon(t) = -a_j x_j^\varepsilon(t) + \left(\displaystyle\sum_{\ell=-n_0+1}^{j-1} u_{\ell,j}^\varepsilon(t) - \sum_{\ell=j+1}^{n} u_{j,\ell}^\varepsilon(t) \right), & 1 \leq j \leq m, \\ \dfrac{d}{dt} x_j^\varepsilon(t) = -a_j x_j^\varepsilon(t) + \left(\displaystyle\sum_{\ell=-n_0+1}^{m} u_{\ell,j}^\varepsilon(t) - z_j \right), & m+1 \leq j \leq n, \end{cases} \tag{9.4}$$

with $\boldsymbol{x}^\varepsilon(0) = (x_1^\varepsilon(0), ..., x_n^\varepsilon(0))' = (x_1, ..., x_n)' = \boldsymbol{x}$, where $a_j > 0$, $j = 1, ..., n$, are constants. The attrition rate a_j has the same physical meaning as in the case of parallel machine systems treated in Section 7.2. Here we assume symmetric deterioration and cancellation rates for finished goods only for convenience in exposition. It is easy to extend our results when the deterioration rates and the order cancellation rates are different (see Chapter 7). On account of (9.2) and (9.3), the control policy $(u_{i,j}(t))$ appearing in (9.4) can be expresed as the vector $\boldsymbol{u}(t)$.

The state constraints are

$$\begin{cases} x_j^\varepsilon(t) \geq 0, \quad t \geq 0, \quad j = 1, ..., m, \\ -\infty < x_j^\varepsilon(t) < +\infty, \quad t \geq 0, \quad j = m+1, ..., n. \end{cases} \tag{9.5}$$

We write the relation (9.4) in the following vector form:

$$\frac{d}{dt} \boldsymbol{x}^\varepsilon(t) = -\text{diag}(\boldsymbol{a}) \boldsymbol{x}^\varepsilon(t) + (A_{-n_0+1}, \ldots, A_{m+1}) \begin{pmatrix} \boldsymbol{u}^\varepsilon(t) \\ \boldsymbol{z} \end{pmatrix}, \tag{9.6}$$

with the initial condition $\boldsymbol{x}^\varepsilon(0) = \boldsymbol{x}$, where $\boldsymbol{a} = (a_1, ..., a_n)'$, $\text{diag}(\boldsymbol{a}) = \text{diag}(a_1, ..., a_n)$, and A_ℓ $(-n_0 + 1 \leq \ell \leq m+1)$ are defined in Section 5.2. Let

$$\mathcal{L}(\boldsymbol{x}^\varepsilon(t), \boldsymbol{u}^\varepsilon(t), \boldsymbol{z}) = -\text{diag}(\boldsymbol{a}) \boldsymbol{x}^\varepsilon(t) + (A_{-n_0+1}, \ldots, A_{m+1}) \begin{pmatrix} \boldsymbol{u}^\varepsilon(t) \\ \boldsymbol{z} \end{pmatrix}.$$

Then (9.6) is changed to

$$\frac{d}{dt} \boldsymbol{x}^\varepsilon(t) = \mathcal{L}(\boldsymbol{x}^\varepsilon(t), \boldsymbol{u}^\varepsilon(t), \boldsymbol{z}), \quad \boldsymbol{x}^\varepsilon(0) = \boldsymbol{x}. \tag{9.7}$$

We are now in a position to formulate our stochastic optimal control problem for the jobshop defined by (9.1), (9.4), and (9.5).

Definition 2.1. We say that a control $\boldsymbol{u}^\varepsilon(\cdot) \in \widehat{\mathcal{U}}$ is admissible with respect to the initial state vector $\boldsymbol{x} = (x_1, \ldots, x_n)' \in \mathcal{Y}$ and $\boldsymbol{k}(\varepsilon, 0) \in \mathcal{M}$, if:

(i) $\boldsymbol{u}^\varepsilon(\cdot)$ is an $\mathcal{F}_t^\varepsilon$-adapted measurable process with $\mathcal{F}_t^\varepsilon = \sigma\{\boldsymbol{k}(\varepsilon, s) : 0 \leq s \leq t\}$;

(ii) $\boldsymbol{u}^\varepsilon(t) \in \mathcal{U}(\boldsymbol{k}(\varepsilon, t))$ for all $t \geq 0$; and

(iii) the solution of

$$\frac{d}{dt}\boldsymbol{x}^\varepsilon(t) = \mathcal{L}(\boldsymbol{x}^\varepsilon(t), \boldsymbol{u}^\varepsilon(t), \boldsymbol{z}), \quad \boldsymbol{x}^\varepsilon(0) = \boldsymbol{x},$$

satisfies $\boldsymbol{x}^\varepsilon(t) = (x_1^\varepsilon(t), \ldots, x_n^\varepsilon(t))' \in \mathcal{Y}$ for all $t \geq 0$. □

Clearly, the admissibility of $\boldsymbol{u}^\varepsilon(\cdot)$ is also dependent on the initial conditions of the corresponding state process $\boldsymbol{x}^\varepsilon(\cdot)$ and the capacity process $\boldsymbol{k}(\varepsilon, \cdot)$. Let $\mathcal{A}^\varepsilon(\boldsymbol{x}, \boldsymbol{k})$ denote the set of all admissible controls with initial conditions $\boldsymbol{x}^\varepsilon(0) = \boldsymbol{x}$ and $\boldsymbol{k}(\varepsilon, 0) = \boldsymbol{k}$.

The problem is to find an admissible control $\boldsymbol{u}(\cdot) \in \mathcal{A}^\varepsilon(\boldsymbol{x}, \boldsymbol{k})$ that minimizes the long-run average cost

$$J^\varepsilon(\boldsymbol{x}, \boldsymbol{k}, \boldsymbol{u}(\cdot)) = \limsup_{T \to \infty} \frac{1}{T} E \int_0^T [h(\boldsymbol{x}(t)) + c(\boldsymbol{u}(t))]\, dt, \tag{9.8}$$

where $h(\cdot)$ defines the cost of inventory/shortage, $c(\cdot)$ is the production cost, $\boldsymbol{x}(\cdot)$ is the surplus process corresponding to $\boldsymbol{u}(\cdot)$ with $\boldsymbol{x}(0) = \boldsymbol{x}$, and $\boldsymbol{k}$ is the initial value of $\boldsymbol{k}(\varepsilon, t)$.

We impose the following assumptions on the random process $\boldsymbol{k}(\varepsilon, t)$ and the cost functions $h(\cdot)$ and $c(\cdot)$ throughout this chapter.

(A1) The capacity process $\boldsymbol{k}(\varepsilon, t) \in \mathcal{M}$ is a finite state Markov chain with the infinitesimal generator

$$Q^\varepsilon = Q^{(1)} + \frac{1}{\varepsilon} Q^{(2)},$$

where $Q^{(1)} = (q^{(1)}_{\boldsymbol{k}^i \boldsymbol{k}^j})$ and $Q^{(2)} = (q^{(2)}_{\boldsymbol{k}^i \boldsymbol{k}^j})$ are matrices such that $q^{(\ell)}_{\boldsymbol{k}^i \boldsymbol{k}^j} \geq 0$ if $\boldsymbol{k}^i \neq \boldsymbol{k}^j$, and $q^{(\ell)}_{\boldsymbol{k}^i \boldsymbol{k}^i} = -\sum_{j \neq i} q^{(\ell)}_{\boldsymbol{k}^i \boldsymbol{k}^j}$ for $\ell = 1, 2$. Assume that $Q^{(2)}$ is strongly irreducible. Let $\boldsymbol{\nu} = (\nu_{\boldsymbol{k}^1}, \ldots, \nu_{\boldsymbol{k}^p})$ denote the equilibrium distribution of $Q^{(2)}$. That is, $\boldsymbol{\nu}$ is the only nonnegative solution to the equation

$$\boldsymbol{\nu} Q^{(2)} = 0 \quad \text{and} \quad \sum_{i=1}^{p} \nu_{\boldsymbol{k}^i} = 1. \tag{9.9}$$

(A2) Let $p_\ell = \sum_{i=1}^{p} k_\ell^i \nu_{\boldsymbol{k}^i}$ and $c(i,j) = \ell$ if $(i,j) \in K_\ell$ for $(i,j) \in A$. Here p_ℓ represents the average capacity of machine ℓ, and $c(i,j)$ is the machine number placed on the arc (i,j). Assume that there exist $\{p_{ij} > 0 : (i,j) \in K_\ell\}$, $\ell = 1, ..., m_c$, such that

$$\begin{aligned} \sum_{(i,j)\in K_\ell} p_{ij} &\le 1, \\ \sum_{i=-n_0+1}^{m} p_{ij} p_{c(i,j)} &> z_j, \quad j = m+1, \ldots, n, \end{aligned} \tag{9.10}$$

and

$$\sum_{i=-n_0+1}^{j-1} p_{ij} p_{c(i,j)} > \sum_{i=j+1}^{n} p_{ji} p_{c(j,i)}, \quad j = 1, \ldots, m. \tag{9.11}$$

(A3) $h(\boldsymbol{x})$ is a nonnegative, convex function with $h(0) = 0$. There are positive constants C_{h1}, C_{h2}, and $\beta_{h1} \ge 1$ such that

$$h(\boldsymbol{x}) \ge C_{h1}|\boldsymbol{x}|^{\beta_{h1}} - C_{h2}, \quad \boldsymbol{x} \in \mathcal{Y}.$$

Moreover, there are constants C_{h3} and $\beta_{h2} \ge \beta_{h1}$ such that

$$|h(\boldsymbol{x}) - h(\boldsymbol{y})| \le C_{h3}(1 + |\boldsymbol{x}|^{\beta_{h2}-1} + |\boldsymbol{y}|^{\beta_{h2}-1})|\boldsymbol{x} - \boldsymbol{y}|, \quad \boldsymbol{x}, \boldsymbol{y} \in \mathcal{Y}.$$

(A4) $c(\boldsymbol{u})$ is a nonnegative convex function.

Let $\lambda^\varepsilon(\boldsymbol{x}, \boldsymbol{k})$ denote the minimal expected cost, i.e.,

$$\lambda^\varepsilon(\boldsymbol{x}, \boldsymbol{k}) = \inf_{\boldsymbol{u}(\cdot)\in\mathcal{A}^\varepsilon(\boldsymbol{x},\boldsymbol{k})} J^\varepsilon(\boldsymbol{x}, \boldsymbol{k}, \boldsymbol{u}(\cdot)). \tag{9.12}$$

Since we are using a long-run average-cost criterion, we know from Chapter 5 and Assumptions (A1) and (A2) that $\lambda^\varepsilon(\boldsymbol{x}, \boldsymbol{k})$ is independent of the initial condition $(\boldsymbol{x}, \boldsymbol{k})$. Thus, we will use the notation λ^ε instead of $\lambda^\varepsilon(\boldsymbol{x}, \boldsymbol{k})$. We use $\mathcal{P}^\varepsilon$ to denote our control problem, i.e.,

$$\mathcal{P}^\varepsilon : \begin{cases} \min\ J^\varepsilon(\boldsymbol{x}, \boldsymbol{k}, \boldsymbol{u}(\cdot)) = \limsup\limits_{T\to\infty} \dfrac{1}{T} E \displaystyle\int_0^T [h(\boldsymbol{x}(t)) + c(\boldsymbol{u}(t))]\, dt, \\ \text{s.t. } \dfrac{d}{dt}\boldsymbol{x}(t) = \mathcal{L}(\boldsymbol{x}(t), \boldsymbol{u}(t), \boldsymbol{z}), \quad \boldsymbol{x}(0) = \boldsymbol{x}, \quad \boldsymbol{u}(\cdot) \in \mathcal{A}^\varepsilon(\boldsymbol{x}, \boldsymbol{k}), \\ \text{minimum average cost } \lambda^\varepsilon = \inf\limits_{\boldsymbol{u}(\cdot)\in\mathcal{A}^\varepsilon(\boldsymbol{x},\boldsymbol{k})} J^\varepsilon(\boldsymbol{x}, \boldsymbol{k}, \boldsymbol{u}(\cdot)). \end{cases} \tag{9.13}$$

As in Section 7.2, the positive attrition rate $\boldsymbol{a}$ implies a uniform bound on the process $\boldsymbol{x}^\varepsilon(t)$, $t \ge 0$. Formally, in view of the fact that the control

$\boldsymbol{u}^\varepsilon(\cdot)$ is bounded between 0 and $(\max\{|\boldsymbol{k}^i|, 1 \le i \le p\})$, it follows that any solution $\boldsymbol{x}^\varepsilon(\cdot)$ of (9.4) with the initial condition $\boldsymbol{x}^\varepsilon(0) = \boldsymbol{x}$ must satisfy, for $j = 1, \ldots, m$,

$$\begin{aligned} |x_j^\varepsilon(t)| &= \left| x_j e^{-a_j t} + e^{-a_j t} \int_0^t e^{a_j s} \left(\sum_{\ell=-n_0+1}^{j-1} u_{\ell,j}^\varepsilon(s) - \sum_{\ell=j+1}^{n} u_{j,\ell}^\varepsilon(s) \right) ds \right| \\ &\le |x_j| e^{-a_j t} + \left(m_c \cdot \max_{1 \le i \le p} \{|\boldsymbol{k}^i|\} \right) \int_0^t e^{-a_j(t-s)}\, ds \\ &\le |x_j| e^{-a_j t} + \frac{m_c \cdot \max_{1 \le i \le p}\{|\boldsymbol{k}^i|\}}{a_j}. \end{aligned} \tag{9.14}$$

Similarly, for $j = m+1, \ldots, n$,

$$\begin{aligned} |x_j^\varepsilon(t)| &= \left| x_j e^{-a_j t} + e^{-a_j t} \int_0^t e^{a_j s} \left(\sum_{\ell=-n_0+1}^{m} u_{\ell j}^\varepsilon(s) - z_j \right) ds \right| \\ &\le |x_j| e^{-a_j t} + \frac{m_c \cdot \max_{1 \le i \le p}\{|\boldsymbol{k}^i|\} + \max_{m+1 \le \ell \le n}\{z_\ell\}}{a_j}. \end{aligned} \tag{9.15}$$

Thus, under the positive deterioration/cancellation rate, the surplus process $\boldsymbol{x}^\varepsilon(t)$, $t \ge 0$, remains bounded.

9.3 The Limiting Control Problem

In this section, we examine elementary properties of the minimum average cost and the potential function, and obtain the limiting control problem as $\varepsilon \to 0$.

The Hamilton-Jacobi-Bellman equation in the directional derivative sense (HJBDD) for the average-cost optimal control problem $\mathcal{P}^\varepsilon$ takes the form

$$\begin{aligned} \lambda^\varepsilon = & \inf_{\boldsymbol{u} \in \mathcal{U}(\boldsymbol{x},\boldsymbol{k})} \left\{ \partial_{\mathcal{L}(\boldsymbol{x},\boldsymbol{u},\boldsymbol{z})} \phi^\varepsilon(\boldsymbol{x},\boldsymbol{k}) + c(\boldsymbol{u}) \right\} \\ & + h(\boldsymbol{x}) + \left(Q^{(1)} + \frac{1}{\varepsilon} Q^{(2)} \right) \phi^\varepsilon(\boldsymbol{x},\cdot)(\boldsymbol{k}), \end{aligned} \tag{9.16}$$

where $\phi^\varepsilon(\cdot,\cdot)$ is a function defined on $\mathcal{Y} \times \mathcal{M}$. Our analysis begins with the proof of the boundedness of λ^ε.

Theorem 3.1. *Let Assumptions* (A1)–(A4) *hold. The minimum long-run average expected cost* λ^ε *of* $\mathcal{P}^\varepsilon$ *is bounded in* ε, *i.e., there exists a constant* $C > 0$ *such that*

$$0 \le \lambda^\varepsilon \le C, \quad \text{for all } \varepsilon > 0.$$

Proof. According to the definition of λ^ε, it suffices to show that

$$\limsup_{T\to\infty} \frac{1}{T} E \int_0^T [h(\boldsymbol{x}^\varepsilon(t)) + c(\boldsymbol{u}^\varepsilon(t))]\, dt \leq C_1, \tag{9.17}$$

for some $C_1 > 0$, where $\boldsymbol{u}^\varepsilon(t) \in \mathcal{A}^\varepsilon(0, \boldsymbol{k})$ and $\boldsymbol{x}^\varepsilon(t)$ is the solution of

$$\frac{d}{dt}\boldsymbol{x}^\varepsilon(t) = \mathcal{L}\left(\boldsymbol{x}^\varepsilon(t), \boldsymbol{u}^\varepsilon(t), \boldsymbol{z}\right), \quad \boldsymbol{x}^\varepsilon(0) = 0.$$

In view of (9.14) and (9.15), for $j = 1, \ldots, n$,

$$|x_j^\varepsilon(t)| \leq \frac{\max\{z_\ell : m+1 \leq \ell \leq n\} + m_c \cdot \max\{|\boldsymbol{k}^i|, 1 \leq i \leq p\}}{a_j}. \tag{9.18}$$

It is clear under Assumptions (A3) and (A4) that the functions $h(\cdot)$ and $c(\cdot)$ are continuous. Consequently, $\boldsymbol{u}^\varepsilon(t) \geq 0$, $|\boldsymbol{u}^\varepsilon(t)| \leq \max\{|\boldsymbol{k}^i|, 1 \leq i \leq p\}$, and inequality (9.18) imply (9.17). □

Next, we derive the limiting control problem as $\varepsilon \to 0$. As the rates of the machine breakdown and repair approach infinity, the problem $\mathcal{P}^\varepsilon$, which is termed the *original control problem*, can be approximated by a simpler problem called the *limiting control problem*, where the stochastic machine capacity process $\boldsymbol{k}(\varepsilon, t)$ is replaced by a weighted form. The limiting control problem can be formulated as follows.

Consider an augmented control

$$U(\cdot) = (\boldsymbol{u}^1(\cdot), \ldots, \boldsymbol{u}^p(\cdot)),$$

where $\boldsymbol{u}^i(t) \in \mathcal{U}(\boldsymbol{k}^i)$ and $\boldsymbol{u}^i(t)$, $t \geq 0$, is a deterministic process.

Definition 3.1. For $\boldsymbol{x} \in \mathcal{Y}$, let $\mathcal{A}^0(\boldsymbol{x})$ denote the set of measurable controls $U(\cdot) = (\boldsymbol{u}^1(\cdot), \ldots, \boldsymbol{u}^p(\cdot))$ such that the solution of

$$\frac{d}{dt}\boldsymbol{x}(t) = \mathcal{L}\left(\boldsymbol{x}(t), \sum_{i=1}^p \nu_{\boldsymbol{k}^i} \boldsymbol{u}^i(t), \boldsymbol{z}\right), \quad \boldsymbol{x}(0) = \boldsymbol{x},$$

satisfies $\boldsymbol{x}(t) \in \mathcal{Y}$ for all $t \geq 0$. □

The objective of the problem is to choose a control $U(\cdot) \in \mathcal{A}^0(\boldsymbol{x})$ that minimizes

$$J(\boldsymbol{x}, U(\cdot)) = \limsup_{T\to\infty} \frac{1}{T} \int_0^T \left[h(\boldsymbol{x}(s)) + \sum_{i=1}^p \nu_{\boldsymbol{k}^i} c(\boldsymbol{u}^i(s)) \right] ds.$$

We use $\mathcal{P}^0$ to denote this problem, known as the limiting control problem, and restate it as follows:

$$\mathcal{P}^0: \begin{cases} \min J(\boldsymbol{x}, U(\cdot)) = \limsup_{T\to\infty} \frac{1}{T}\int_0^T [h(\boldsymbol{x}(s)) + \sum_{i=1}^p \nu_{\boldsymbol{k}^i} c(\boldsymbol{u}^i(s))]\, ds, \\ \text{s.t. } \frac{d}{dt}\boldsymbol{x}(t) = \mathcal{L}\left(\boldsymbol{x}(t), \sum_{i=1}^p \nu_{\boldsymbol{k}^i} \boldsymbol{u}^i(t), \boldsymbol{z}\right), \\ \boldsymbol{x}(0) = \boldsymbol{x}, \quad U(\cdot) \in \mathcal{A}^0(\boldsymbol{x}), \\ \text{minimum average cost } \lambda = \inf_{U(\cdot)\in\mathcal{A}^0(\boldsymbol{x})} J(\boldsymbol{x}, U(\cdot)). \end{cases} \tag{9.19}$$

The HJBDD equation associated with $\mathcal{P}^0$ is

$$\lambda = \inf_{U\in\mathcal{U}^0(\boldsymbol{x})} \left\{ \partial_{\mathcal{L}(\boldsymbol{x}, \sum_{i=1}^p \nu_{\boldsymbol{k}^i}\boldsymbol{u}^i, \boldsymbol{z})}\phi(\boldsymbol{x}) + \sum_{i=1}^p \nu_{\boldsymbol{k}^i} c(\boldsymbol{u}^i) \right\} + h(\boldsymbol{x}), \tag{9.20}$$

where $\phi(\cdot)$ is a function defined on $\mathcal{Y}$, and

$$\mathcal{U}^0(\boldsymbol{x}) = \Big\{ (\boldsymbol{u}^1, \ldots, \boldsymbol{u}^p) \; : \; \boldsymbol{u}^i \in \mathcal{U}(\boldsymbol{k}^i) \text{ and } x_j = 0 \ (1 \le j \le m) \Rightarrow \sum_{i=1}^p \sum_{\ell=-n_0+1}^{j-1} u^i_{\ell,j} - \sum_{i=1}^p \sum_{\ell=j+1}^{n} u^i_{j,\ell} \ge 0 \Big\}.$$

9.4 Convergence of the Minimum Average Expected Cost

In this section we consider the convergence of the minimum average expected cost λ^ε as ε goes to zero, and establish its convergence rate. In order to get the required convergence result, we need the following auxiliary result, which is a key to obtaining our main result.

Lemma 4.1. *Let Assumptions* (A3) *and* (A4) *hold. For* $\delta \in [0,\ 1/2)$, *there exist an* $\varepsilon_0 > 0$, $C > 0$, $\boldsymbol{x} = (x_1, ..., x_n)' \in \mathcal{Y}$, *such that for each given* $\varepsilon \in (0,\ \varepsilon_0]$, *we can find a control*

$$(\overline{\boldsymbol{u}}^1(\cdot), ..., \overline{\boldsymbol{u}}^p(\cdot)) \in \mathcal{A}^0(\boldsymbol{x})$$

satisfying, for $j = 1, ..., m_c$ *and* $i = 1, ..., p$,

$$\sum_{(\ell_1,\ell_2)\in K_j} \bar{u}^i_{\ell_1,\ell_2}(\cdot) \le k^i_j - \varepsilon^\delta \quad \text{when } k^i_j \ne 0, \tag{9.21}$$

and

$$\lambda + C\varepsilon^{\delta} \geq \limsup_{T\to\infty} \frac{1}{T} \int_0^T \left[h(\overline{\boldsymbol{x}}(t)) + \sum_{i=1}^{p} \nu_{\boldsymbol{k}^i} c(\overline{\boldsymbol{u}}^i(t)) \right] dt, \tag{9.22}$$

where $\overline{\boldsymbol{x}}(t)$ *is the solution of*

$$\frac{d}{dt}\boldsymbol{x}(t) = \mathcal{L}\left(\boldsymbol{x}(t), \sum_{i=1}^{p} \nu_{\boldsymbol{k}^i} \overline{\boldsymbol{u}}^i(t), \boldsymbol{z} \right), \quad \boldsymbol{x}(0) = \boldsymbol{x}.$$

Proof. For each fixed $\varepsilon > 0$, we select $\widetilde{\boldsymbol{x}} \in \mathcal{Y}$ and

$$(\widetilde{\boldsymbol{u}}^1(\cdot), ..., \widetilde{\boldsymbol{u}}^p(\cdot)) \in \mathcal{A}^0(\widetilde{\boldsymbol{x}}) \tag{9.23}$$

such that

$$\lambda + \varepsilon > \limsup_{T\to\infty} \frac{1}{T} \int_0^T \left[h(\widetilde{\boldsymbol{x}}(t)) + \sum_{i=1}^{p} \nu_{\boldsymbol{k}^i} c(\widetilde{\boldsymbol{u}}^i(t)) \right] dt, \tag{9.24}$$

where $\widetilde{\boldsymbol{x}}(t)$ is the solution of

$$\frac{d}{dt}\boldsymbol{x}(t) = \mathcal{L}\left(\boldsymbol{x}(t), \sum_{i=1}^{p} \nu_{\boldsymbol{k}^i} \widetilde{\boldsymbol{u}}^i(t), \boldsymbol{z} \right), \quad \boldsymbol{x}(0) = \widetilde{\boldsymbol{x}}.$$

Define

$$\mathcal{C}(j) = \{i : k_j^i \neq 0\}, \quad j = 1, ..., m_c.$$

Based on (9.23) and (9.24), we show that there exist $\boldsymbol{x} \in \mathcal{Y}$ and

$$(\overline{\boldsymbol{u}}^1(\cdot), \ldots, \overline{\boldsymbol{u}}^p(\cdot)) \in \mathcal{A}^0(\boldsymbol{x})$$

such that

$$\sum_{(\ell_1,\ell_2)\in K_j} \bar{u}^i_{\ell_1,\ell_2}(t) \leq k_j^i - \varepsilon^{\delta}, \quad i \in \mathcal{C}(j) \quad \text{and} \quad j = 1, \ldots, m_c, \tag{9.25}$$

$$\left| \overline{\boldsymbol{u}}^i(t) - \widetilde{\boldsymbol{u}}^i(t) \right| \leq C_1 \varepsilon^{\delta}, \quad i = 1, ..., p, \tag{9.26}$$

and

$$|\overline{\boldsymbol{x}}(t) - \widetilde{\boldsymbol{x}}(t)| \leq C_1 \varepsilon^{\delta}, \tag{9.27}$$

for some $C_1 > 0$, where

$$\frac{d}{dt}\overline{\boldsymbol{x}}(t) = \mathcal{L}\left(\overline{\boldsymbol{x}}(t), \sum_{i=1}^{p} \nu_{\boldsymbol{k}^i} \overline{\boldsymbol{u}}^i(t), \boldsymbol{z} \right), \quad \overline{\boldsymbol{x}}(0) = \boldsymbol{x}.$$

Recall that Assumptions (A3) and (A4) and the boundedness of $\widetilde{x}(t)$ and $\overline{x}(t)$ imply that there is a constant $C_2 > 0$ such that

$$\begin{aligned}
&\Bigg| \limsup_{T\to\infty} \frac{1}{T} \int_0^T \left[h(\overline{x}(t)) + \sum_{i=1}^p \nu_{k^i} c(\overline{u}^i(t)) \right] dt \\
&\qquad - \limsup_{T\to\infty} \frac{1}{T} \int_0^T \left[h(\widetilde{x}(t)) + \sum_{i=1}^p \nu_{k^i} c(\widetilde{u}^i(t)) \right] dt \Bigg| \\
&\quad \le \limsup_{T\to\infty} \frac{1}{T} \int_0^T \Bigg| \left[h(\overline{x}(t)) + \sum_{i=1}^p \nu_{k^i} c(\overline{u}^i(t)) \right] \qquad\qquad (9.28) \\
&\qquad\qquad - \left[h(\widetilde{x}(t)) + \sum_{i=1}^p \nu_{k^i} c(\widetilde{u}^i(t)) \right] \Bigg| dt \\
&\quad \le \limsup_{T\to\infty} \frac{1}{T} \int_0^T C_2 \varepsilon^\delta \, dt \le C_2 \varepsilon^\delta.
\end{aligned}$$

Thus, (9.24) implies that

$$\lambda + C_3 \varepsilon^\delta \ge \limsup_{T\to\infty} \frac{1}{T} \int_0^T \left[h(\overline{x}(t)) + \sum_{i=1}^p \nu_{k^i} c(\overline{u}^i(t)) \right] dt, \qquad (9.29)$$

for some $C_3 > 0$. In order to complete the proof of the lemma, it suffices to show the existence of $(\overline{u}^1(\cdot), \ldots, \overline{u}^p(\cdot))$ satisfying (9.25)–(9.27). To do this, let

$$C_4 = \min_{1\le j\le m_c} \min_{i\in C(j)} \left\{ k_j^i \right\}.$$

For $i = 1, ..., p$, and $(\ell_1, \ell_2) \in A$ with $\ell_2 \ne n+1$, define

$$\overline{u}^i_{\ell_1,\ell_2}(t) = \frac{\widetilde{u}^i_{\ell_1,\ell_2}(t)}{1 + 2\varepsilon^\delta/(C_4 - 2\varepsilon^\delta)}.$$

Then, for $j = 1, \ldots, m$,

$$\begin{aligned}
\overline{x}_j(t) &= \frac{\widetilde{x}_j}{1 + 2\varepsilon^\delta/(C_4 - 2\varepsilon^\delta)} e^{-a_j t} \\
&\quad + e^{-a_j t} \int_0^t e^{a_j s} \left[\sum_{\ell=-n_0+1}^{j-1} \sum_{i=1}^p \nu_{k^i} \overline{u}^i_{\ell,j}(s) - \sum_{\ell=j+1}^{n} \sum_{i=1}^p \nu_{k^i} \overline{u}^i_{j,\ell}(s) \right] ds \\
&= \frac{1}{1 + 2\varepsilon^\delta/(C_4 - 2\varepsilon^\delta)} \widetilde{x}_j(t) \ge 0 \quad \text{(by (9.23))}.
\end{aligned} \qquad (9.30)$$

This implies that

$$(\overline{\boldsymbol{u}}^1(\cdot), ..., \overline{\boldsymbol{u}}^p(\cdot)) \in \mathcal{A}^0 \left(\frac{\widetilde{\boldsymbol{x}}}{1 + 2\varepsilon^\delta/(C_4 - 2\varepsilon^\delta)} \right).$$

Furthermore, for $j = m+1, ..., n$,

$$\begin{aligned} \bar{x}_j(t) &= \frac{\tilde{x}_j}{1 + 2\varepsilon^\delta/(C_4 - 2\varepsilon^\delta)} e^{-a_j t} \\ &\quad + e^{-a_j t} \int_0^t e^{a_j s} \left[\sum_{i=1}^p \nu_{\boldsymbol{k}^i} \sum_{\ell=-n_0+1}^m \bar{u}^i_{\ell,j}(s) - z_j \right] ds \\ &= \frac{1}{1 + 2\varepsilon^\delta/(C_4 - 2\varepsilon^\delta)} \tilde{x}_j(t) - \frac{2 z_j \varepsilon^\delta}{a_j C_4} (1 - e^{-a_j t}). \end{aligned} \tag{9.31}$$

Therefore, in view of the boundedness of $\widetilde{\boldsymbol{x}}(t)$ and $\widetilde{\boldsymbol{u}}^i(t)$, (9.30) and (9.31) imply

$$\begin{aligned} &|\overline{\boldsymbol{x}}(t) - \widetilde{\boldsymbol{x}}(t)| \\ &= \left(\left(1 - \frac{1}{1 + 2\varepsilon^\delta/(C_4 - 2\varepsilon^\delta)} \right) \max_{1 \le j \le m} \{ \widetilde{x}_j(t) \} \right) \\ &\quad \vee \left(\max_{m+1 \le j \le n} \left\{ \left| \frac{2 z_j \varepsilon^\delta}{a_j C_4} (1 - e^{-a_j t}) + \left(1 - \frac{1}{1 + 2\varepsilon^\delta/(C_4 - 2\varepsilon^\delta)} \right) \widetilde{x}_j(t) \right| \right\} \right) \\ &\le C_5 \varepsilon^\delta, \end{aligned} \tag{9.32}$$

and

$$\left| \overline{\boldsymbol{u}}^i(t) - \widetilde{\boldsymbol{u}}^i(t) \right| \le C_5 \varepsilon^\delta, \quad i = 1, \ldots, p, \tag{9.33}$$

for some $C_5 > 0$, which give (9.26)–(9.27).

On the other hand, for $i \in \mathcal{C}(j)$ and $j = 1, \ldots, m_c$,

$$\begin{aligned} \sum_{(\ell_1,\ell_2) \in K_j} \bar{u}^i_{\ell_1,\ell_2}(t) &= \frac{\sum_{(\ell_1,\ell_2) \in K_j} \widetilde{u}^i_{\ell_1,\ell_2}(t)}{1 + 2\varepsilon^\delta/(C_4 - 2\varepsilon^\delta)} \\ &\le \frac{k^i_j}{1 + 2\varepsilon^\delta/(C_4 - 2\varepsilon^\delta)} \\ &\le k^i_j - 2\varepsilon^\delta, \end{aligned} \tag{9.34}$$

which implies (9.25). Thus the proof of the result is completed. □

In view of Lemma 4.1, we have the following lemma.

Lemma 4.2. *Let Assumptions* (A3) *and* (A4) *hold. For* $\delta \in [0,\ 1/2)$*, there exist* $C > 0$, $\widehat{C} > 0$, $\boldsymbol{x} = (x_1, ..., x_n)' \in \mathcal{Y}$*, and an* $\varepsilon_0 > 0$ *such that for each* *given* $\varepsilon \in (0, \varepsilon_0]$, *we can find a control*

$$(\overline{\boldsymbol{u}}^1(\cdot), \ldots, \overline{\boldsymbol{u}}^p(\cdot)) \in \mathcal{A}^0(\boldsymbol{x})$$

satisfying

$$\min_{1 \le j \le m} \inf_{0 \le t < \infty} \bar{x}_j(t) \ge C\varepsilon^\delta, \tag{9.35}$$

and

$$\lambda + \widehat{C}\varepsilon^\delta \ge \limsup_{T \to \infty} \frac{1}{T} \int_0^T \left[h(\overline{\boldsymbol{x}}(t)) + \sum_{i=1}^p \nu_{\boldsymbol{k}^i} c(\overline{\boldsymbol{u}}^i(t)) \right] dt, \tag{9.36}$$

where $\overline{\boldsymbol{x}}(t)$ *is the solution of*

$$\frac{d}{dt}\boldsymbol{x}(t) = \mathcal{L}\left(\boldsymbol{x}(t), \sum_{i=1}^p \nu_{\boldsymbol{k}^i} c(\overline{\boldsymbol{u}}^i(t)), \boldsymbol{z} \right), \quad \boldsymbol{x}(0) = \boldsymbol{x}.$$

Proof. First, we introduce the concept of the parameter distribution to be used in the proof. The set $\{w_{i,j} \ge 0 : -n_0 + 1 \le i \le n \text{ and } 1 \le j \le n\}$ is called *a parameter distribution* if:

(i) $w_{i,j} \ge 0$;

(ii) $w_{i,j} = 0$ for $(i,j) \notin A$;

(iii) $\sum_{(i,j) \in K_\ell} w_{i,j} \le 1$ for $\ell = 1, ..., m_c$.

We show that there is a parameter distribution $\{w_{i,j} \ge 0 : -n_0 + 1 \le i \le m \text{ and } 1 \le j \le n\}$ such that

$$\min_{1 \le j \le m} \left\{ \sum_{\ell=-n_0+1}^{j-1} w_{\ell,j} - \sum_{\ell=j+1}^{n} w_{j,\ell} \right\} > 0. \tag{9.37}$$

We use the induction on the internal buffer number m to prove (9.37). First for $m = 1$, let

$$n(1) = \min\{j \le 0 \ : \ (j,1) \in A\},$$

and

$$w_{i,j} = \begin{cases} 1, & \text{if } (i,j) = (-n_0 + 1, 1), \\ 0, & \text{otherwise.} \end{cases}$$

We know that this $\{w_{i,j}\}$ is a parameter distribution, and

$$\sum_{\ell=-n_0+1}^{0} w_{\ell,1} - \sum_{\ell=1}^{n} w_{1,\ell} = 1 - 0 > 0.$$

Thus (9.37) holds for $m = 1$. In order to apply an induction-based proof, let us now suppose that (9.37) is true for $m-1$ ($m \geq 2$). Our task is to show that (9.37) holds for m. Consider now the system $(\widehat{V}, \widehat{A}, \widehat{\mathcal{K}})$ with buffer 1 removed, that is,

$$\widehat{A} = A \setminus A(1),$$

$$\widehat{\mathcal{K}} = \{\widehat{K}_1, \ldots, \widehat{K}_{m_c}\} \quad \text{with } \widehat{K}_j = K_j \setminus A(1),$$

$$\widehat{V} = \{-n_0 + 1, \ldots, 0, 2, \ldots, n, n+1\},$$

where

$$A(1) = \{(\ell, 1) : \ell = -n_0 + 1, \ldots, 0\} \cup \{(1, \ell) : \ell = 2, \ldots, n\}.$$

Applying the induction hypothesis, there is a parameter distribution $\{\widehat{w}_{i,j} \geq 0 : (i,j) \in \widehat{A} \text{ and } j \neq n+1\}$ such that

$$\min_{2 \leq j \leq m} \left\{ \sum_{\ell \neq 1, \ell = -n_0+1}^{j-1} \widehat{w}_{\ell,j} - \sum_{\ell=j+1}^{n} \widehat{w}_{j,\ell} \right\} > 0. \tag{9.38}$$

Now we select $C_1 > 1$ such that

$$\sum_{\ell_2=1}^{n} \sum_{(\ell_1,\ell_2) \in \widehat{K}_{c(1,\ell_2)}} \widehat{w}_{\ell_1,\ell_2} + \sum_{\ell_1=-n_0+1}^{0} \sum_{(\ell_1,\ell_2) \in \widehat{K}_{c(\ell_1,1)}} \widehat{w}_{\ell_1,\ell_2} < \frac{C_1}{3}.$$

Let

$$w_{i,j} = \begin{cases} \dfrac{2}{3}, & \text{if } (i,j) = (n(1), 1), \\ 0, & \text{if } j = 1 \text{ and } i \neq n(1) \text{ or } i = 1, \\ \dfrac{\widehat{w}_{i,j}}{C_1}, & \text{otherwise.} \end{cases} \tag{9.39}$$

This implies that, for $\ell = 1, \ldots, m_c$, if $(n(1), 1) \in K_\ell$, then

$$\sum_{(i,j) \in K_\ell} w_{i,j} = \frac{2}{3} + \sum_{(i,j) \in \widehat{K}_\ell} \frac{\widehat{w}_{i,j}}{C_1} < 1,$$

and if $(n(1), 1) \notin K_\ell$, then

$$\sum_{(i,j) \in K_\ell} w_{i,j} = \sum_{(i,j) \in \widehat{K}_\ell} \frac{\widehat{w}_{i,j}}{C_1} < 1.$$

We know that $\{w_{i,j}\}$ defined in (9.39) is also a parameter distribution for the system $(V, A, \mathcal{K})$. Furthermore,

$$\min_{2\le j\le m}\left\{\sum_{\ell=-n_0+1}^{j-1} w_{\ell,j} - \sum_{\ell=j+1}^{n} w_{j,\ell}\right\} = \min_{2\le j\le m}\left\{\sum_{\ell=-n_0+1}^{j-1} \frac{\widehat{w}_{\ell,j}}{C_1} - \sum_{\ell=j+1}^{n} \frac{\widehat{w}_{j,\ell}}{C_1}\right\}$$
$$> 0 \quad \text{(by (9.38))},$$

and

$$\sum_{\ell=-n_0+1}^{0} w_{\ell,1} - \sum_{\ell=2}^{n} w_{1,\ell} = \frac{2}{3} - 0 = \frac{2}{3} > 0.$$

Thus, we know that (9.37) holds for m. Let

$$\min_{1\le j\le m}\left\{\sum_{\ell=-n_0+1}^{j-1} w_{\ell,j} - \sum_{\ell=j+1}^{n} w_{j,\ell}\right\} := C_2 > 0.$$

Next, let

$$(\widehat{\boldsymbol{u}}^1(t),\ldots,\widehat{\boldsymbol{u}}^p(t)) \in \mathcal{A}^0(\widehat{\boldsymbol{x}})$$

satisfy (9.21) and (9.22) in Lemma 4.1. Furthermore, let

$$\boldsymbol{x} = \widehat{\boldsymbol{x}} + \mathbf{1}\varepsilon^{\delta},$$

and for $(\ell_1,\ell_2) \in A$ with $\ell_2 \ne n+1$, $i \in \mathcal{C}(j)$, $j = 1,\ldots,m_c$, let

$$\bar{u}^i_{\ell_1,\ell_2}(t) = w_{\ell_1,\ell_2}\varepsilon^{\delta} + \widehat{u}^i_{\ell_1,\ell_2}(t).$$

Then (9.37) implies that, for $i \in \mathcal{C}(j)$ and $j = 1,\ldots,m_c$,

$$\sum_{(\ell_1,\ell_2)\in K_j} \bar{u}^i_{\ell_1,\ell_2}(t) = \sum_{(\ell_1,\ell_2)\in K_j} w_{\ell_1,\ell_2}\varepsilon^{\delta} + \sum_{(\ell_1,\ell_2)\in K_j} \widehat{u}^i_{\ell_1,\ell_2}(t)$$
$$\le \varepsilon^{\delta} + \sum_{(\ell_1,\ell_2)\in K_j} \widehat{u}^i_{\ell_1,\ell_2}(t) < k^i_j.$$

Let $\overline{\boldsymbol{x}}(t)$ be the solution of

$$\frac{d}{dt}\boldsymbol{x}(t) = \mathcal{L}\left(\boldsymbol{x}(t), \sum_{i=1}^{p} \nu_{\boldsymbol{k}^i}\overline{\boldsymbol{u}}^i(t), \boldsymbol{z}\right), \quad \boldsymbol{x}(0) = \widehat{\boldsymbol{x}} + \mathbf{1}\varepsilon^{\delta}.$$

For $j = 1, \ldots, m$,

$$\begin{aligned}
\bar{x}_j(t) &= (\widehat{x}_j + \varepsilon^\delta)e^{-a_j t} \\
&\quad + \int_0^t e^{-a_j(t-s)} \left[\sum_{i=1}^p \nu_{\boldsymbol{k}^i} \sum_{\ell=-n_0+1}^{j-1} \bar{u}^i_{\ell,j}(s) - \sum_{i=1}^p \nu_{\boldsymbol{k}^i} \sum_{\ell=j+1}^{n} \bar{u}^i_{j,\ell}(s) \right] ds \\
&\geq \widehat{x}_j(t) + \varepsilon^\delta e^{-a_j t} \\
&\quad + \int_0^t e^{-a_j(t-s)} \left[\sum_{\ell=-n_0+1}^{j-1} w_{\ell,j}\varepsilon^\delta - \sum_{\ell=j+1}^{n} w_{j,\ell}\varepsilon^\delta \right] ds \\
&\geq \widehat{x}_j(t) + \varepsilon^\delta e^{-a_j t} + \frac{1}{a_j}(1 - e^{-a_j t})C_2 C_3 \varepsilon^\delta \\
&> \widehat{x}_j(t) + C_3 \varepsilon^\delta,
\end{aligned} \tag{9.40}$$

for some $C_3 > 0$. Consequently,

$$(\overline{\boldsymbol{u}}^1(t), \ldots, \overline{\boldsymbol{u}}^p(t)) \in \mathcal{A}^0 \left(\widehat{\boldsymbol{x}} + \mathbf{1}\varepsilon^\delta \right).$$

Furthermore, there is a $C_5 > 0$ such that

$$\left| \widehat{\boldsymbol{u}}^i(t) - \overline{\boldsymbol{u}}^i(t) \right| \leq C_5 \varepsilon^\delta, \quad i = 1, \ldots, p. \tag{9.41}$$

This implies that

$$|\widehat{\boldsymbol{x}}(t) - \overline{\boldsymbol{x}}(t)| \leq C_6 \varepsilon^\delta, \tag{9.42}$$

for some $C_6 > 0$. It follows from (9.40)–(9.42) that $\overline{\boldsymbol{x}}(t)$ satisfies (9.35) and (9.36). □

Using Lemmas 4.1 and 4.2, we can show our main result that says that the problem $\mathcal{P}^0$ is indeed a limiting problem in the sense that the minimum average cost λ^ε of $\mathcal{P}^\varepsilon$ converges to the minimum average cost λ of the limiting problem $\mathcal{P}^0$. Moreover, we can prove the following corresponding convergence rate.

Theorem 4.1. *Let Assumptions* (A1)–(A4) *hold. Then for any* $\delta \in [0, 1/2)$, *there exist a constant* $C > 0$ *and* $\varepsilon_0 > 0$ *such that, for* $\varepsilon \in (0, \varepsilon_0]$,

$$|\lambda^\varepsilon - \lambda| \leq C\varepsilon^\delta. \tag{9.43}$$

This implies in particular that $\lim_{\varepsilon \to 0} \lambda^\varepsilon = \lambda$.

Proof. First, we prove $\lambda^\varepsilon < \lambda + C\varepsilon^\delta$ by constructing an admissible control $\boldsymbol{u}^\varepsilon(t)$ of $\mathcal{P}^\varepsilon$ from the near-optimal control of the limiting problem $\mathcal{P}^0$ and by estimating the difference between the state trajectories corresponding to these two controls. Then we establish the opposite inequality, namely, $\lambda^\varepsilon > \lambda - C\varepsilon^\delta$, by constructing a control of the limiting problem $\mathcal{P}^0$ from a near-optimal control of $\mathcal{P}^\varepsilon$, and then using Assumptions (A3) and (A4).

In order to show that

$$\lambda^\varepsilon \leq \lambda + C\varepsilon^\delta, \tag{9.44}$$

we can choose, in view of Lemma 4.2, $\boldsymbol{x} \in \mathcal{Y}$ and

$$(\overline{\boldsymbol{u}}^1(t), \ldots, \overline{\boldsymbol{u}}^p(t)) \in \mathcal{A}^0(\boldsymbol{x}),$$

such that

$$\min_{1 \leq j \leq m} \inf_{0 \leq t < \infty} \bar{x}_j(t) \geq C_1 \varepsilon^\delta \tag{9.45}$$

and

$$\lambda + C_2\varepsilon^\delta \geq \limsup_{T \to \infty} \frac{1}{T} \int_0^T \left[h(\overline{\boldsymbol{x}}(t)) + \sum_{i=1}^p \nu_{\boldsymbol{k}^i} c(\overline{\boldsymbol{u}}^i(t)) \right] dt, \tag{9.46}$$

for some constants $C_1 > 0$ and $C_2 > 0$, where

$$\frac{d}{dt}\overline{\boldsymbol{x}}(t) = \mathcal{L}\left(\overline{\boldsymbol{x}}(t), \sum_{i=1}^p \nu_{\boldsymbol{k}^i} \overline{\boldsymbol{u}}^i(t), \boldsymbol{z} \right), \quad \overline{\boldsymbol{x}}(0) = \boldsymbol{x}.$$

We construct the control

$$\overline{\boldsymbol{u}}^\varepsilon(t) = \sum_{i=1}^p I_{\{\boldsymbol{k}(\varepsilon,t) = \boldsymbol{k}^i\}} \overline{\boldsymbol{u}}^i(t),$$

and let

$$\frac{d}{dt}\overline{\boldsymbol{x}}^\varepsilon(t) = \mathcal{L}(\overline{\boldsymbol{x}}^\varepsilon(t), \overline{\boldsymbol{u}}^\varepsilon(t), \boldsymbol{z}), \quad \overline{\boldsymbol{x}}^\varepsilon(0) = \boldsymbol{x}.$$

Then, for $j = 1, \ldots, m$,

$$\begin{aligned}
&E\left|\bar{x}_j^\varepsilon(t) - \bar{x}_j(t)\right|^2 \\
&= E\left| e^{-a_j t} \int_0^t e^{a_j s} \left[\left(\sum_{\ell=-n_0+1}^{j-1} \bar{u}_{\ell,j}^\varepsilon(s) - \sum_{\ell=-n_0+1}^{j-1} \sum_{i=1}^p \nu_{\boldsymbol{k}^i} \bar{u}_{\ell,j}^i(s) \right) \right.\right. \\
&\qquad \left.\left. - \left(\sum_{\ell=j+1}^n \bar{u}_{j,\ell}^\varepsilon(s) - \sum_{\ell=j+1}^n \sum_{i=1}^p \nu_{\boldsymbol{k}^i} \bar{u}_{j,\ell}^i(s) \right) \right] ds \right|^2 \\
&\leq 2E\left[e^{-a_j t} \int_0^t e^{a_j s} \left(\sum_{\ell=-n+1}^{j-1} \bar{u}_{\ell,j}^\varepsilon(s) - \sum_{\ell=-n+1}^{j-1} \sum_{i=1}^p \nu_{\boldsymbol{k}^i} \bar{u}_{\ell,j}^i(s) \right) ds \right]^2 \\
&\quad + 2E\left[e^{-a_j t} \int_0^t e^{a_j s} \left(\sum_{\ell=j+1}^n \bar{u}_{j,\ell}^\varepsilon(s) - \sum_{\ell=j+1}^n \sum_{i=1}^p \nu_{\boldsymbol{k}^i} \bar{u}_{j,\ell}^i(s) \right) ds \right]^2,
\end{aligned}$$

and for $j = m+1, ..., n$,

$$E\left|\bar{x}_j^\varepsilon(t) - \bar{x}_j(t)\right|^2$$

$$= E\left[e^{-a_j t}\int_0^t e^{a_j s}\left(\sum_{\ell=-n_0+1}^{m} \bar{u}_{\ell,j}^\varepsilon(s) - \sum_{\ell=-n_0+1}^{m}\sum_{i=1}^{p} \nu_{\boldsymbol{k}^i}\bar{u}_{\ell,j}^i(s)\right) ds\right]^2.$$

Similar to the proof of (7.21), we have

$$E\left|\bar{x}_j^\varepsilon(t) - \bar{x}_j(t)\right| \le C_2\varepsilon^{1/2}, \quad j = 1, ..., n, \tag{9.47}$$

for some $C_2 > 0$. Consequently, by the boundedness of $\overline{\boldsymbol{x}}^\varepsilon(t)$ and $\overline{\boldsymbol{x}}(t)$ and Assumption (A3), we obtain

$$\begin{aligned}
&\left|\limsup_{T\to\infty}\frac{1}{T}E\left[\int_0^T h(\overline{\boldsymbol{x}}^\varepsilon(t))\,dt - \limsup_{T\to\infty}\frac{1}{T}\int_0^T h(\overline{\boldsymbol{x}}(t))\right] dt\right| \\
&\le \limsup_{T\to\infty}\frac{C_{h3}}{T}\int_0^T E\left[\left(1 + |\overline{\boldsymbol{x}}^\varepsilon(t)|^{\beta_{h2}-1} + |\overline{\boldsymbol{x}}(t)|^{\beta_{h2}-1}\right)|\overline{\boldsymbol{x}}^\varepsilon(t) - \overline{\boldsymbol{x}}(t)|\right] dt \\
&\le C_3\varepsilon^{1/2},
\end{aligned} \tag{9.48}$$

for some $C_3 > 0$. It follows from the boundedness of $c(\cdot)$ that

$$\begin{aligned}
&\left|\limsup_{T\to\infty}\frac{1}{T}E\int_0^T c(\overline{\boldsymbol{u}}^\varepsilon(t))\,dt - \limsup_{T\to\infty}\frac{1}{T}\int_0^T\sum_{i=1}^p \nu_{\boldsymbol{k}^i}c(\overline{\boldsymbol{u}}^i(t))\,dt\right| \\
&= \left|\limsup_{T\to\infty}\frac{1}{T}E\int_0^T\sum_{i=1}^p I_{\{\boldsymbol{k}(\varepsilon,t)=\boldsymbol{k}^i\}}c(\overline{\boldsymbol{u}}^i(t))\,dt\right. \\
&\qquad \left. - \limsup_{T\to\infty}\frac{1}{T}\int_0^T\sum_{i=1}^p \nu_{\boldsymbol{k}^i}c(\overline{\boldsymbol{u}}^i(t))\,dt\right| \\
&\le \limsup_{T\to\infty}\frac{1}{T}\int_0^T\sum_{i=1}^p \left|P(\boldsymbol{k}(\varepsilon,t) = \boldsymbol{k}^i) - \nu_{\boldsymbol{k}^i}\right| c(\overline{\boldsymbol{u}}^i(t))\,dt \\
&\le C_4\varepsilon,
\end{aligned} \tag{9.49}$$

for some $C_4 > 0$. Thus, combining (9.46), (9.48), and (9.49), we have that there is a constant $C_5 > 0$ such that

$$\limsup_{T\to\infty}\frac{1}{T}E\int_0^T [h(\overline{\boldsymbol{x}}^\varepsilon(t)) + c(\overline{\boldsymbol{u}}^\varepsilon(t))]\,dt \le \lambda + C_5(\varepsilon^\delta + \varepsilon^{1/2} + \varepsilon). \tag{9.50}$$

If $\overline{\boldsymbol{u}}^\varepsilon(t) \in \mathcal{A}^\varepsilon(\boldsymbol{x}, \boldsymbol{k}(\varepsilon, 0))$, then from (9.50) and

$$\lambda^\varepsilon \le \limsup_{T\to\infty}\frac{1}{T}E\int_0^T [h(\overline{\boldsymbol{x}}^\varepsilon(t)) + c(\overline{\boldsymbol{u}}^\varepsilon(t))]\,dt,$$

we have (9.44). But generally, $\overline{\boldsymbol{u}}^\varepsilon(t) \notin \mathcal{A}^\varepsilon(\boldsymbol{x}, \boldsymbol{k}(\varepsilon, 0))$. Thus, based on $\overline{\boldsymbol{u}}^\varepsilon(t)$ we next construct $\boldsymbol{u}^\varepsilon(t) \in \mathcal{A}^\varepsilon(\boldsymbol{x}, \boldsymbol{k}(\varepsilon, 0))$ such that it and the solution $\boldsymbol{x}^\varepsilon(t)$ of

$$\frac{d}{dt}\boldsymbol{x}(t) = \mathcal{L}(\boldsymbol{x}(t), \boldsymbol{u}^\varepsilon(t), \boldsymbol{z}), \quad \boldsymbol{x}(0) = \boldsymbol{x},$$

satisfy (9.50). Consequently, we get (9.44). To do this, let

$$M_1 = \max_{1 \le i \le p} \{k_j^i : 1 \le j \le m_c\}$$

and

$$M_2 = \max_{1 \le j \le n} \left\{ \frac{1}{a_j} \ln \left(\frac{C_1 a_j \varepsilon^\delta}{5(n + n_0) M_1} \right)^{-1} \right\},$$

where C_1 is given in (9.45). We show momentarily that there exists a control $\boldsymbol{u}^\varepsilon(t) \in \mathcal{A}^\varepsilon(\boldsymbol{x}, \boldsymbol{k}(\varepsilon, 0))$ such that, for $t > 2M_2$ and $(i, j) \in A$ with $j \ne n+1$,

$$E\left|u_{i,j}^\varepsilon(t) - \overline{u}_{i,j}^\varepsilon(t)\right| \le C_6 \exp\left\{ -M_3 \varepsilon^{-(1-2\delta)/2}(1 + M_2)^{-3} \right\}, \tag{9.51}$$

for some $C_6 > 0$ and $M_3 > 0$. Hence, Assumptions (A3) and (A4) imply that there is a constant $C_7 > 0$ such that

$$\begin{aligned}
&E\int_0^T |h(\boldsymbol{x}^\varepsilon(t)) - h(\overline{\boldsymbol{x}}^\varepsilon(t))|\, dt \\
&\le C_7 \int_0^T E\,|\boldsymbol{x}^\varepsilon(t) - \overline{\boldsymbol{x}}^\varepsilon(t)|\, dt \\
&\le C_7 \sum_{j=1}^m \int_0^T \int_0^t e^{-a_j(t-s)} E\left[\sum_{\ell=-n_0+1}^{j-1} \left|u_{\ell,j}^\varepsilon(s) - \bar{u}_{\ell,j}^\varepsilon(s)\right| \right. \\
&\qquad\qquad \left. + \sum_{\ell=j+1}^{n} \left|u_{j,\ell}^\varepsilon(s) - \bar{u}_{j,\ell}^\varepsilon(s)\right| \right] ds\, dt \\
&\quad + C_7 \sum_{j=m+1}^{n} \int_0^T \int_0^t e^{-a_j(t-s)} E\left[\sum_{\ell=-n_0+1}^{m} \left|u_{\ell,j}^\varepsilon(s) - \bar{u}_{\ell,j}^\varepsilon(s)\right| \right] ds\, dt \\
&\le 2C_7 \sum_{j=1}^n (1/a_j) \int_0^T \sum_{(\ell,j)\in A} E\left|u_{\ell,j}^\varepsilon(s) - \bar{u}_{\ell,j}^\varepsilon(s)\right| ds
\end{aligned} \tag{9.52}$$

and

$$\begin{aligned}
E\int_0^T |c(\boldsymbol{u}^\varepsilon(t)) - c(\overline{\boldsymbol{u}}^\varepsilon(t))|\, dt &\le C_7 \int_0^T E\,|\boldsymbol{u}^\varepsilon(t) - \overline{\boldsymbol{u}}^\varepsilon(t)|\, dt \\
&\le C_7 \int_0^T \sum_{(\ell,j)\in A} E\left|u_{\ell,j}^\varepsilon(t) - \bar{u}_{\ell,j}^\varepsilon(t)\right| dt.
\end{aligned} \tag{9.53}$$

Therefore, we know, in view of (9.50) and (9.51)–(9.53), that

$$\limsup_{T\to\infty} \frac{1}{T} E \int_0^T [h(\boldsymbol{x}^\varepsilon(t)) + c(\boldsymbol{u}^\varepsilon(t))]\, dt \le \lambda + C_8 \varepsilon^\delta,$$

for some $C_8 > 0$, which implies that (9.44) holds.

Thus, it suffices to show that there is a control $\boldsymbol{u}^\varepsilon(\cdot) \in \mathcal{A}^\varepsilon(\boldsymbol{x}, \boldsymbol{k}(\varepsilon, 0))$ satisfying (9.51). We will modify $\overline{\boldsymbol{u}}^\varepsilon(\cdot)$ to $\boldsymbol{u}^\varepsilon(t)$ such that (9.51) holds and $\boldsymbol{u}^\varepsilon(\cdot) \in \mathcal{A}^\varepsilon(\boldsymbol{x}, \boldsymbol{k}(\varepsilon, 0))$. This modification is based on the estimation of $P(\bar{x}_j^\varepsilon(t) < 0)$ for $t \ge 2M_2$. Thus, we first establish the following inequality by (9.45). For $t \ge 2M_2$ and $j = 1, \ldots, m$,

$$\begin{aligned}
& P\left(\bar{x}_j^\varepsilon(s) \le C_1 \varepsilon^\delta / 2\right) \\
& \le P\left(\bar{x}_j^\varepsilon(t) \le \bar{x}_j(t) - C_1 \varepsilon^\delta / 2\right) \quad \text{(by (9.45))} \\
& = P\left(\bar{x}_j(t) - \bar{x}_j^\varepsilon(t) \ge C_1 \varepsilon^\delta / 2\right) \\
& \le P\left(\left|\bar{x}_j(t) - \bar{x}_j^\varepsilon(t)\right| \ge C_1 \varepsilon^\delta / 2\right) \\
& = P\Bigg(\Bigg|\int_0^t e^{a_j(s-t)} \Bigg(\sum_{\ell=-n_0+1}^{j-1} \sum_{i=1}^{p} I_{\{\boldsymbol{k}(\varepsilon,s)=\boldsymbol{k}^i\}} \bar{u}_{\ell,j}^i(s) \\
& \qquad\qquad - \sum_{\ell=j+1}^{n} \sum_{i=1}^{p} I_{\{\boldsymbol{k}(\varepsilon,s)=\boldsymbol{k}^i\}} \bar{u}_{j,\ell}^i(s) \Bigg) ds \\
& \qquad - \int_0^t e^{a_j(s-t)} \Bigg(\sum_{\ell=-n_0+1}^{j-1} \sum_{i=1}^{p} \nu_{\boldsymbol{k}^i} \bar{u}_{\ell,j}^i(s) \\
& \qquad\qquad - \sum_{\ell=j+1}^{n} \sum_{i=1}^{p} \nu_{\boldsymbol{k}^i} \bar{u}_{j,\ell}^i(s) \Bigg) ds \Bigg| \ge C_1 \varepsilon^\delta / 2 \Bigg) \\
& \le \sum_{\ell=-n_0+1}^{j-1} P\Bigg(\Bigg|\int_0^t e^{a_j(s-t)} \\
& \qquad \times \left[\sum_{i=1}^{p} \left(I_{\{\boldsymbol{k}(\varepsilon,s)=\boldsymbol{k}^i\}} - \nu_{\boldsymbol{k}^i}\right) \bar{u}_{\ell,j}^i(s)\right] ds \Bigg| \ge \frac{C_1 \varepsilon^\delta}{4(n+n_0)} \Bigg) \\
& \quad + \sum_{\ell=j+1}^{n} P\Bigg(\Bigg|\int_0^t e^{a_j(s-t)} \\
& \qquad \times \left[\sum_{i=1}^{p} \left(I_{\{\boldsymbol{k}(\varepsilon,s)=\boldsymbol{k}^i\}} - \nu_{\boldsymbol{k}^i}\right) \bar{u}_{j,\ell}^i(s)\right] ds \Bigg| \ge \frac{C_1 \varepsilon^\delta}{4(n+n_0)} \Bigg).
\end{aligned} \tag{9.54}$$

Note that, for $t > 2M_2$,

$$\begin{aligned}
&\left| e^{-a_j t} \int_0^t e^{a_j s} \left(\sum_{i=1}^p \left[I_{\{\boldsymbol{k}(\varepsilon,s)=\boldsymbol{k}^i\}} - \nu_{\boldsymbol{k}^i} \right] \bar{u}^i_{\ell,j}(s) \right) ds \right| \\
&\quad \le \left| e^{-a_j t} \int_0^{t-M_2} e^{a_j s} \left(\sum_{i=1}^p \left[I_{\{\boldsymbol{k}(\varepsilon,s)=\boldsymbol{k}^i\}} - \nu_{\boldsymbol{k}^i} \right] \bar{u}^i_{\ell,j}(s) \right) ds \right| \\
&\qquad + \left| e^{-a_j t} \int_{t-M_2}^t e^{a_j s} \left(\sum_{i=1}^p \left[I_{\{\boldsymbol{k}(\varepsilon,s)=\boldsymbol{k}^i\}} - \nu_{\boldsymbol{k}^i} \right] \bar{u}^i_{\ell,j}(s) \right) ds \right| \\
&\quad \le e^{-a_j t} \int_0^{t-M_2} e^{a_j s} M_1 \, ds \\
&\qquad + \left| e^{-a_j t} \int_{t-M_2}^t e^{a_j s} \left(\sum_{i=1}^p \left[I_{\{\boldsymbol{k}(\varepsilon,s)=\boldsymbol{k}^i\}} - \nu_{\boldsymbol{k}^i} \right] \bar{u}^i_{\ell,j}(s) \right) ds \right| \\
&\quad \le \left| e^{-a_j t} \int_{t-M_2}^t e^{a_j s} \left(\sum_{i=1}^p \left[I_{\{\boldsymbol{k}(\varepsilon,s)=\boldsymbol{k}^i\}} - \nu_{\boldsymbol{k}^i} \right] \bar{u}^i_{\ell,j}(s) \right) ds \right| \\
&\qquad + \frac{C_1 \varepsilon^\delta}{5(n+n_0)}.
\end{aligned} \tag{9.55}$$

Therefore, it follows from Corollary B.3, (9.54), and (9.55) that, for $j = 1,...,m$ and $t \ge 2M_2$,

$$P\left(\bar{x}^\varepsilon_j(t) \le C_1\varepsilon^\delta/2\right) \le C_9 \exp\left\{ - M_3 \varepsilon^{-(1-2\delta)/2}(1+M_2)^{-3/2} \right\}, \tag{9.56}$$

for some $C_9 > 0$ and $M_3 > 0$.

Based on (9.56), we use the induction on the internal buffer number m to construct the desired $\boldsymbol{u}^\varepsilon(t)$ from $\overline{\boldsymbol{u}}^\varepsilon(t)$. First, for $m = 1$, define

$$\mathcal{B}^\varepsilon_1 = \left\{ t : \bar{x}^\varepsilon_1(t) - \inf_{0 \le s \le t} \bar{x}^\varepsilon_1(s) = 0 \text{ and } \bar{x}^\varepsilon_1(t) < 0 \right\}. \tag{9.57}$$

We choose $\{\widetilde{u}^\varepsilon_{\ell,1}(t) : -n_0 + 1 \le \ell \le 0\}$ and $\{\widetilde{u}^\varepsilon_{1,\ell}(t) : 2 \le \ell \le n\}$ such that

$$\begin{aligned}
\widetilde{u}^\varepsilon_{\ell,1}(t) &\le \bar{u}^\varepsilon_{\ell,1}(t), \quad \ell = -n_0+1, \ldots, 0, \\
\widetilde{u}^\varepsilon_{1,\ell}(t) &\le \bar{u}^\varepsilon_{1,\ell}(t), \quad \ell = 2, \ldots, n,
\end{aligned}$$

and

$$\sum_{\ell=-n_0+1}^{0} \widetilde{u}^\varepsilon_{\ell,1}(t) = \sum_{\ell=2}^{n} \widetilde{u}^\varepsilon_{1,\ell}(t) = \left(\sum_{\ell=2}^{n} \bar{u}^\varepsilon_{1,\ell}(t) \right) \wedge \left(\sum_{\ell=-n_0+1}^{0} \bar{u}^\varepsilon_{\ell,1}(t) \right). \tag{9.58}$$

For $(i,j) \in \{(\ell,1) : -n_0 + 1 \le \ell \le 0\} \cup \{(1,\ell) : 2 \le \ell \le n\}$, define

$$u^\varepsilon_{i,j}(t) = \begin{cases} \bar{u}^\varepsilon_{i,j}(t), & \text{if } t \notin \mathcal{B}^\varepsilon_1, \\ \widetilde{u}^\varepsilon_{i,j}(t), & \text{if } t \in \mathcal{B}^\varepsilon_1. \end{cases} \tag{9.59}$$

For $(i,j) \in A$ with $j \neq 1$ and $i \neq 1$, let

$$u^{\varepsilon}_{i,j}(t) = \bar{u}^{\varepsilon}_{i,j}(t).$$

Then, for $(i,j) \in A$,

$$u^{\varepsilon}_{i,j}(t) \leq \bar{u}^{\varepsilon}_{i,j}(t).$$

Define

$$\begin{cases} x^{\varepsilon}_1(t) = x_1 e^{-a_1 t} + \displaystyle\int_0^t e^{-a_1(t-s)} \left(\sum_{\ell=-n_0+1}^{0} u^{\varepsilon}_{\ell,1}(s) - \sum_{\ell=2}^{n} u^{\varepsilon}_{2,\ell}(s) \right) ds, \\ x^{\varepsilon}_j(t) = x_j e^{-a_j t} + \displaystyle\int_0^t e^{-a_j(t-s)} \left(\sum_{\ell=-n_0+1}^{1} u^{\varepsilon}_{\ell,j}(s) - z_j \right) ds, \\ \qquad\qquad\qquad\qquad\qquad\qquad\qquad\qquad j = 2, ..., n. \end{cases}$$

We know that $\boldsymbol{x}^{\varepsilon}(t) \in \mathcal{Y}$ and $\boldsymbol{u}^{\varepsilon}(t) \in \mathcal{A}^{\varepsilon}(\boldsymbol{x}, \boldsymbol{k}(\varepsilon, 0))$. Furthermore,

$$\begin{aligned} E|\boldsymbol{u}^{\varepsilon}(t) - \bar{\boldsymbol{u}}^{\varepsilon}(t)| &= E\left[|\boldsymbol{u}^{\varepsilon}(t) - \bar{\boldsymbol{u}}^{\varepsilon}(t)| I_{\{\bar{x}^{\varepsilon}_1(t)<0\}} \right] \\ &\leq \widehat{C}_1 P\left(\bar{x}^{\varepsilon}_1(t) < 0\right), \end{aligned}$$

for some $\widehat{C}_1 > 0$. Thus, (9.56) implies that there is a control $\boldsymbol{u}^{\varepsilon}(\cdot)$ such that (9.51) is true and $\boldsymbol{u}^{\varepsilon}(\cdot) \in \mathcal{A}^{\varepsilon}(\boldsymbol{x}, \boldsymbol{k}(\varepsilon, 0))$ for $m = 1$. In order to apply an induction-based proof, let us now suppose that there exists a $\boldsymbol{u}^{\varepsilon}(\cdot)$ such that (9.51) is true and $\boldsymbol{u}^{\varepsilon}(\cdot) \in \mathcal{A}^{\varepsilon}(\boldsymbol{x}, \boldsymbol{k}(\varepsilon, 0))$ for $m - 1$. Then our task is to show that this is also true for m.

The same argument, as in the case $m = 1$ above, yields that there exists a $\hat{u}^{\varepsilon}_{i,j}(t)$ with $(i,j) \in \{(\ell,1) : -n_0 + 1 \leq \ell \leq 0\} \cup \{(1,\ell) : 2 \leq \ell \leq n\}$ such that, for $(i,j) \in \{(\ell,1) : -n_0 + 1 \leq \ell \leq 0\} \cup \{(1,\ell) : 2 \leq \ell \leq n\}$,

$$\hat{u}^{\varepsilon}_{i,j}(t) \leq \bar{u}^{\varepsilon}_{i,j}(t), \tag{9.60}$$

$$E|\bar{u}^{\varepsilon}_{i,j}(t) - \hat{u}^{\varepsilon}_{i,j}(t)| \leq \widehat{C}_1 P\left(\bar{x}^{\varepsilon}_1(t) < 0\right), \tag{9.61}$$

and

$$\hat{x}^{\varepsilon}_1(t) = x_1 e^{-a_1 t} + \int_0^t e^{-a_1(t-s)} \left(\sum_{\ell=-n_0+1}^{0} \hat{u}^{\varepsilon}_{\ell,1}(s) - \sum_{\ell=2}^{n} \hat{u}^{\varepsilon}_{1,\ell}(s) \right) ds \geq 0. \tag{9.62}$$

Now we consider the system

$$\begin{cases} \hat{x}_j^\varepsilon(t) = x_j e^{-a_j t} + \displaystyle\int_0^t e^{-a_j(t-s)} \Bigg(\sum_{\ell\neq 1, \ell=-n_0+1}^{j-1} \bar{u}_{\ell j}^\varepsilon(s) + \hat{u}_{1,j}^\varepsilon(s) \\ \qquad\qquad - \displaystyle\sum_{\ell=j+1}^{n} \bar{u}_{j,\ell}^\varepsilon(s) \Bigg) ds, \quad j = 2, ..., m, \\ \hat{x}_j(t) = x_j e^{-a_j t} + \displaystyle\int_0^t e^{-a_j(t-s)} \Bigg(\sum_{\ell\neq 1, \ell=-n_0+1}^{j-1} \bar{u}_{\ell,j}^\varepsilon(s) + \hat{u}_{1,j}^\varepsilon - z_j \Bigg) ds, \\ \qquad\qquad j = m+1, ..., n. \end{cases} \tag{9.63}$$

In order to apply the induction hypothesis, we also have to show that an inequality like (9.56) is also satisfied for $(\hat{x}_2^\varepsilon(t), ..., \hat{x}_m^\varepsilon(t))$. Note that, for $j = 2, ..., m$, we have from (9.63) that, for $t \geq 2M_2$,

$$\begin{aligned} \hat{x}_j^\varepsilon(t) &= \bar{x}_j^\varepsilon(t) + \int_0^t e^{-a_j(t-s)} \left[\hat{u}_{1,j}^\varepsilon(s) - \bar{u}_{1,j}^\varepsilon(s)\right] ds \\ &= \bar{x}_j^\varepsilon(t) + \int_0^t e^{-a_j(t-s)} \left[\hat{u}_{1,j}^\varepsilon(s) - \bar{u}_{1,j}^\varepsilon(s)\right] I_{\{\bar{x}_1^\varepsilon(s)<0\}} ds \\ &= \bar{x}_j^\varepsilon(t) + \int_0^{t-M_2} e^{-a_j(t-s)} \left[\hat{u}_{1,j}^\varepsilon(s) - \bar{u}_{1,j}^\varepsilon(s)\right] I_{\{\bar{x}_1^\varepsilon(s)<0\}} ds \\ &\quad + \int_{t-M_2}^t e^{-a_j(t-s)} \left[\hat{u}_{1,j}^\varepsilon(s) - \bar{u}_{1,j}^\varepsilon(s)\right] I_{\{\bar{x}_1^\varepsilon(s)<0\}} ds \\ &\geq \bar{x}_j^\varepsilon(t) - M_1 \int_{t-M_2}^t e^{-a_j(t-s)} I_{\{\bar{x}_1^\varepsilon(s)<0\}} ds - \frac{C_1\varepsilon^\delta}{5(n+n_0)}. \end{aligned}$$

Hence, for $t \geq 2M_2$,

$$\begin{aligned} &P\left(\hat{x}_j^\varepsilon(t) < \frac{C_1\varepsilon^\delta}{5}\right) \\ &\leq P\left(\bar{x}_j^\varepsilon(t) < \frac{C_1\varepsilon^\delta}{2}\right) \\ &\quad + P\left(\bar{x}_j^\varepsilon(t) \geq \frac{C_1\varepsilon^\delta}{2}, \bar{x}_j^\varepsilon(t) - M_1 \int_{t-M_2}^t e^{-a_1(t-s)} I_{\{\bar{x}_1^\varepsilon(s)<0\}} ds < \frac{2C_1\varepsilon^\delta}{5}\right) \\ &\leq P\left(\bar{x}_j^\varepsilon(t) < \frac{C_1\varepsilon^\delta}{2}\right) + P\left(M_1 \int_{t-M_2}^t e^{-a_1(t-s)} I_{\{\bar{x}_1^\varepsilon(s)<0\}} ds \geq \frac{C_1\varepsilon^\delta}{10}\right). \end{aligned}$$

Therefore, similar to (8.66),

$$\begin{aligned} P\left(\hat{x}_j^\varepsilon(t) \leq C_1\varepsilon^\delta/4\right) \leq \widehat{C}_2 \exp\left\{ - M_3 \varepsilon^{-(1-2\bar{\delta})/2} (1+M_2)^{-3/2} \right\}, \\ j = 2, ..., m, \end{aligned} \tag{9.64}$$

for some $\widetilde{C}_2 > 0$ and $\overline{\delta} \in (0,\ 1/2)$. Applying the induction hypothesis to system (9.63), there is a $u_{i,j}^{\varepsilon}(t)$ ($(i,j) \in A \setminus \{(\ell,1) : -n_0+1 \le \ell \le 0\}$) such that

$$u_{i,j}^{\varepsilon}(t) \le \hat{u}_{i,j}^{\varepsilon}(t), \tag{9.65}$$

$$\begin{aligned} x_j^{\varepsilon}(t) &= x_j e^{-a_j t} + \int_0^t e^{-a_j(t-s)} \left(\sum_{\ell=n_0+1}^{j-1} u_{\ell,j}^{\varepsilon}(s) - \sum_{\ell=j+1}^{n} u_{j,\ell}^{\varepsilon}(s) \right) ds \\ &\ge 0, \quad j = 2, \dots, m, \end{aligned} \tag{9.66}$$

$$E|u_{i,j}^{\varepsilon}(t) - \hat{u}_{i,j}^{\varepsilon}(t)| \le \widehat{C}_3 \sum_{j=2}^{m} P\left(\hat{x}_j^{\varepsilon}(t) < 0\right), \tag{9.67}$$

for some $\widehat{C}_3 > 0$. Consequently, let

$$u_{\ell,1}^{\varepsilon}(t) = \hat{u}_{\ell,1}^{\varepsilon}(t), \quad \ell = -n_0+1, \dots, 0.$$

Then, (9.62) and (9.65) yield

$$x_1^{\varepsilon}(t) = x_1 e^{-a_1 t} + \int_0^t e^{-a_1(t-s)} \left(\sum_{\ell=-n_0+1}^{0} u_{\ell,1}^{\varepsilon}(s) - \sum_{\ell=2}^{n} u_{1,\ell}^{\varepsilon}(s) \right) ds \ge 0.$$

Thus, it follows from (9.66) that $\boldsymbol{u}^{\varepsilon}(\cdot) \in \mathcal{A}^{\varepsilon}(\boldsymbol{x}, \boldsymbol{k}(\varepsilon, 0))$. By (9.61), (9.64), (9.66), and (9.67), we know that (9.51) is true.

We now show

$$\lambda^{\varepsilon} > \lambda + C\varepsilon^{\delta}. \tag{9.68}$$

First we show that for any control $u^{\varepsilon}(\cdot) \in \mathcal{A}^{\varepsilon}(\boldsymbol{x}, \boldsymbol{k})$, there exists a control $U(\cdot) = (\boldsymbol{u}^1(\cdot), \dots, \boldsymbol{u}^p(\cdot)) \in \mathcal{A}^0(\boldsymbol{x})$ such that $E[\boldsymbol{x}^{\varepsilon}(t)] - \boldsymbol{x}(t)$ is small when ε is small enough, where $\boldsymbol{x}^{\varepsilon}(\cdot)$ and $\boldsymbol{x}(\cdot)$ are the respective state trajectories under controls $\boldsymbol{u}^{\varepsilon}(\cdot)$ and $U(\cdot)$ with the same initial condition $\boldsymbol{x}$. Now we choose $U(\cdot)$ defined by

$$\boldsymbol{u}^i(t) = E\left[\boldsymbol{u}^{\varepsilon}(t) | \boldsymbol{k}(\varepsilon, t) = \boldsymbol{k}^i\right], \quad i = 1, \dots, p.$$

Then, for $j = 1, \dots, m$,

$$\begin{aligned} E[x_j^{\varepsilon}(t)] = \ & x_j e^{-a_j t} + \int_0^t e^{-a_j(t-s)} \left(\sum_{i=1}^{p} P(\boldsymbol{k}(\varepsilon, s) = \boldsymbol{k}^i) \sum_{\ell=-n_0+1}^{j-1} u_{\ell,j}^i(s) \right. \\ & \left. - \sum_{i=1}^{p} P(\boldsymbol{k}(\varepsilon, s) = \boldsymbol{k}^i) \sum_{\ell=j+1}^{n} u_{j,\ell}^i(s) \right) ds, \end{aligned}$$

$$\begin{aligned} x_j(t) = \ & x_j e^{-a_j t} + \int_0^t e^{-a_j(t-s)} \\ & \times \left(\sum_{\ell=-n_0+1}^{j-1} \sum_{i=1}^{p} \nu_{\boldsymbol{k}^i} u_{\ell,j}^i(s) - \sum_{\ell=j+1}^{n} \sum_{i=1}^{p} \nu_{\boldsymbol{k}^i} u_{j,\ell}^i(s) \right) ds, \end{aligned}$$

and, for $j = m+1, ..., n$,

$$E[x_j^\varepsilon(t)] = x_j e^{-a_j t} + \int_0^t e^{-a_j(t-s)} \times \left(\sum_{i=1}^p P(\boldsymbol{k}(\varepsilon, s) = \boldsymbol{k}^i) \sum_{\ell=-n_0+1}^m u_{\ell,j}^i(s) - z_j \right) ds,$$

$$x_j(t) = x_j e^{-a_j t} + \int_0^t e^{-a_j(t-s)} \left(\sum_{i=1}^p \nu_{\boldsymbol{k}^i} \sum_{\ell=-n_0+1}^m u_{\ell,j}^i(s) - z_j \right) ds.$$

Similar to Lemma 4.2, we can prove that there exists a control $\boldsymbol{u}^\varepsilon(\cdot) \in \mathcal{A}^\varepsilon(\boldsymbol{x}, \boldsymbol{k})$ such that

$$\min_{1 \le j \le m} \inf_{0 \le t < \infty} E[x_j^\varepsilon(t)] > \widehat{C}_4 \varepsilon^\delta \tag{9.69}$$

and

$$\limsup_{T \to \infty} \frac{1}{T} E \int_0^T [h(\boldsymbol{x}^\varepsilon(t)) + c(\boldsymbol{u}^\varepsilon(t))]\, dt \le \lambda^\varepsilon + \widehat{C}_5 \varepsilon^\delta, \tag{9.70}$$

for some $\widehat{C}_4 > 0$ and $\widehat{C}_5 > 0$, where $\boldsymbol{x}^\varepsilon(t)$ is the state trajectory under the control $\boldsymbol{u}^\varepsilon(t)$. Similar to (9.47), we have

$$\left| E[x_j^\varepsilon(t)] - x_j(t) \right| \le \widehat{C}_6 \varepsilon^{1/2}, \quad j = 1, ..., n.$$

Consequently, it follows from (9.69) that, for sufficiently small ε,

$$(\boldsymbol{u}^1(\cdot), ..., \boldsymbol{u}^p(\cdot)) \in \mathcal{A}^0(\boldsymbol{x}).$$

In view of the convexity and the local Lipschitz continuity of $h(\cdot)$, Jensen's inequality yields

$$\begin{aligned} E[h(\boldsymbol{x}^\varepsilon(t))] &\ge h(E[\boldsymbol{x}^\varepsilon(t)]) \\ &= h(\boldsymbol{x}(t)) + [h(E[\boldsymbol{x}^\varepsilon(t)]) - h(\boldsymbol{x}(t))] \\ &\ge h(\boldsymbol{x}(t)) - C_{h3} \left(1 + |E[\boldsymbol{x}^\varepsilon(t))]|^{\beta_{h2}} + |\boldsymbol{x}(t)|^{\beta_{h2}}\right) \\ &\quad \times |E[\boldsymbol{x}^\varepsilon(t)] - \boldsymbol{x}(t)| \\ &\ge h(\boldsymbol{x}(t)) - \widehat{C}_7 \varepsilon, \end{aligned} \tag{9.71}$$

for some $\widehat{C}_7 > 0$. In the same way, using Lemma B.3, we can establish

$$\begin{aligned} E[c(\boldsymbol{u}^\varepsilon(t))] &= \sum_{i=1}^p P(\boldsymbol{k}(\varepsilon, t) = \boldsymbol{k}^i) E[c(\boldsymbol{u}^\varepsilon(t)) | \boldsymbol{k}(\varepsilon, t) = \boldsymbol{k}^i] \\ &\ge \sum_{i=1}^p P(\boldsymbol{k}(\varepsilon, t) = \boldsymbol{k}^i) c(\boldsymbol{u}^i(t)) \\ &\ge \sum_{i=1}^p \nu_{\boldsymbol{k}^i} c(\boldsymbol{u}^i(t)) - \widehat{C}_8(\varepsilon + e^{-\beta_0 t/\varepsilon}), \end{aligned} \tag{9.72}$$

for some positive $\widehat{C}_8$, where β_0 is specified in Lemma B.3. By combining (9.71) and (9.72), we obtain

$$\begin{aligned}&\limsup_{T\to\infty}\frac{1}{T}E\int_0^T [h(\boldsymbol{x}^\varepsilon(t))+c(\boldsymbol{u}^\varepsilon(t))]\,dt\\&\qquad\geq \limsup_{T\to\infty}\frac{1}{T}\int_0^T\left[h(\boldsymbol{x}(t))+\sum_{i=1}^p \nu_{\boldsymbol{k}^i}c(\boldsymbol{u}^i(t))\right]dt-\widehat{C}_9\varepsilon,\end{aligned}$$

for some positive constant $\widehat{C}_9$. Inequalities (9.70)–(9.72) imply (9.68). □

9.5 Construction of a Near-Optimal Control

In this section, based on the proofs of Lemmas 4.1 and 4.2 and Theorem 4.1, we present a procedure to construct a near-optimal control from a near-optimal control of the limiting problem $\mathcal{P}^0$.

Construction of a Near-Optimal Control:

Step 1. In view of Lemma 4.2, choose

$$(\overline{\boldsymbol{u}}^1(t),\ldots,\overline{\boldsymbol{u}}^p(t))\in\mathcal{A}^0(\boldsymbol{x})$$

so that

$$\min_{1\leq j\leq m}\ \inf_{0\leq t<\infty}\bar{x}_j(t)\geq C_1\varepsilon^\delta \tag{9.73}$$

and

$$\lambda+\overline{C}_1\varepsilon^\delta\geq\limsup_{T\to\infty}\frac{1}{T}\int_0^T\left[h(\overline{\boldsymbol{x}}(t))+\sum_{i=1}^p\nu_{\boldsymbol{k}^i}c(\overline{\boldsymbol{u}}^i(t))\right]dt, \tag{9.74}$$

where $\overline{\boldsymbol{x}}(t)$ is the solution of

$$\frac{d}{dt}\boldsymbol{x}(t)=\mathcal{L}\left(\boldsymbol{x}(t),\sum_{i=1}^p\nu_{\boldsymbol{k}^i}\overline{\boldsymbol{u}}^i(t),\boldsymbol{z}\right),\quad \boldsymbol{x}(0)=\boldsymbol{x}.$$

Step 2. We construct the control

$$\overline{\boldsymbol{u}}^\varepsilon(t)=\sum_{i=1}^p I_{\{\boldsymbol{k}(\varepsilon,t)=\boldsymbol{k}^i\}}\overline{\boldsymbol{u}}^i(t),$$

and let $\overline{\boldsymbol{x}}^\varepsilon(t)$ be the solution of

$$\frac{d}{dt}\boldsymbol{x}(t)=\mathcal{L}\left(\boldsymbol{x}(t),\overline{\boldsymbol{u}}^\varepsilon(t),\boldsymbol{z}\right),\quad \boldsymbol{x}(0)=\boldsymbol{x}.$$

Step 3. *Sub-step* 1. Define

$$\mathcal{B}_1^\varepsilon = \left\{ t : \bar{x}_1^\varepsilon(t) - \inf_{0\le s\le t} \bar{x}_1^\varepsilon(s) = 0 \text{ and } x_1^\varepsilon(t) < 0 \right\}.$$

We choose $\{\widetilde{u}_{\ell,1}^\varepsilon(t) : -n_0+1 \le \ell \le 0\}$ and $\{\widetilde{u}_{1,\ell}^\varepsilon(t) : 2 \le \ell \le n\}$ such that

$$\begin{aligned} \widetilde{u}_{\ell,1}^\varepsilon(t) &\le \bar{u}_{\ell,1}^\varepsilon(t), \quad \ell = -n_0+1,\dots,0, \\ \widetilde{u}_{1,\ell}^\varepsilon(t) &\le \bar{u}_{1,\ell}^\varepsilon(t), \quad \ell = 2,\dots,n, \end{aligned}$$

and

$$\sum_{\ell=-n_0+1}^{0} \widetilde{u}_{\ell,1}^{\varepsilon,1}(t) = \sum_{\ell=2}^{n} \widetilde{u}_{1,\ell}^{\varepsilon,1}(t) = \left(\sum_{\ell=2}^{n} \bar{u}_{1,\ell}^\varepsilon(t) \right) \wedge \left(\sum_{\ell=-n_0+1}^{0} \bar{u}_{\ell,1}^\varepsilon(t) \right). \tag{9.75}$$

For $(i,j) \in \{(\ell,1) : -n_0+1 \le \ell \le 0\} \cup \{(1,\ell) : 2 \le \ell \le n\}$, define

$$u_{i,j}^{\varepsilon,1}(t) = \begin{cases} \bar{u}_{i,j}^\varepsilon(t), & \text{if } t \notin \mathcal{B}_1^\varepsilon, \\ \widetilde{u}_{i,j}^\varepsilon(t), & \text{if } t \in \mathcal{B}_1^\varepsilon. \end{cases} \tag{9.76}$$

Sub-step 2. Let

$$\widetilde{x}_2^{\varepsilon,2} = x_2 e^{-a_2 t} + \int_0^t e^{-a_2(t-s)} \left[\sum_{\ell=-n_0+1}^{0} \bar{u}_{\ell,2}^\varepsilon(s) + u_{1,2}^{\varepsilon,1}(s) - \sum_{\ell=3}^{n} \bar{u}_{2,\ell}^\varepsilon(s) \right] ds$$

and define

$$\mathcal{B}_2^\varepsilon = \left\{ t : \widetilde{x}_2^{\varepsilon,2}(t) - \inf_{0\le s\le t} \widetilde{x}_2^{\varepsilon,2}(s) = 0 \text{ and } x_2^{\varepsilon,2}(t) < 0 \right\}.$$

We choose $\{\widetilde{u}_{\ell,2}^{\varepsilon,2}(t) : -n_0+1 \le \ell \le 1\}$ and $\{\widetilde{u}_{2,\ell}^{\varepsilon,2}(t) : 3 \le \ell \le n\}$ such that

$$\begin{aligned} &\widetilde{u}_{\ell,2}^{\varepsilon,2}(t) \le \bar{u}_{\ell,2}^\varepsilon(t), \quad -n_0+1 \le \ell \le 1, \quad \widetilde{u}_{1,2}^{\varepsilon,2}(t) \le u_{1,2}^{\varepsilon,1}(t), \\ &\widetilde{u}_{2,\ell}^{\varepsilon,2}(t) \le \bar{u}_{2,\ell}^\varepsilon(t), \quad 3 \le \ell \le n, \end{aligned}$$

and

$$\begin{aligned} \sum_{\ell=-n_0+1}^{1} \widetilde{u}_{\ell,2}^{\varepsilon,2}(t) &= \sum_{\ell=3}^{n} \widetilde{u}_{2,\ell}^{\varepsilon,2}(t) \\ &= \left(\sum_{\ell=3}^{n} \bar{u}_{2,\ell}^\varepsilon(t) \right) \wedge \left(\sum_{\ell=-n_0+1}^{0} \bar{u}_{\ell,2}^\varepsilon(t) + u_{1,2}^{\varepsilon,1}(t) \right). \end{aligned} \tag{9.77}$$

For $(i,j) \in \{(\ell,2) : -n_0+1 \le \ell \le 0\} \cup \{(2,\ell) : 3 \le \ell \le n\}$, define

$$u_{i,j}^{\varepsilon,2}(t) = \begin{cases} \bar{u}_{i,j}^{\varepsilon}(t), & \text{if } t \notin \mathcal{B}_2^{\varepsilon}, \\ \widetilde{u}_{i,j}^{\varepsilon,2}(t), & \text{if } t \in \mathcal{B}_2^{\varepsilon}, \end{cases} \tag{9.78}$$

and

$$u_{1,2}^{\varepsilon,2}(t) = \begin{cases} u_{1,2}^{\varepsilon,1}(t), & \text{if } t \notin \mathcal{B}_2^{\varepsilon}, \\ \widetilde{u}_{1,2}^{\varepsilon,2}(t), & \text{if } t \in \mathcal{B}_2^{\varepsilon}. \end{cases}$$

Sub-step ℓ $(3 \le \ell \le m)$:
Let

$$\widetilde{x}_{\ell}^{\varepsilon,\ell}(t) = x_{\ell}e^{-a_{\ell}t} + \int_0^t e^{-a_{\ell}(t-s)} \left[\sum_{\ell_1=-n_0+1}^{0} \bar{u}_{\ell_1,\ell}^{\varepsilon}(s) + \sum_{\ell_1=1}^{\ell-1} u_{\ell_1,\ell}^{\varepsilon,\ell_1}(s) - \sum_{\ell_1=\ell+1}^{n} \bar{u}_{\ell,\ell_1}^{\varepsilon}(s) \right] ds,$$

and define

$$\mathcal{B}_{\ell}^{\varepsilon} = \left\{ t : \widetilde{x}_{\ell}^{\varepsilon,\ell}(t) - \inf_{0 \le s \le t} \widetilde{x}_{\ell}^{\varepsilon,\ell}(s) = 0 \text{ and } \widetilde{x}_{\ell}^{\varepsilon,\ell}(t) < 0 \right\}.$$

We choose $\{\widetilde{u}_{\ell_1,\ell}^{\varepsilon,\ell}(t) : -n_0+1 \le \ell_1 \le \ell-1\}$ and $\{\widetilde{u}_{\ell,\ell_1}^{\varepsilon,\ell}(t) : \ell+1 \le \ell_1 \le n\}$ such that

$$\begin{aligned} &\widetilde{u}_{\ell_1,\ell}^{\varepsilon,\ell}(t) \le \bar{u}_{\ell_1,\ell}^{\varepsilon}(t), \quad -n_0+1 \le \ell_1 \le 0, \\ &\widetilde{u}_{\ell_1,\ell}^{\varepsilon,\ell}(t) \le u_{\ell_1,\ell}^{\varepsilon,\ell_1}(t), \quad \ell_1 = 1,\ldots,\ell-1, \\ &\widetilde{u}_{\ell,\ell_1}^{\varepsilon,\ell}(t) \le \bar{u}_{\ell,\ell_1}^{\varepsilon}(t), \quad \ell+1 \le \ell_1 \le n, \end{aligned}$$

and

$$\begin{aligned} \sum_{\ell_1=-n_0+1}^{\ell-1} \widetilde{u}_{\ell_1,\ell}^{\varepsilon,\ell}(t) &= \sum_{\ell_1=\ell+1}^{n} \widetilde{u}_{\ell,\ell_1}^{\varepsilon,\ell}(t) \\ &= \left(\sum_{\ell_1=\ell+1}^{n} \bar{u}_{\ell,\ell_1}^{\varepsilon}(t) \right) \wedge \left(\sum_{\ell_1=-n_0+1}^{0} \bar{u}_{\ell_1,\ell}^{\varepsilon}(t) + \sum_{\ell_1=1}^{\ell-1} u_{\ell_1,\ell}^{\varepsilon,\ell_1}(t) \right). \end{aligned} \tag{9.79}$$

For $(i,j) \in \{(\ell_1,\ell) : -n_0+1 \le \ell_1 \le 0\} \cup \{(\ell,\ell_1) : \ell+1 \le \ell_1 \le n\}$, define

$$u_{i,j}^{\varepsilon,\ell}(t) = \begin{cases} \bar{u}_{i,j}^{\varepsilon}(t), & \text{if } t \notin \mathcal{B}_{\ell}^{\varepsilon}, \\ \widetilde{u}_{i,j}^{\varepsilon,\ell}(t), & \text{if } t \in \mathcal{B}_{\ell}^{\varepsilon}, \end{cases} \tag{9.80}$$

and for $(i,j) \in \{(\ell_1, \ell) : 1 \le \ell_1 \le \ell - 1\}$, define

$$u_{i,j}^{\varepsilon,\ell}(t) = \begin{cases} u_{i,j}^{\varepsilon,\ell-1}(t), & \text{if } t \notin \mathcal{B}_\ell^\varepsilon, \\ \widetilde{u}_{i,j}^{\varepsilon,\ell}(t), & \text{if } t \in \mathcal{B}_\ell^\varepsilon. \end{cases} \tag{9.81}$$

Finally, set

$$u_{i,j}^\varepsilon = u_{i,j}^{\varepsilon,j}(t).$$

We get the desired control $\boldsymbol{u}^\varepsilon(\cdot)$.

9.6 Notes

This chapter is based on Sethi, Zhang, and Zhang [123]. The averaging approach used here requires computation of equilibrium measures associated with the generator of the capacity process. Lasserre [86] has suggested how this might be efficiently carried out for processes with a large number of states. Furthermore, it should be noted that the limiting problem can be solved by the software MISER3 developed by Jennings, Fisher, Teo, and Goh [74]. Indeed, the method has been applied by Jennings, Sethi, and Teo [75] to solve the limiting problem associated with the example of Figure 5.1 and with other more complicated jobshops.

10

Near-Optimal Risk-Sensitive Control

10.1 Introduction

In this chapter we consider near-optimal control with a risk-sensitive cost criterion. The machine capacity process is assumed to be a finite state Markov chain with generator Q/ε, where $\varepsilon > 0$ is a parameter related to the frequency of machine breakdown and repair relative to an underlying production time scale. For simplicity, as in Chapter 6, we consider a one-product manufacturing system with constant demand. The control is the production rate, which is subject to a random machine capacity constraint. For a fixed $\varepsilon > 0$, the goal is to find a control policy which minimizes the long-term growth rate of an expected exponential-of-integral criterion. We discuss the asymptotic property of the problem as the rate of fluctuation of the production capacity process goes to infinity ($\varepsilon \to 0$). We show that the risk-sensitive control problem can be approximated by a limiting problem in which the stochastic capacity process can be averaged out and replaced by its average. This procedure is analogous to passing in limit the disturbance attenuation problem from the risk-sensitive model with small noise intensity to the deterministic robust control.

The plan of this chapter is as follows. In Section 10.2 we formulate the model. In Section 10.3 we derive the limiting problem, and investigate the asymptotic property of the problem as the rate of fluctuation of the production capacity process goes to infinity ($\varepsilon \to 0$). The chapter is concluded in Section 10.4.

10.2 Problem Formulation

Let us consider a one-product, parallel-machine manufacturing system with stochastic production capacity facing a constant demand for its production. For $t \geq 0$, let $x(t)$, $u(t)$, and z denote the surplus level, the production rate, and the constant demand rate, respectively. We assume $x(t) \in \Re$, $u(t) \in \Re_+$, $t \geq 0$, and $z > 0$ a constant. They satisfy the differential equation

$$\frac{d}{dt}x(t) = -ax(t) + u(t) - z, \quad x(0) = x, \tag{10.1}$$

where $a > 0$ is a constant representing the deterioration rate (or spoilage rate) of the finished product.

Let $k(\varepsilon, t) \in \mathcal{M} = \{0, 1, 2, \ldots, m\}$, $t \geq 0$, denote a Markov chain generated by $(1/\varepsilon)Q$, where $\varepsilon > 0$ is a small parameter and $Q = (q_{ij})$, $i, j \in \mathcal{M}$, is an $(m+1)\times(m+1)$ matrix such that $q_{ij} \geq 0$ for $i \neq j$ and $q_{ii} = -\sum_{j\neq i} q_{ij}$. We let $k(\varepsilon, t)$ represent the maximum production capacity of the system at time t. The representation for $\mathcal{M}$ usually stands for the case of m identical machines, each with a unit capacity and having two states: up and down.

The production constraints are given by the inequalities

$$0 \leq u(t) \leq k(\varepsilon, t), \quad t \geq 0.$$

Definition 2.1. A production control process $u(\cdot) = \{u(t) : t \geq 0\}$ is *admissible* if: (i) $u(t)$ is $\sigma\{k(\varepsilon, s) : 0 \leq s \leq t\}$ progressively measurable; and (ii) $0 \leq u(t) \leq k(\varepsilon, t)$ for all $t \geq 0$. □

Let $\mathcal{A}^\varepsilon(k)$ denote the class of admissible controls with the initial condition $k(\varepsilon, 0) = k$. Let $g(x, u)$ denote a cost function of the surplus and the production. For each $\varepsilon > 0$, the objective of the problem is to choose $u(\cdot) \in \mathcal{A}^\varepsilon(k)$ to minimize

$$J^\varepsilon(x, k, u(\cdot)) = \limsup_{T\to\infty} \frac{\varepsilon}{T} \log E\left[\exp\left(\frac{1}{\varepsilon}\int_0^T g(x(t), u(t))\,dt\right)\right], \tag{10.2}$$

where $x(\cdot)$ is the surplus process corresponding to the production process $u(\cdot)$. We summarize our control problem as follows:

$$\mathcal{P}^\varepsilon: \begin{cases} \min\ J^\varepsilon(x, k, u(\cdot)) \\ \qquad = \displaystyle\limsup_{T\to\infty} \frac{\varepsilon}{T} \log E \exp\left(\frac{1}{\varepsilon}\int_0^T g(x(t), u(t))\,dt\right), \\ \text{s.t. } \displaystyle\frac{d}{dt}x(t) = -ax(t) + u(t) - z, \quad x(0) = x, \quad u(\cdot) \in \mathcal{A}^\varepsilon(k), \\ \text{minimum average cost } \ \lambda^\varepsilon = \displaystyle\inf_{u(\cdot)\in\mathcal{A}^\varepsilon(k)} J^\varepsilon(x, k, u(\cdot)). \end{cases} \tag{10.3}$$

Remark 2.1. The positive spoilage rate a implies a uniform bound for $x(t)$; see Remark 6.2.2. □

We assume the cost function $g(x,u)$ and the production capacity process $k(\varepsilon,\cdot)$ to satisfy the following:

(A1) $g(x,u) \geq 0$ is continuous, bounded, and uniformly Lipschitz in x.

(A2) Q is strongly irreducible.

Let $\boldsymbol{\nu} = (\nu_0, \nu_1, \ldots, \nu_m)$ denote the equilibrium distribution of $k(\varepsilon,\cdot)$. Formally, we can write the associated Hamilton-Jacobi-Bellman (HJB) equation as follows:

$$\begin{aligned}\frac{\lambda^\varepsilon}{\varepsilon} = \inf_{0\le u\le k}\Bigg\{&\frac{-ax+u-z}{\varepsilon}\frac{\partial\psi^\varepsilon(x,k)}{\partial x}\\ &+\exp\left(-\frac{\psi^\varepsilon(x,k)}{\varepsilon}\right)\frac{Q}{\varepsilon}\exp\left(\frac{\psi^\varepsilon(x,\cdot)}{\varepsilon}\right)(k)+\frac{g(x,u)}{\varepsilon}\Bigg\},\end{aligned}$$

where $\psi^\varepsilon(x,k)$ is the potential function defined on $\Re \times \mathcal{M}$. By multiplying both sides of this equation with ε, we have

$$\begin{aligned}\lambda^\varepsilon = \inf_{0\le u\le k}\Bigg\{&(-ax+u-z)\frac{\partial\psi^\varepsilon(x,k)}{\partial x}\\ &+\exp\left(-\frac{\psi^\varepsilon(x,k)}{\varepsilon}\right)Q\exp\left(\frac{\psi^\varepsilon(x,\cdot)}{\varepsilon}\right)(k)+g(x,u)\Bigg\}.\end{aligned} \tag{10.4}$$

10.3 The Limiting Problem

In this section, we analyze the asymptotic property of the HJB equation (10.4) as $\varepsilon \to 0$. We note that this HJB equation has an additional term involving the exponential functions, when compared to the HJB equation associated with an ordinary long-run average-cost problem. In order to get rid of the exponential term, we make use of the logarithmic transformation, as in Fleming and Soner [56, p. 275].

Let

$$\mathcal{V} = \{v = (v(0),\ldots,v(m)) \in \Re^{m+1} : \ v(i) > 0,\ i = 0,1,\ldots,m\}.$$

Define

$$Q^v = (q^v_{ij}) \quad \text{such that} \quad q^v_{ij} = q_{ij}\frac{v(j)}{v(i)} \quad \text{for} \quad i \neq j \text{ and } q^v_{ii} = -\sum_{j\neq i} q^v_{ij}.$$

Then, in view of the logarithmic transformation, we have, for each $k \in \mathcal{M}$,

$$\begin{aligned}
&\exp\left(\frac{-\psi^\varepsilon(x,k)}{\varepsilon}\right) Q \exp\left(\frac{-\psi^\varepsilon(x,\cdot)}{\varepsilon}\right)(k) \\
&\quad = \sup_{v\in\mathcal{V}} \left\{ \frac{Q^v}{\varepsilon}\psi^\varepsilon(x,\cdot)(k) + \frac{Qv(\cdot)(k)}{v(k)} - Q^v(\log v(\cdot))(k) \right\}.
\end{aligned}$$

The supremum is obtained at $v(k) = \exp(-\psi^\varepsilon(x,k)/\varepsilon)$.

The logarithmic transformation suggests that the HJB equation is equivalent to an Isaacs equation of a two-player zero-sum dynamic stochastic game. The Isaacs equation is given as follows:

$$\lambda^\varepsilon = \inf_{0\le u\le k}\sup_{v\in\mathcal{V}} \left\{ (-ax+u-z)\frac{\partial\psi^\varepsilon(x,k)}{\partial x} + \widetilde{g}(x,u,v,k) + \frac{Q^v}{\varepsilon}\psi^\varepsilon(x,\cdot)(k) \right\} \tag{10.5}$$

where

$$\widetilde{g}(x,u,v,k) = g(x,u) + \frac{Qv(\cdot)(k)}{v(k)} - Q^v(\log v(\cdot))(k), \tag{10.6}$$

for $k \in \mathcal{M}$.

Remark 3.1. Note that if $v = (1,\dots,1)$, then $\widetilde{g}(x,u,v,k) = g(x,u)$ and $Q^v = Q$. □

Remark 3.2. In the results to follow, we will not give a precise description of the stochastic dynamic game with the Isaacs equation (10.5), since this interpretation will not be used in proving our results about the deterministic limit as $\varepsilon \to 0$. In the game, $u(t)$ and $v(t)$ represent minimizing and maximizing controls, based on information available at time t. Note that the maximizing control v produces a change in transition rates, from q_{ij} to q_{ij}^v. This idea can be made precise using Elliott-Kalton type strategies; see Fleming and Souganidis [57]. Since the order in (10.5) is $\inf(\sup(\cdots))$ rather than $\sup(\inf(\cdots))$, λ^ε turns out to be the upper game value for the game payoff

$$\limsup_{T\to\infty} \frac{1}{T} E \int_0^T \widetilde{g}(x(t),u(t),v(t),k(\varepsilon,t))\,dt. \qquad \square$$

We consider the limit of the problem as $\varepsilon \to 0$. We first define the control sets for the limiting problem. Let

$$\Gamma_u = \{U = (u^0,\dots,u^m);\ 0 \le u^i \le i,\ i = 0,\dots,m\}$$

and

$$\Gamma_v = \{V = (v^0,\dots,v^m);\ v^i = (v^i(0),\dots,v^i(m)) \in \mathcal{V},\ i = 0,\dots,m\}.$$

For each $V \in \Gamma_v$, let $\overline{Q}^V := (q_{ij}^V)$ such that

$$q_{ij}^{v^i} = q_{ij}^V = \frac{q_{ij} v^i(j)}{v^i(i)} \quad \text{for} \quad i \neq j \text{ and } q_{ii}^V = -\sum_{j \neq i} q_{ij}^V,$$

and let $\nu^V = (\nu_0^V, \ldots, \nu_m^V)$ denote the equilibrium distribution of $\overline{Q}^V$. The next lemma says that $\overline{Q}^V$ is irreducible. Therefore, there exists a unique positive ν^V for each $V \in \Gamma_v$. Moreover, ν^V depends continuously on V.

Lemma 3.1. *Assume that Assumptions* (A1) *and* (A2) *hold. For each* $V \in \Gamma_v$, $\overline{Q}^V$ *is irreducible.*

Proof. We divide the proof into three steps.

Step 1. $\text{rank}(\overline{Q}^V) = m$.

First of all, it is easy to see that the irreducibility of Q implies $q_{kk}^V < 0$, for $k = 0, 1, \ldots, m$. We multiply the first row of $\overline{Q}^V$ by $-q_{k0}^V / q_{00}^V$ and add to the kth row, $k = 1, \ldots, m$, to make the first component of that row vanish. Let $Q^{V,1} = (q_{ij}^{V,1})$ denote the resulting matrix. Then, $Q^{V,1}$ must satisfy

$$\begin{aligned} q_{0j}^{V,1} &= q_{0j}^V, \quad j = 0, 1, \ldots, m, \\ q_{k0}^{V,1} &= 0, \quad\quad k = 1, \ldots, m, \\ q_{kk}^{V,1} &\leq 0, \quad\quad k = 1, \ldots, m, \quad \text{and} \\ \sum_{j=0}^{m} q_{kj}^{V,1} &= 0, \quad\quad k = 0, 1, \ldots, m. \end{aligned}$$

We now show that $q_{kk}^{V,1} < 0$ for $k = 1, \ldots, m$. For $k = 1$, if $q_{11}^{V,1} \geq 0$, then it must be equal to 0, which implies

$$(q_{12}^V, \ldots, q_{1m}^V) - \left(\frac{q_{10}^V}{q_{00}^V} \right) (q_{02}^V, \ldots, q_{0m}^V) = 0. \tag{10.7}$$

Recall that $q_{11}^V \neq 0$. One must have $q_{10}^V > 0$, since otherwise $q_{10}^V = 0$ implies $q_{11}^V = q_{11}^{V,1} = 0$, which contradicts the fact that $q_{kk}^V < 0$, for $k = 0, 1, \ldots, m$. Thus, $-q_{10}^V / q_{00}^V > 0$. This together with the nonnegativity of q_{ij}^V, $i \neq j$, implies that both of the vectors in (10.7) must be equal to 0, i.e.,

$$(q_{12}^V, \ldots, q_{1m}^V) = 0 \quad \text{and} \quad (q_{02}^V, \ldots, q_{0m}^V) = 0.$$

These equations imply that a state in $\{2, 3, \ldots, m\}$ is not accessible from a state in $\{0, 1\}$. They contradict the irreducibility of Q. Therefore, one must have $q_{11}^{V,1} < 0$. Similarly, we can show that $q_{kk}^{V,1} < 0$ for $k = 2, \ldots, m$.

We repeat this procedure in a similar way by multiplying the second row of $Q^{V,1}$ by $-q_{k1}^{V,1}/q_{11}^{V,1}$, $k = 2, \ldots, m$, and adding to the kth row. Let $Q^{V,2)} = (q_{ij}^{V,2})$ denote the resulting matrix. Then one has

$$\begin{aligned} q_{ij}^{V,2} &= q_{ij}^{V,1}, \quad i = 0, 1, \quad j = 0, 1, \ldots, m, \\ q_{ij}^{V,2} &= 0, \qquad i = 2, \ldots, m, \quad j = 0, 1, \ldots, m, \\ q_{kk}^{V,2} &\leq 0, \qquad k = 2, \ldots, m, \quad \text{and} \\ \sum_{j=0}^{m} q_{kj}^{V,2} &= 0, \qquad k = 0, 1, \ldots, m. \end{aligned}$$

Similarly, we can show that $q_{kk}^{V,2} < 0$ for $k = 2, \ldots, m$.

We continue this procedure and transform $Q \to Q^{V,1} \to \cdots \to Q^{V,m-1}$ with $Q^{V,m-1} = (q_{ij}^{V,m-1})$, such that

$$\begin{aligned} q_{ij}^{V,m-1} &= 0, \quad i > j, \\ q_{kk}^{V,m-1} &< 0, \quad k = 0, 1, \ldots, m-1, \\ \sum_{j=0}^{m} q_{kj}^{V,m-1} &= 0, \quad k \in \mathcal{M}, \quad \text{and} \\ q_{mm}^{V,m-1} &= 0. \end{aligned}$$

Notice that the prescribed transformations do not change the rank of the original matrix. Thus,

$$\operatorname{rank}(\overline{Q}^V) = \operatorname{rank}(Q^{V,1}) = \cdots = \operatorname{rank}(Q^{V,m-1}) = m.$$

Step 2. $\overline{Q}^V$ is weakly irreducible.
Consider an $(m+1)$ row vector $b = (b_0, \ldots, b_m)$ such that

$$b\overline{Q}^V = 0 \quad \text{and} \quad b_0 + \cdots + b_m = 1.$$

It follows from Lemma B.3 that

$$\lim_{t \to \infty} \exp(\overline{Q}^V t) = \begin{pmatrix} \nu_0^V & \nu_0^V & \cdots & \nu_0^V \\ \nu_1^V & \nu_1^V & \cdots & \nu_1^V \\ \vdots & \vdots & \cdots & \vdots \\ \nu_m^V & \nu_m^V & \cdots & \nu_m^V \end{pmatrix}.$$

Since $\exp(\overline{Q}^V t)$ represents the transition probabilities, the limit b must be nonnegative. Thus, $b = (\nu_0^V, \ldots, \nu_m^V)$ is an equilibrium distribution of $\overline{Q}^V$. Note that the kernel$(\overline{Q}^V)' = \operatorname{span}\{(\nu_0^V, \ldots, \nu_m^V)\}$, since $\operatorname{rank}(\overline{Q}^V)' =$

$\text{rank}(\overline{Q}^V) = m$. Then, $c = (\nu_0^V, \ldots, \nu_m^V)$ is the unique nonnegative solution to $b\overline{Q}^V = 0$ and $b_0 + \cdots + b_m = 1$. Hence, $\overline{Q}^V$ is weakly irreducible.

Step 3. $\overline{Q}^V$ is irreducible, i.e., $(\nu_0^V, \ldots, \nu_m^V) > 0$.

If not, then without loss of generality we may assume $\nu_0^V > 0, \ldots, \nu_{k_0}^V > 0$ and $\nu_{k_0+1}^V = 0, \ldots, \nu_m^V = 0$, for some $k_0 = 0, 1, \ldots, m$. Note that the equation $(\nu_0^V, \ldots, \nu_m^V)\overline{Q}^V = 0$ implies that $q_{ij}^V = 0$, and thus $q_{ij} = 0$, for $i = 0, \ldots, k_0$ and $j = k_0 + 1, \ldots, m$. This in turn implies that Q is not irreducible, since the process $k(\varepsilon, \cdot)$ cannot jump from a state in $\{0, 1, \ldots, k_0\}$ to a state in $\{k_0 + 1, \ldots, m\}$. The contradiction yields the irreducibility of $\overline{Q}^V$. □

Let

$$\widehat{g}(x, U, V) = \sum_{i=0}^{m} \nu_i^V g(x, u^i) + \sum_{i=0}^{m} \nu_i^V \frac{Qv^i(\cdot)(i)}{v^i(i)} - \sum_{i=0}^{m} \nu_i^V \overline{Q}^V (\log v^i(\cdot))(i).$$

Note that $\widehat{g}(x, U, V) \le ||g||$, where $||\cdot||$ is the sup norm. Moreover, since $g(x, u) \ge 0$, $\widehat{g}(x, U, 1) \ge 0$, where $V = 1$ means $v^i(j) = 1$ for all i, j.

Theorem 3.1. *Let Assumptions* (A1) *and* (A2) *hold. Let* $\varepsilon_n \to 0$ *be a sequence such that* $\lambda^{\varepsilon_n} \to \lambda^0$ *and* $\psi^{\varepsilon_n}(x, k) \to \psi^0(x, k)$*. Then:*

(i) $\psi^0(x, k)$ *is independent of* k, *i.e.,* $\psi^0(x, k) = \psi^0(x)$;

(ii) $\psi^0(x)$ *is Lipschitz;*

(iii) $(\lambda^0, w^0(x))$ *is a viscosity solution to the following Isaacs equation:*

$$\lambda^0 = \inf_{U \in \Gamma_u} \sup_{V \in \Gamma_v} \left\{ \left(-ax + \sum_{i=0}^{m} \nu_i^V u^i - z \right) \frac{\partial \psi^0(x)}{\partial x} + \widehat{g}(x, U, V) \right\}; \tag{10.8}$$

(iv) *with*

$$J^0(U(\cdot), V(\cdot)) = \limsup_{T \to \infty} \frac{1}{T} \int_0^T \widehat{g}(x(t), U(t), V(t))\, dt,$$

$$\lambda^0 = \inf_{U(\cdot)} \left(\sup_{V(\cdot)} J^0(U(\cdot), V(\cdot)) \right),$$

subject to

$$\frac{d}{dt} x(t) = -ax(t) + \sum_{i=0}^{m} \nu_i^{V(t)} u^i(t) - z, \quad x(0) = x,$$

where $U(\cdot)$ *and* $V(\cdot)$ *are Borel measurable functions and* $U(t) \in \Gamma_u$ *and* $V(t) \in \Gamma_v$ *for* $t \ge 0$.

Note that the equation in (10.8) is an Isaacs equation associated with a two-player, zero-sum dynamic game with the objective J^0.

Proof of Theorem 3.1. Lemma 6.3.2(iii) implies that

$$|\psi_\rho^\varepsilon(x,k) - \psi_\rho^\varepsilon(x,\tilde{k})| \le \varepsilon \log C_3,$$

for x in any finite interval. Thus, the limit of $\psi_\rho^\varepsilon(x,k)$ must be independent of k, i.e.,

$$\psi^0(x,0) = \cdots = \psi^0(x,m) =: \psi^0(x).$$

The Lipschitz property of $\psi^0(x)$ follows from the Lipschitz property of $\psi^\varepsilon(x,k)$.

Note that

$$\nu^V \overline{Q}^V = (\nu_0^V, \ldots, \nu_m^V)\overline{Q}^V = 0.$$

It follows that

$$\sum_{i=0}^{m} \nu_i^V Q^{v^i} \psi^\varepsilon(x,\cdot)(i) = 0. \tag{10.9}$$

The remaining proof of (iii) is standard and can be carried out in the manner of Chapter 7.

The proof of (iv) involves the theory of viscosity solutions of differential games. We only sketch the proof below, and refer to Fleming and Zhang [58] for details.

Let

$$\begin{aligned} H(x,p) = \inf_{U\in\Gamma_u} \sup_{V\in\Gamma_v} \Bigg\{ &\left(-ax + \sum_{i=0}^{m} \nu^{V_i} u^i - z\right)p + \sum_{i=0}^{m} \nu_i^V g(x,u^i) \\ &+ \left(\sum_{i=0}^{m} \nu_i^V \frac{Qv^i(\cdot)(i)}{v^i(i)} - \sum_{i=0}^{m} \nu^{V_i} \overline{Q}^V (\log v^i(\cdot))(i)\right) \Bigg\}. \end{aligned}$$

Then,

$$|H(\tilde{x},p) - H(x,p)| \le \left(a|p| + \left\|\frac{\partial g(x,u)}{\partial x}\right\|\right) |\tilde{x} - x|,$$

$$|H(x,\tilde{p}) - H(x,p)| \le (a|x| + m + z)|\tilde{p} - p|.$$

These conditions imply the uniqueness of a viscosity solution to the following finite-time problem:

$$\begin{cases} \dfrac{\partial \Psi}{\partial T} = H\left(x, \dfrac{\partial \Psi(x,T)}{\partial x}\right) - \lambda^0, \quad T > 0, \\ \Psi(0,x) = w^0(x). \end{cases}$$

Uniqueness is in the class of continuous viscosity solutions $\Psi(x,T)$ such that $\Psi(\cdot,T)$ satisfies a uniform Lipschitz condition on every finite time interval $0 \le T \le T_1$; see Crandall and Lions [38] and Ishii [73].

The method of Evans and Souganidis [49] shows that the

$$\text{upper value}\left\{\int_0^T \left(\widehat{g}(x(t),U(t),V(t))-\lambda^0\right)dt+\psi^0(x(T))\right\} \tag{10.10}$$

is such a viscosity solution, and $\psi^0(x)$ is also a viscosity solution. So $\psi^0(x) = \Psi(T,x)$. Namely,

$$\psi^0(x) = \text{upper value}\left\{\int_0^T \left(\widehat{g}(x(t),U(t),V(t))-\lambda^0\right)dt+\psi^0(x(T))\right\}.$$

Using the above equality, one can show as in Fleming and McEneaney [52] that

$$\lambda^0 = \inf_{U(\cdot)}\sup_{V(\cdot)} J^0(U(\cdot),V(\cdot)). \qquad \square$$

We would like to comment on how to use the solution to the limiting problem to obtain a control for the original problem. Typically, an explicit solution is not available to either of the problems. A numerical scheme has to be used to obtain an approximate solution. The advantage of the limiting problem is its dimensionality, which is much smaller than that of the original problem if the number of states in $\mathcal{M}$ is large.

Let $(U^*(x),V^*(x))$ denote a solution to the upper value problem. The control

$$u(x,k(\varepsilon,t)) = \sum_{i=0}^m I_{\{k(\varepsilon,t)=i\}}u^{i*}(x)$$

is expected to be nearly optimal for the original problem.

Remark 3.3. Note that the last term in (10.8) is nonpositive. In fact, for each $v\in\mathcal{V}$ and $i\in\mathcal{M}$, we have

$$\begin{aligned}\frac{Qv(\cdot)(i)}{v(i)} - Q^v(\log v(\cdot))(i) &= \sum_{j\neq i} q_{ij}\left(\frac{v(j)}{v(i)}-1\right) - \sum_{j\neq i} q_{ij}\frac{v(j)}{v(i)}\log\frac{v(j)}{v(i)}\\ &= \sum_{j\neq i} q_{ij}\left(\frac{v(j)}{v(i)}-1-\frac{v(j)}{v(i)}\log\frac{v(j)}{v(i)}\right)\le 0,\end{aligned}$$

because the function $(x-1-x\log x)$ is nonpositive on $(0,\infty)$. It follows that

$$\sum_{i=0}^m \nu_i^V\frac{Qv^i(\cdot)(i)}{v^i(i)} - \sum_{i=0}^m \nu_i^V Q^{v^i}(\log v^i(\cdot))(i)\le 0. \qquad \square$$

Remark 3.4. Let $I(\mu)$ be the Donsker-Varadhan function, defined for any probability vector $\mu = (\mu_0,\ldots,\mu_m) > 0$. That is, for $\mu_i > 0$ and $\sum_{i=0}^m \mu_i = 1$,

$$I(\mu) = \sup_{\beta\in\mathcal{V}}[-\langle\mu,\beta^{-1}Q\beta\rangle],$$

see Fleming, Sheu, and Soner [54]. Then we have

$$-I(\mu) = \sup_{V,\nu^V=\mu} \left\{ \sum_{i=0}^{m} \nu_i^V \frac{Qv^i(\cdot)(i)}{v^i(i)} - \sum_{i=0}^{m} \nu_i^V \overline{Q}^V (\log v^i(\cdot))(i) \right\}. \tag{10.11}$$

To prove this, for each $V \in \Gamma_v$, let

$$K^V(i) = \overline{Q}^V (\log v^i(\cdot))(i) - \frac{1}{v^i(i)} Qv^i(\cdot)(i).$$

Then (10.11) can be written as $I(\mu) = \inf_{\nu^V=\mu} \langle \mu, K^V \rangle$. We first show that

$$I(\mu) \geq \inf_{\nu^V=\mu} \langle \mu, K^V \rangle. \tag{10.12}$$

It is elementary to show that there exists $\beta^* \in \mathcal{V}$ such that $I(\mu) = \langle \mu, (\beta^*)^{-1} Q\beta^* \rangle$. Then, in view of Lemma 3.2 in [54], we have

$$\mu = \nu^{V^*}, \quad \text{where } v^{i*}(j) = \beta^*(i)/\beta^*(j).$$

It follows that $\langle \mu, K^{V^*} \rangle = -\langle \mu, (\beta^*)^{-1} Q\beta^* \rangle$, because

$$\langle \mu, \overline{Q}^{V^*}(\phi) \rangle = \langle \nu^{V^*}, \overline{Q}^{V^*} \phi \rangle = 0.$$

This implies (10.12).

To show the opposite inequality, note that the logarithmic transformation

$$e^{-\phi(i)} Q(e^{\phi(\cdot)})(i) = \sup_{V \in \Gamma_v} [\overline{Q}^V \phi(\cdot)(i) - K^V(i)],$$

for all ϕ. Let $\phi = \beta^*$. Then, for each V such that $\nu^V = \mu$, we have

$$\frac{1}{\beta^*} Q\beta^* \geq \overline{Q}^V \phi - K^V.$$

Hence, for $\nu^V = \mu$, we obtain

$$\left\langle \mu, \frac{1}{\beta^*} Q\beta^* \right\rangle \geq \langle \mu, \overline{Q}^V \phi \rangle - \langle \mu, K^V \rangle = -\langle \mu, K^V \rangle.$$

This implies that $I(\mu) \leq \inf_{\nu^V=\mu} \langle \mu, K^V \rangle$. Thus, (10.11) is established.

It follows from (10.11) that (10.8) is equivalent to

$$\lambda^0 = \inf_{U \in \Gamma_u} \sup_{\mu} \left\{ \big(-ax + \langle \mu, u \rangle - z \big) \frac{\partial w^0(x)}{\partial x} + \langle \mu, g(x,u) \rangle - I(\mu) \right\}.$$

Similarly, the dynamics of $x(t)$ can be written as

$$\frac{d}{dt} x(t) = -ax(t) + \langle \mu(t), u(t) \rangle - z. \qquad \square$$

10.4 Notes

This chapter is based on Fleming and Zhang [58]. Only a single machine single product model is considered. It would be interesting to generalize the results to more general manufacturing systems such as flowshops and jobshops in Chapters 8 and 9.

In addition to the long-run average version of the risk-sensitive cost considered in this chapter, Zhang [151] studied the risk-sensitive discounted cost criterion

$$J^{\varepsilon,\rho}(x,k,u(\cdot)) = \sqrt{\varepsilon}\log E\exp\left(\frac{1}{\sqrt{\varepsilon}}\int_0^\infty e^{-\rho t}g(x(t),u(t))\,dt\right).$$

The scale parameter in the cost is $\sqrt{\varepsilon}$ instead of ε as in (6.2). This is because the convergence involving a discounted cost is mainly affected by the convergence rate of $k(\varepsilon,\cdot)$ to its equilibrium distribution, which is of order $\sqrt{\varepsilon}$.

Part IV:

Conclusions

11

Further Extensions and Open Research Problems

11.1 Introduction

In this book we have considered average-cost manufacturing systems in which deterministic as well as stochastic events occur at different time scales. In cases where rates at which production machines break down and get repaired are large, we have presented theoretical developments devoted to showing that properly designed decision-making procedures can lead to near optimization of its overall objective. In other words, we have shown that it is possible to base longer-term decisions (that respond to slower frequency events) on the average existing production capacity and an appropriately modified objective, and we expect these decisions to be nearly optimal even though the shorter-term capacity fluctuations, because of machine breakdowns and repairs, are ignored. Furthermore, having the longer-term decision in hand, one can then solve the simpler problem for obtaining optimal or near-optimal production rates.

In this chapter we summarize some specific results that have been obtained and related open problems. In Chapter 7 we have constructed feedback policies or controls for parallel-machine manufacturing systems and have shown these policies to be asymptotically optimal as the rates of occurrence of machine breakdown and repair events become arbitrarily large. Also obtained are the error estimates associated with some of the constructed controls. For more general manufacturing systems such as flowshops and jobshops, *only* asymptotic optimal open-loop controls have been constructed in Chapters 8 and 9. This is because of the difficulty posed by the inherent presence of the state constraints requiring that the inventories in the internal buffers (between any two machines) remain nonnegative.

The remainder of this chapter is devoted to indicating some important open problems and to concluding remarks. In Section 11.2 we discuss possible extensions of our results to multilevel hierarchical systems. In Section 11.3, we discuss some difficulties associated with obtaining asymptotic optimal feedback controls for flowshops and jobshops. In Section 11.4 we mention possibilities of relaxing the deterioration rate used in Chapter 10 in the context of risk-sensitive controls. In Section 11.5 we briefly discuss more general systems of interest. Section 11.6 concludes the book with some final thoughts.

11.2 Multilevel Systems

In this book, we have only considered production planning problems. It would be interesting to extend these results to multilevel hierarchical systems. For example, one could incorporate the capacity expansion decisions, as in Sethi and Zhang [125], into these models.

Our approach considers manufacturing firms in which events occur at different time scales. For example, changes in demand may occur far more slowly than breakdowns and repairs of production machines. This suggests that capital expansion decisions that respond to demand are relatively longer-term decisions than decisions regarding production. It is then possible to base capital expansion decisions on the average existing production capacity and to expect these decisions to be nearly optimal, even though the rapid capacity fluctuations are ignored. Having the longer-term decisions in hand, one can then solve the simpler problem of obtaining production rates. More specifically, we shall show that the two-level decisions constructed in this manner are asymptotically optimal as the rate of fluctuation in the production capacity becomes large in comparison with the rates at which other events are taking place.

11.3 Asymptotic Optimal Feedback Controls

As mentioned above, asymptotic optimal open-loop controls have been constructed for flowshops and jobshops, whereas it is only for parallel machine systems that asymptotic optimal *feedback* controls have been obtained. In single- or parallel-machine systems, either the Lipschitz property of optimal control for the corresponding deterministic systems, or the monotonicity property of optimal control with respect to the state variables, makes the proof of asymptotic optimality hold. Unfortunately, neither of these properties is available in the case of flowshops or jobshops.

However, it is an important task to construct asymptotic optimal feedback policies for flowshops and jobshops and their multilevel extensions, possibly with stochastic demand. Such control policies need to be constructed from optimal or near-optimal feedback controls of the appropriate limiting problems as $\varepsilon \to 0$. Furthermore, it is of interest to estimate how close the cost associated with the constructed feedback policies would be to the (theoretically) optimal cost.

There are two major obstacles in achieving this task. First, optimal feedback controls may not be Lipschitz in the presence of state constraints, and the ordinary differential equation describing the evolution of the state of the system may not have a unique solution. As Davis [42] has mentioned, this puts a severe restriction on the class of control policies that can be considered. Second, the existing proof of asymptotic optimality of constructed feedback controls (Theorem 7.5.2) requires the Lipschitz property.

In Theorem 7.5.2 for the single product case, the availability of the monotonicity property eliminates the need for the Lipschitz property, and allows us to obtain an asymptotic optimal feedback control.

Other possible research directions for constructing asymptotic optimal feedback policies are: (i) to make a smooth approximation of a non-Lipschitz feedback control; (ii) to use barrier or penalty methods to transform a state-constrained problem to an unconstrained problem and obtain a Lipschitz control; and (iii) to make a diffusion approximation for improving the regularity of the optimal policy. Once a candidate control policy is obtained by any of these means, new methods still need to be developed for proving the asymptotic optimality of the policy.

11.4 Robust Controls

In the risk-sensitive control considered in Chapter 10, it would be of interest to study the stability without the deterioration condition used in Lemma 6.3.1. One possible direction for attacking the problem is to use a "diminishing deterioration" approach by sending the deterioration rate $a \to 0$. In order to obtain the desired convergence of the potential function $\psi^\varepsilon(x, k)$ as $a \to 0$, it is necessary to have the uniform equicontinuity property that is typically guaranteed by the Lipschitz condition uniform with respect to $a > 0$. A major difficulty, however, is the absence of such a uniform Lipschitz property.

In our model, we assume a positive deterioration rate a for items in storage (formula (10.1)). This corresponds to a stability condition typically imposed for disturbance attenuation problems on an infinite time horizon (see Fleming and McEneaney [52]). Nevertheless, it would be interesting to weaken the assumption that $a > 0$.

11.5 General Systems

Our main concern in this book has been the construction of near-optimal decisions in observable manufacturing systems. In particular, we have assumed system dynamics to be linear and that the running cost or profit function be separable in the relevant variables, such as surplus levels and production rates.

For dealing with nonmanufacturing systems such as urban design systems, traffic flow systems, environmental systems, or even general systems for that matter, one might need to consider nonlinear dynamics, nonseparable cost functions, and partially observed stochastic processes; see Alexander [4] and Auger [7] for some nonmanufacturing systems of interest. Systems may also involve several players competing against each other. Such systems would require stochastic differential game formulations. While some progress has been made in the literature along these lines, much remains to be done. We shall now describe briefly the progress made to date.

Zhou and Sethi [153] consider systems with nonlinear dynamics and nonseparable costs and use a maximum principle approach in order to construct asymptotic optimal open-loop controls. In Sethi and Zhang [125], we have also considered nonseparable costs of surplus and production and have obtained asymptotic optimal feedback controls.

Finally, we would like to mention that it would be interesting to consider models with Markov chains involving weak and strong interactions. Quite often, the states of the underlying Markov chain are either subject to rather frequent changes or are naturally divisible to a number of groups such that the chain fluctuates very rapidly from one state to another within a single group, but jumps less rapidly from one group to another. In Yin and Zhang [149], such Markov chains are studied extensively. The results include asymptotic expansions of the probability distributions for the singularly perturbed Markov chains, the asymptotic normality of occupation measures, exponential error bounds, and structural properties of the underlying Markov chains with weak and strong interactions. These results can be used when dealing with more general manufacturing systems with much larger state space in $k(\varepsilon, t)$.

11.6 Final Thoughts

In conclusion, we have considered a variety of different manufacturing systems in this book and have constructed nearly optimal hierarchical controls for them. Additional features can be incorporated into the models considered. These include control-dependent generators, setups and capacity expansion, and marketing; see Sethi and Zhang [125].

In real-life systems many of these features may be simultaneously present.

Nevertheless, it should be possible without too much difficulty to construct candidate feedback controls that are likely to be asymptotically optimal. Moreover, in some cases, it may be possible, although tedious, to apply the methods presented in this book for proving asymptotic optimality of the candidate controls. In other cases, the existing methods may need to be supplemented with new ideas for obtaining a proof of asymptotic optimality.

While we should endeavor to develop these new ideas, we hope that the research presented in this book and in Sethi and Zhang [125] has already provided sufficient (at least heuristic) justification for using the candidate feedback policies in the management of realistically complex manufacturing systems.

Part V:

Appendices

Appendix A
Finite State Markov Chains

Let $k(\cdot) = \{k(t) : t \geq 0\}$ denote a stochastic process defined on a probability space $(\Omega, \mathcal{F}, P)$ with values in $\mathcal{M} = \{0, 1, 2, \ldots, m\}$. Then $\{k(t) : t \geq 0\}$ is a *Markov chain* if

$$P(k(t+s) = i | k(r) : r \leq s) = P(k(t+s) = i | k(s)),$$

for all $s, t \geq 0$ and $i \in \mathcal{M}$. We shall also write $k(\cdot)$ as $k(t)$, $t \geq 0$.

Let us assume that the transition probability $P(k(t+s) = j | k(s) = i)$ is stationary, i.e., it is independent of s. This allows us to introduce the notation $P_{ij}(t) = P(k(t+s) = j | k(s) = i)$. Then,

$$\begin{cases} P_{ij}(t) \geq 0, \quad i, j \in \mathcal{M}, \\ \displaystyle\sum_{j=0}^{m} P_{ij}(t) = 1, \quad i \in \mathcal{M}, \\ P_{ij}(t+s) = \displaystyle\sum_{l=0}^{m} P_{il}(s) P_{lj}(t), \quad t, s \geq 0, \quad i, j \in \mathcal{M}, \\ \text{(the Chapman-Kolmogorov relation)}. \end{cases}$$

Let $P(t)$ denote the $(m+1) \times (m+1)$ matrix $(P_{ij}(t))$ of stationary transition probabilities. We shall refer to $P(t)$ as the transition matrix of the Markov chain $k(\cdot)$. We postulate that

$$\lim_{t \to 0} P(t) = I_{m+1},$$

where I_{m+1} denotes the $(m+1) \times (m+1)$ identity matrix.

Let Q denote an $(m+1) \times (m+1)$ matrix such that $Q = (q_{ij})$ with $q_{ij} \geq 0$ for $j \neq i$ and $q_{ii} = -\sum_{j \neq i} q_{ij}$. Then one can constract a Markov chain $k(\cdot)$ as in Lemma A.1.

The following notation is used throughout the book:

$$Q\phi(\cdot)(i) = \sum_{j \neq i} q_{ij}(\phi(j) - \phi(i)),$$

for any function $\phi(\cdot)$ on $\mathcal{M}$. The matrix Q is called the *infinitesimal generator* (or simply *generator*) of $k(\cdot)$.

The transition matrix $P(t)$ is determined uniquely by the generator Q according to the following differential equation (see Karlin and Taylor [78]):

$$\frac{d}{dt}P(t) = P(t)Q = QP(t), \quad P(0) = I_{m+1}.$$

Thus,

$$q_{ij} = \begin{cases} \lim_{t \to 0^+} \dfrac{P_{ii}(t) - 1}{t}, & \text{if } j = i, \\ \lim_{t \to 0^+} \dfrac{P_{ij}(t)}{t}, & \text{if } j \neq i, \end{cases}$$

can be interpreted as the transition rate from state i to state j when $i \neq j$, and as the (negative of the) transition rate out of state i when $j = i$.

Let $\{\eta_k : k = 0, 1, 2, \ldots\}$ be a discrete-time Markov chain in $\mathcal{M}$ with initial distribution $(p_0, p_1, \ldots, p_m)$ and transition matrix $(p_{ij})_{(m+1)\times(m+1)}$ such that $p_{ii} = 0$ and $p_{ij} = q_{ij}/(\sum_{\ell \neq i} q_{i\ell})$. Let τ_0, τ_1, τ_2,... be independent and exponentially distributed random variables with unity density parameter.

Lemma A.1 (Construction of Markov Chains). *Let* $\lambda(i) = \sum_{j \neq i} q_{ij}$. *Then*

$$k(t) = \begin{cases} \eta_0, & \textit{if} \quad 0 \leq t < \dfrac{\tau_0}{\lambda(\eta_0)}, \\ \eta_\ell, & \textit{if} \quad \displaystyle\sum_{j=0}^{\ell-1} \frac{\tau_j}{\lambda(\eta_j)} \leq t < \sum_{j=0}^{\ell} \frac{\tau_j}{\lambda(\eta_j)}, \end{cases}$$

defines a Markov chain in $\mathcal{M}$ *with initial distribution* $(p_0, p_1, \ldots, p_m)$ *and generator* Q.

Proof. See Ethier and Kurtz [48] for a proof. □

Definition A.1 (Irreducibility). (i) A $(m+1) \times (m+1)$ matrix Q is said to be *weakly irreducible* if the equations

$$\boldsymbol{x}'Q = 0 \quad \text{and} \quad \sum_{i=0}^{m} x_i = 1 \tag{A.1}$$

have a unique solution $\boldsymbol{x}$ and $\boldsymbol{x} \geq \mathbf{0}$.

(ii) A $(m+1) \times (m+1)$ matrix Q is said to be *strongly irreducible*, or simply *irreducible*, if equations (A.1) have a unique solution $\boldsymbol{x} > \mathbf{0}$. □

The solution $\boldsymbol{x}$ to equations (A.1) is termed an *equilibrium distribution.*

Note that the rank of a weakly irreducible matrix Q is m. The difference between weak and strong irreducibility is that the former only requires the unique solution $\boldsymbol{x}$ to be nonnegative and the latter requires $\boldsymbol{x}$ to be strictly positive. In fact, the nonnegativity requirement in the weak irreducibility is superfluous. It is shown in Yin and Zhang [149] that if $\boldsymbol{x}$ is a solution to (A.1), then $\boldsymbol{x} \geq 0$. We keep the nonnegativity requirement in the definition for the sake of clarity.

Appendix B

Convergence and Error Estimates of Markov Chains

In this appendix we consider a Markov chain $k(t) = k(\varepsilon, t)$, $t \geq 0$, that has a generator $Q = Q^{(1)} + \varepsilon^{-1} Q^{(2)}$, where $\varepsilon > 0$ is a small parameter. We discuss several technical results that concern the asymptotic properties of the Markov chain $k(\varepsilon, \cdot)$ as ε tends to zero.

Let us assume that $Q^{(2)}$ is weakly irreducible throughout this appendix. Let $\boldsymbol{\nu} = (\nu_0, \nu_1, \ldots, \nu_m)$ denote the equilibrium distribution of $Q^{(2)}$. Then $\boldsymbol{\nu}$ is the only solution to the equations

$$\boldsymbol{\nu} Q^{(2)} = 0 \quad \text{and} \quad \sum_{i=0}^{m} \nu_i = 1.$$

Lemma B.1. *Let*

$$\overline{P} = \begin{pmatrix} \nu_0 & \nu_1 & \cdots & \nu_m \\ \vdots & \vdots & \cdots & \vdots \\ \nu_0 & \nu_1 & \cdots & \nu_m \end{pmatrix},$$

and let $P^0(t) = \exp(Q^{(2)} t)$. If $Q^{(2)}$ is weakly irreducible, then there exist constants C and $\beta_0 > 0$ such that

$$|P^0(t) - \overline{P}| \leq C e^{-\beta_0 t}.$$

Proof. By Chung [35, Theorem II.10.1], there exists an $(m+1) \times (m+1)$ matrix $\overline{P}_0$ such that $|P^0(t) - \overline{P}_0| \to 0$ as $t \to \infty$. Since $P^0(t)$ is a finite-dimensional matrix, the convergence must be exponential, i.e., $|P^0(t) - \overline{P}_0| \leq C e^{-\beta_0 t}$ for some $\beta_0 > 0$.

It remains to show that $\overline{P}_0 = \overline{P}$. Note that $dP^0(t)/dt \to 0$ as $t \to \infty$ (see Chung [35, Theorem II.12.8]. This implies $\overline{P}_0 Q^{(2)} = 0$. By the weak irreducibility of $Q^{(2)}$, we conclude that $\overline{P}_0 = \overline{P}$. □

Lemma B.2. *Let $P^\varepsilon(t) = \exp((\varepsilon Q^{(1)} + Q^{(2)})t)$ and $P^0(t) = \exp(Q^{(2)}t)$. Then a constant $C > 0$ exists such that, for $t \geq 0$,*

$$|P^\varepsilon(t) - P^0(t)| \leq C\varepsilon.$$

Proof. Let $Y(t) = P^\varepsilon(t) - P^0(t)$. Then,

$$\frac{d}{dt}Y(t) = (\varepsilon Q^{(1)} + Q^{(2)})Y(t) + \varepsilon Q^{(1)} P^0(t), \quad Y(0) = 0.$$

By solving this ordinary differential equation, we obtain

$$Y(t) = \varepsilon \int_0^t P^\varepsilon(t-s) Q^{(1)} P^0(s)\, ds.$$

Note that $Q^{(1)}\overline{P} = 0$. This yields

$$Y(t) = \varepsilon \int_0^t P^\varepsilon(t-s) Q^{(1)} (P^0(s) - \overline{P})\, ds.$$

By Lemma B.1, $|P^0(s) - \overline{P}| \leq C_1 e^{-\beta_0 s}$, where C_1 is a positive constant. Since $P^\varepsilon(t)$ is a transition probability matrix, there exists a positive constant C_2 such that $|P^\varepsilon(t-s)Q^{(1)}| \leq C_2$ for all $t \geq s \geq 0$. Thus,

$$\begin{aligned}
|Y(t)| &\leq \varepsilon(m+1) \int_0^t |P^\varepsilon(t-s)Q^{(1)}| \cdot |P^0(s) - \overline{P}|\, ds \\
&\leq \varepsilon C_1 C_2 (m+1) \int_0^t e^{-\beta_0 s}\, ds \\
&\leq \varepsilon C_1 C_2 (m+1) \beta_0^{-1}.
\end{aligned}$$

□

Lemma B.3. *Let $P(t)$ denote the transition matrix of the Markov chain $k(\varepsilon, \cdot)$. Then $P(t) = P^\varepsilon(t/\varepsilon)$ and*

$$|P(t) - \overline{P}| \leq C(\varepsilon + e^{-\beta_0 t/\varepsilon}), \tag{B.1}$$

for some constant C, where β_0 is given in Lemma B.1. *Moreover, for all $i \in \mathcal{M}$ and $t \geq 0$,*

$$|P(k(\varepsilon, t) = i) - \nu_i| \leq C(\varepsilon + e^{-\beta_0 t/\varepsilon}).$$

Proof. By Lemmas B.1 and B.2, we have

$$\begin{aligned}
|P(t) - \overline{P}| &= |P^\varepsilon(t/\varepsilon) - \overline{P}| \\
&\leq |P^\varepsilon(t/\varepsilon) - P^0(t/\varepsilon)| + |P^0(t/\varepsilon) - \overline{P}| \\
&\leq C_1(\varepsilon + e^{-\beta_0 t/\varepsilon}),
\end{aligned}$$

for some $C_1 > 0$. Therefore, by the Markov properties of $k(\varepsilon, t)$, $t \geq 0$,

$$\begin{aligned} |P(k(\varepsilon,t)=i)-\nu_i| &= \left| \sum_{j=0}^{m} P(k(\varepsilon,t)=i|k(\varepsilon,0)=j)P(k(\varepsilon,0)=j)-\nu_i \right| \\ &\leq \sum_{j=0}^{m} |P(k(\varepsilon,t)=i|k(\varepsilon,0)=j)-\nu_i|P(k(\varepsilon,0)=j) \\ &\leq C_1(\varepsilon + e^{-\beta_0 t/\varepsilon}). \end{aligned}$$

□

Lemma B.4. *Positive constants ε_0, θ, and C exist such that for $0 < \varepsilon \leq \varepsilon_0$, $i \in \mathcal{M}$, and for any uniformly bounded deterministic process $\beta(t)$, $t \geq 0$, on $[0, \infty)$, we have*

$$E\left[\exp\left\{\frac{\theta}{\sqrt{\varepsilon}(T+1)^{3/2}} \sup_{0\leq t\leq T}\left|\int_0^t \left(I_{\{k(\varepsilon,s)=i\}}-\nu_i\right)\beta(s)\,ds\right|\right\}\right] \leq C. \quad \text{(B.2)}$$

Proof. The proof is divided into several steps.

Step 1. In the first step, we prove (B.2) when the "sup" is absent.

Without loss of generality, we assume the uniform bound of $\beta(\cdot)$ is one. We let

$$\begin{aligned} \boldsymbol{\lambda}(t) &= \left(1_{\{k(\varepsilon,t)=0\}}, \ldots, 1_{\{k(\varepsilon,t)=m\}}\right)' \quad \text{and} \\ \boldsymbol{w}(t) &= \boldsymbol{\lambda}(t) - \boldsymbol{\lambda}(0) - \int_0^t Q'\boldsymbol{\lambda}(s)\,ds, \end{aligned}$$

where $Q = Q^{(1)} + \varepsilon^{-1}Q^{(2)}$ is the generator of the process $k(\varepsilon, \cdot)$. Then, it is well known (see Elliott [47]) that $\boldsymbol{w}(t) = (w_0(t), \ldots, w_m(t))'$, $t \geq 0$, is an $\{\mathcal{F}_t\}$-martingale, where $\mathcal{F}_t = \sigma\{k(\varepsilon, s) : s \leq t\}$, and

$$\boldsymbol{\lambda}(t) = \exp(Q't)\boldsymbol{\lambda}(0) + \int_0^t \exp(Q'(t-s))\,d\boldsymbol{w}(s).$$

By Lemma B.3,

$$\exp(Q't) - \overline{P}' = O(\varepsilon + e^{-\beta_0 t/\varepsilon}) \quad \text{and} \quad \overline{P}'\boldsymbol{\lambda}(t) = \boldsymbol{\nu},$$

where $\overline{P}$ is given in Lemma B.3. Hence,

$$\begin{aligned} \boldsymbol{\lambda}(t) - \boldsymbol{\nu} &= (\exp(Q't) - \overline{P}')\boldsymbol{\lambda}(0) \\ &\quad + \int_0^t [(\exp(Q'(t-s)) - \overline{P}') + \overline{P}']\,d\boldsymbol{w}(s) \\ &= O(\varepsilon + e^{-\beta_0 t/\varepsilon}) + \int_0^t [O(\varepsilon + e^{-\beta_0 (t-s)/\varepsilon}) + \overline{P}']\,d\boldsymbol{w}(s) \\ &= O(\varepsilon + e^{-\beta_0 t/\varepsilon}) + \int_0^t [O(\varepsilon + e^{-\beta_0 (t-s)/\varepsilon})]\,d\boldsymbol{w}(s), \end{aligned} \quad \text{(B.3)}$$

where the last equality follows from the observation that

$$\begin{aligned}\overline{P}'\boldsymbol{w}(t) &= \overline{P}'\left[\boldsymbol{\lambda}(t) - \boldsymbol{\lambda}(0) - \int_0^t Q'\boldsymbol{\lambda}(s)\,ds\right]\\ &= \boldsymbol{\nu} - \boldsymbol{\nu} - \int_0^t \overline{P}'Q'\boldsymbol{\lambda}(s)\,ds = 0.\end{aligned}$$

Consequently,

$$\begin{aligned}&\int_0^t (\boldsymbol{\lambda}(s) - \boldsymbol{\nu})\beta(s)\,ds\\ &= O(\varepsilon(t+1)) + \int_0^t \left[\int_0^s O(\varepsilon + e^{-\beta_0(s-r)/\varepsilon})\,d\boldsymbol{w}(r)\right]\beta(s)\,ds\\ &= O(\varepsilon(t+1)) + \int_0^t \left(\int_r^t O(\varepsilon + e^{-\beta_0(s-r)/\varepsilon})\beta(s)\,ds\right)d\boldsymbol{w}(r)\\ &= O(\varepsilon(t+1)) + O(\varepsilon)\int_0^t \left[(t-s) + \beta_0^{-1}(1 - e^{-\beta_0(t-s)/\varepsilon})\right]d\boldsymbol{w}(s).\end{aligned}$$

Dividing both sides by $(T+1)$, we have

$$\frac{1}{T+1}\left|\int_0^t (\boldsymbol{\lambda}(s) - \boldsymbol{\nu})\beta(s)\,ds\right| = O(\varepsilon) + O(\varepsilon)\left|\int_0^t b(s,t)\,d\boldsymbol{w}(s)\right|, \tag{B.4}$$

where $b(s,t)$ is measurable and $|b(s,t)| \le 1$ for all t and s. Therefore,

$$\begin{aligned}&E\exp\left\{\frac{\theta}{\sqrt{\varepsilon}(T+1)^{3/2}}\left|\int_0^t (\boldsymbol{\lambda}(s) - \boldsymbol{\nu})\beta(s)\,ds\right|\right\}\\ &\quad\le E\exp\left\{\frac{\theta}{\sqrt{\varepsilon}\sqrt{T+1}}\left[O(\varepsilon) + O(\varepsilon)\left|\int_0^t b(s,t)\,d\boldsymbol{w}(s)\right|\right]\right\}.\end{aligned}$$

We assume that $\theta\sqrt{\varepsilon_0} \le 1$. Then, for $0 < \varepsilon \le \varepsilon_0$,

$$\begin{aligned}&E\exp\left\{\frac{\theta}{\sqrt{\varepsilon}(T+1)^{3/2}}\left|\int_0^t (\boldsymbol{\lambda}(s) - \boldsymbol{\nu})\beta(s)\,ds\right|\right\}\\ &\quad\le \exp\frac{\theta\sqrt{\varepsilon}}{\sqrt{T+1}}E\exp\left\{\frac{\theta\sqrt{\varepsilon}}{\sqrt{T+1}}\left|\int_0^t b(s,t)\,d\boldsymbol{w}(s)\right|\right\}\\ &\quad\le eE\exp\left\{\frac{\theta\sqrt{\varepsilon}}{\sqrt{T+1}}\left|\int_0^t b(s,t)\,d\boldsymbol{w}(s)\right|\right\}.\end{aligned} \tag{B.5}$$

Recall that $\boldsymbol{w}(t) = (w_0(t), \ldots, w_m(t))'$. It suffices to show that for θ small enough and for each $i \in \mathcal{M}$,

$$E\exp\left\{\frac{\theta\sqrt{\varepsilon}}{\sqrt{T+1}}\left|\int_0^t b(s,t)\,dw_i(s)\right|\right\} \le C. \tag{B.6}$$

Note that for any nonnegative random variable ξ,

$$Ee^{\xi} \le e + (e-1) \sum_{j=1}^{\infty} e^j P(\xi \ge j).$$

For each $t_0 \ge 0$, let $b_0(s) = b(s, t_0)$. Then,

$$\begin{aligned} & E \exp \left\{ \frac{\theta\sqrt{\varepsilon}}{\sqrt{T+1}} \left| \int_0^t b_0(s)\, dw_i(s) \right| \right\} \\ & \qquad \le e + (e-1) \sum_{j=1}^{\infty} e^j P \left(\frac{\theta\sqrt{\varepsilon}}{\sqrt{T+1}} \left| \int_0^t b_0(s)\, dw_i(s) \right| \ge j \right). \end{aligned} \tag{B.7}$$

Now we estimate

$$P \left(\frac{\theta\sqrt{\varepsilon}}{\sqrt{T+1}} \left| \int_0^t b_0(s)\, dw_i(s) \right| \ge j \right).$$

Let $p(t) = \int_0^t b_0(s)\, dw_i(s)$. Then $p(t)$, $t \ge 0$, is a local martingale. Let $q(\cdot)$ denote the only solution to the equation (see Elliott [46])

$$q(t) = 1 + \zeta \int_0^t q(s^-)\, dp(s),$$

where $q(s^-)$ is the left-hand limit of $q(\cdot)$ at s and ζ is a positive constant to be determined later. Since $\zeta \int_0^t q(s^-)\, dp(s)$, $t \ge 0$, is a local martingale, we have $Eq(t) \le 1$ for all $t \ge 0$. Moreover, $q(t)$ can be written as follows (see Elliott [46]):

$$q(t) = e^{\zeta p(t)} \prod_{s \le t} (1 + \zeta \Delta p(s)) e^{-\zeta \Delta p(s)}, \tag{B.8}$$

where $\Delta p(s) := p(s) - p(s^-), |\Delta p(s)| \le 1$.

Let us now observe that there exist positive constants ζ_0 and β_1 such that, for $0 < \zeta \le \zeta_0$ and for all $s > 0$,

$$(1 + \zeta \Delta p(s)) e^{-\zeta \Delta p(s)} \ge e^{-\beta_1 \zeta^2}. \tag{B.9}$$

Combining (B.8) and (B.9), we conclude

$$q(t) \ge \exp\{\zeta p(t) - \beta_1 \zeta^2 N(t)\}, \quad \text{for } 0 < \zeta \le \zeta_0,\ t > 0,$$

where $N(t)$ is the number of jumps of $p(s)$ in $s \in [0, t]$. Since $N(\cdot)$ is a monotone increasing process, we have

$$q(t) \ge \exp\left\{\zeta p(t) - \beta_1 \zeta^2 N(T)\right\}, \quad \text{for } 0 < \zeta \le \zeta_0.$$

Note also that

$$\begin{aligned}
&P\left(\frac{\theta\sqrt{\varepsilon}}{\sqrt{T+1}}\left|\int_0^t b_0(s)dw_i(s)\right| \geq j\right) \\
&\quad = P\left(|p(t)| \geq \frac{j\sqrt{T+1}}{\theta\sqrt{\varepsilon}}\right) \\
&\quad \leq P\left(p(t) \geq \frac{j\sqrt{T+1}}{\theta\sqrt{\varepsilon}}\right) + P\left(-p(t) \geq \frac{j\sqrt{T+1}}{\theta\sqrt{\varepsilon}}\right).
\end{aligned}$$

We consider the first term

$$P\left(p(t) \geq \frac{j\sqrt{T+1}}{\theta\sqrt{\varepsilon}}\right).$$

Let $a_j = j(T+1)/(8\beta_1\theta^2\varepsilon)$. Then,

$$\begin{aligned}
&P\left(p(t) \geq \frac{j\sqrt{T+1}}{\theta\sqrt{\varepsilon}}\right) \\
&\quad \leq P\left(q(t) \geq \exp\left\{\frac{j\zeta\sqrt{T+1}}{\theta\sqrt{\varepsilon}} - \beta_1\zeta^2 N(T)\right\}\right) \\
&\quad \leq P\left(q(t) \geq \exp\left\{\frac{j\zeta\sqrt{T+1}}{\theta\sqrt{\varepsilon}} - \beta_1\zeta^2 N(T)\right\}, N(T) \leq a_j\right) \\
&\qquad + P(N(T) \geq a_j) \\
&\quad \leq P\left(q(t) \geq \exp\left(\frac{j\zeta\sqrt{T+1}}{\theta\sqrt{\varepsilon}} - \beta_1\zeta^2 a_j\right)\right) + P(N(T) \geq a_j) \\
&\quad \leq 2\exp\left(-\frac{j\zeta\sqrt{T+1}}{\theta\sqrt{\varepsilon}} + \beta_1\zeta^2 a_j\right) + P(N(T) \geq a_j).
\end{aligned}$$

The last inequality is because of the local martingale property (see Elliott [46, Theorem 4.2]).

Now if we take $\zeta = 4\theta\sqrt{\varepsilon}/\sqrt{T+1}$, then

$$\exp\left(-\frac{j\zeta\sqrt{T+1}}{\theta\sqrt{\varepsilon}} + \beta_1\zeta^2 a_j\right) = e^{-2j}.$$

In view of the construction of the Markov chain in Appendix A, there exists a Poisson process $N_0(\cdot)$ with parameter $a\varepsilon^{-1}$ for some $a > 0$, such that $N(t) \leq N_0(t)$. We may assume $a = 1$, otherwise we may take ε as εa^{-1}. By using the Poisson distribution of $N_0(t)$ and Stirling's formula, we can show that, for ε small enough,

$$P(N(T) \geq a_j) \leq 2\gamma^{a_j - 1},$$

where $\gamma = 8e\beta_1\theta^2 \in (0,1)$ for θ small enough.

Thus,

$$P\left(\frac{\theta\sqrt{\varepsilon}}{\sqrt{T+1}}\int_0^t b_0(s)\,dw_i(s)\geq j\right)\leq 2e^{-2j}+2\gamma^{a_j-1}.$$

Repeating the same argument for the martingale $(-p(\cdot))$, we get

$$P\left(\frac{\theta\sqrt{\varepsilon}}{\sqrt{T+1}}\left(-\int_0^t b_0(s)\,dw_i(s)\right)\geq j\right)\leq 2e^{-2j}+2\gamma^{a_j-1}.$$

Combining the above two inequalities, we obtain

$$P\left(\frac{\theta\sqrt{\varepsilon}}{\sqrt{T+1}}\left|\int_0^t b_0(s)\,dw_i(s)\right|\geq j\right)\leq 4(e^{-2j}+\gamma^{a_j-1}).$$

Then, by (B.7),

$$E\exp\left\{\frac{\theta\sqrt{\varepsilon}}{\sqrt{T+1}}\left|\int_0^t b(s,t)\,dw_i(s)\right|\right\}\leq e+4(e-1)\sum_{j=1}^{\infty}e^j(e^{-2j}+\gamma^{a_j-1}).$$

Now we choose ε small enough so that $e\gamma^{1/(8\beta_1\theta^2\varepsilon)}\leq 1/2$. Then,

$$E\exp\left\{\frac{\theta\sqrt{\varepsilon}}{\sqrt{T+1}}\left|\int_0^t b(s,t_0)\,dw_i(s)\right|\right\}\leq e+4e\gamma^{-1}.$$

Since t_0 is arbitrary, we may take $t_0=t$ in the above inequality. Then,

$$E\exp\left\{\frac{\theta\sqrt{\varepsilon}}{\sqrt{T+1}}\left|\int_0^t b(s,t)\,dw_i(s)\right|\right\}\leq e+4e\gamma^{-1}.$$

Combining this inequality with inequality (B.5), we obtain

$$E\exp\left\{\frac{\theta}{\sqrt{\varepsilon}(T+1)^{3/2}}\left|\int_0^t (\boldsymbol{\lambda}(s)-\boldsymbol{\nu})\beta(s)\,ds\right|\right\}\leq e(e+4e\gamma^{-1}).$$

Step 2. Let

$$n^\varepsilon(t,i)=\frac{1}{\sqrt{\varepsilon}}\int_0^t\left(I_{\{k(\varepsilon,s)=i\}}-\nu_i\right)\beta(s)\,ds.$$

Then, for each $i\in\mathcal{M}$, $n^\varepsilon(t,i)$ is nearly a martingale, i.e., for ε small enough,

$$|E[n^\varepsilon(t,i)|\mathcal{F}_s]-n^\varepsilon(s,i)|\leq O(\sqrt{\varepsilon}),\quad \text{for all } \omega\in\Omega \text{ and } 0\leq s\leq t\leq T. \tag{B.10}$$

Here $O(\sqrt{\varepsilon})$ is deterministic.

To see this, note that, for all $i_0 \in \mathcal{M}$,

$$\begin{aligned} E&\left[\int_s^t (I_{\{k(\varepsilon,r)=i\}} - \nu_i)\beta_i(r)\,dr \Big| k(\varepsilon,s)=i_0\right] \\ &= \int_s^t (E[I_{\{k(\varepsilon,r)=i\}}|k(\varepsilon,s)=i_0] - \nu_i)\beta_i(r)\,dr \\ &= \int_s^t [P(k(\varepsilon,r)=i|k(\varepsilon,s)=i_0) - \nu_i)]\beta_i(r)\,dr \\ &= \int_s^t O(\varepsilon + \exp(-\kappa_0(r-s)/\varepsilon)\,dr = O(\varepsilon). \end{aligned}$$

So, (B.10) follows.

Step 3. We show that, for each $a > 0$,

$$E[\exp\{a|n^\varepsilon(t,i)|\}|\mathcal{F}_s] \geq \exp\{a|n^\varepsilon(s,i)|\}(1+O(\sqrt{\varepsilon})).$$

First of all, note that $\phi(x) = |x|$ is a convex function. We have, noting that $O(\sqrt{\varepsilon}) = -O(\sqrt{\varepsilon})$,

$$\begin{aligned} E[|n^\varepsilon(t,i)|\,|\mathcal{F}_s] &\geq |n^\varepsilon(s,i)| + \phi'_+(n^\varepsilon(s,i)) \cdot E[n^\varepsilon(t,i) - n^\varepsilon(s,i)|\mathcal{F}_s] \\ &\geq |n^\varepsilon(s,i)| + O(\sqrt{\varepsilon}), \end{aligned}$$

where ϕ'_+ is the right-hand derivative which is bounded by 1. Moreover, note that e^{ax} is also convex. It follows that

$$\begin{aligned} &E[\exp(a|n^\varepsilon(t,i)|)|\mathcal{F}_s] \\ &\quad\geq \exp(a|n^\varepsilon(s,i)|) + a\exp\{a|n^\varepsilon(s,i)|\}E[|n^\varepsilon(t,i)| - |n^\varepsilon(s,i)|\,|\mathcal{F}_s] \\ &\quad\geq \exp(a|n^\varepsilon(s,i)|)(1+O(\sqrt{\varepsilon})). \end{aligned}$$

Step 4. Let $x^\varepsilon(t) = \exp(a|n^\varepsilon(t,i)|)$ for $a > 0$. Then, for any $\mathcal{F}_t$ stopping time $\tau \leq T$,

$$E[x^\varepsilon(T)|\mathcal{F}_\tau] \geq x^\varepsilon(\tau)(1+O(\sqrt{\varepsilon})). \tag{B.11}$$

Note that $x^\varepsilon(t)$ is continuous. Therefore, it suffices to show the above inequality when τ takes values in a countable set $\{t_1, t_2, \ldots\}$. To this end, note that, for each t_i,

$$E[x^\varepsilon(T)|\mathcal{F}_{t_i}] \geq x^\varepsilon(t_i)(1+O(\sqrt{\varepsilon})).$$

For all $A \in \mathcal{F}_\tau$, we have $A \cap \{\tau = t_i\} \in \mathcal{F}_{t_i}$. Therefore,

$$\int_{A\cap\{\tau=t_i\}} x^\varepsilon(T)\,dP \geq \left(\int_{A\cap\{\tau=t_i\}} x^\varepsilon(\tau)\,dP\right)(1+O(\sqrt{\varepsilon})).$$

Thus

$$\int_A x^\varepsilon(T)\,dP \geq \left(\int_A x^\varepsilon(\tau)\,dP\right)(1+O(\sqrt{\varepsilon})),$$

and (B.11) follows.

Step 5. Let $a = \theta/\sqrt{(T+1)^3}$ in Step 3. Then, for ε small enough, there exists K such that

$$P\left(\sup_{t\le T} x^{\varepsilon}(t) \ge x\right) \le \frac{K}{x}, \tag{B.12}$$

for all $x > 0$.

In fact, let $\tau = \inf\{t > 0 \,:\, x^{\varepsilon}(t) \ge x\}$. We adopt the convention and take $\tau = \infty$ if $\{t > 0 \,:\, x^{\varepsilon}(t) \ge x\} = \emptyset$. Then we have

$$E[x^{\varepsilon}(T)] \ge (E[x^{\varepsilon}(T\wedge\tau)])(1 + O(\sqrt{\varepsilon})),$$

and we can write

$$E[x^{\varepsilon}(T\wedge\tau)] = E[x^{\varepsilon}(\tau)I_{\{\tau<T\}}] + E[x^{\varepsilon}(T)I_{\{\tau\ge T\}}] \ge E[x^{\varepsilon}(\tau)I_{\{\tau<T\}}].$$

Moreover, in view of the definition of τ, we have

$$E\left[x^{\varepsilon}(\tau)I_{\{\tau<T\}}\right] \ge xP(\tau < T) \le xP\left(\sup_{t\le T} x^{\varepsilon}(t) \ge x\right).$$

It follows that

$$P\left(\sup_{t\le T} x^{\varepsilon}(t) \ge x\right) \le \frac{E[x^{\varepsilon}(T)]}{(1+O(\sqrt{\varepsilon}))x} \le \frac{K}{x}.$$

Thus, (B.12) follows.

Finally, to complete the proof of (B.2), note that, for $0 < \kappa < 1$,

$$E\exp\left(\frac{\kappa\theta}{\sqrt{(1+T)^3}}\sup_{t\le T}|n^{\varepsilon}(t,i)|\right) = E\left[\sup_{t\le T}(x^{\varepsilon}(t))^{\kappa}\right].$$

It follows that

$$\begin{aligned} E\left[\sup\nolimits_{t\le T}(x^{\varepsilon}(t))^{\kappa}\right] &= \int_0^{\infty} P\left(\sup_{t\le T}(x^{\varepsilon}(t))^{\kappa} \ge x\right)dx \\ &\le 1 + \int_1^{\infty} P\left(\sup_{t\le T}(x^{\varepsilon}(t))^{\kappa} \ge x\right)dx \\ &\le 1 + \int_1^{\infty} P\left(\sup_{t\le T} x^{\varepsilon}(t) \ge x^{1/\kappa}\right)dx \\ &\le 1 + \int_1^{\infty} Kx^{-1/\kappa}\,dx < \infty. \end{aligned}$$

This completes the proof. □

Corollary B.1. *For any Markov time τ and uniformly bounded deterministic process $\beta(t)$, $t \ge 0$, on $[0,\ \infty)$, there exist positive constants ε_0, θ, and C such that for $0 < \varepsilon \le \varepsilon_0$ and $i \in \mathcal{M}$, we have*

$$E\left[\exp\left\{\frac{\theta}{\sqrt{\varepsilon}(T+1)^{3/2}}\sup_{0\le t\le T}\left|\int_{\tau}^{\tau+t}\left(I_{\{k(\varepsilon,s)=i\}} - \nu_i\right)\beta(s)\,ds\right|\right\}\right] \le C.$$

Proof. Using the strong Markov property of $k(\varepsilon,\cdot)$ and going along the same lines of the proof of the lemma, the corollary can be proved. □

Lemma B.5. *In Lemma* B.4, *if* $Q^{(1)}=0$, *i.e.,* $Q=\varepsilon^{-1}Q^{(2)}$, *then we have the following stronger estimate:*

$$E\left[\exp\left\{\frac{\theta}{\sqrt{\varepsilon}\sqrt{T+1}}\sup_{0\le t\le T}\left|\int_0^t (1_{\{k(\varepsilon,s)=i\}}-\nu_i)\beta(s)\,ds\right|\right\}\right]\le C.$$

Proof. If $Q^{(1)}=0$, then equation (B.4) can be replaced by

$$\left|\int_0^t (\boldsymbol{\lambda}(s)-\boldsymbol{\nu})\beta(s)\,ds\right|=\varepsilon C_1+\varepsilon C_2\left|\int_0^t b(s,t)\,d\boldsymbol{w}(s)\right|.$$

The proof of Lemma B.5 follows in the same way as the proof of Lemma B.4 from equation (B.4) on. □

In a same way, we have the following corollary.

Corollary B.2. *In Corollary* B.1, *if* $Q^{(1)}=0$, *i.e.,* $Q=\varepsilon^{-1}Q^{(2)}$, *then we have the following stronger estimate:*

$$E\left[\exp\left\{\frac{\theta}{\sqrt{\varepsilon}\sqrt{T+1}}\sup_{0\le t\le T}\left|\int_\tau^{\tau+t} (1_{\{k(\varepsilon,s)=i\}}-\nu_i)\beta(s)\,ds\right|\right\}\right]\le C.$$

Lemma B.6. *For each* $0<\delta<1/2$, *and any deterministic process* $\beta(\cdot)$ *with* $|\beta(t)|\le B_0$, *for some constant* B_0, *positive constants* ε_0, θ, *and* C *exist such that for* $0<\varepsilon\le\varepsilon_0$ *and* $i\in\mathcal{M}$, *we have*

$$P\left(\sup_{0\le t\le T}\left|\int_0^t (1_{\{k(\varepsilon,s)=i\}}-\nu_i)\beta(s)\,ds\right|\ge\varepsilon^{1/2-\delta}\right)\le C\exp\left\{-\frac{\theta}{\varepsilon^{\delta}\sqrt{(1+T)^3}}\right\}. \tag{B.13}$$

Moreover, if $Q^{(1)}=0$, *then*

$$P\left(\sup_{0\le t\le T}\left|\int_0^t (1_{\{k(\varepsilon,s)=i\}}-\nu_i)\beta(s)\,ds\right|\ge\varepsilon^{1/2-\delta}\right)\le C\exp\left\{-\frac{\theta}{\varepsilon^{\delta}\sqrt{1+T}}\right\}. \tag{B.14}$$

Proof. Using Lemma B.4, we see

$$P\left(\sup_{0\le t\le T}\left|\int_0^t (1_{\{k(\varepsilon,s)=i\}}-\nu_i)\beta(s)\,ds\right|\ge \varepsilon^{1/2-\delta}\right)$$

$$=P\left(\exp\left\{\frac{\theta}{\sqrt{\varepsilon}\sqrt{(1+T)^3}}\sup_{0\le t\le T}\left|\int_0^t (I_{\{k(\varepsilon,s)=i\}}-\nu_i)\beta(s)\,ds\right|\right\}\right.$$

$$\left.\ge \exp\left\{\frac{\theta\varepsilon^{1/2-\delta}}{\sqrt{\varepsilon}\sqrt{(1+T)^3}}\right\}\right)$$

$$\le C\exp\left\{-\frac{\theta}{\varepsilon^{\delta}\sqrt{(1+T)^3}}\right\}.$$

This proves (B.13). Similarly, (B.14) follows from Lemma B.5. □

Corollary B.3. *In Lemma* B.6, *for any Markov time* τ, *we have*

$$P\left(\sup_{0\le t\le T}\left|\int_\tau^{\tau+t} (1_{\{k(\varepsilon,s)=i\}}-\nu_i)\beta(s)\,ds\right|\ge \varepsilon^{1/2-\delta}\right)\le C\exp\left\{-\frac{\theta}{\varepsilon^{\delta}\sqrt{(1+T)^3}}\right\}. \tag{B.15}$$

Moreover, if $Q^{(1)}=0$, then

$$P\left(\sup_{0\le t\le T}\left|\int_\tau^{\tau+t} (1_{\{k(\varepsilon,s)=i\}}-\nu_i)\beta(s)\,ds\right|\ge \varepsilon^{1/2-\delta}\right)\le C\exp\left\{-\frac{\theta}{\varepsilon^{\delta}\sqrt{1+T}}\right\}. \tag{B.16}$$

Proof. The proof is along the same lines as the proof of Corollary B.1. □

Lemma B.7. *If* $Q^{(2)}$ *is irreducible (resp., weakly irreducible), then there exists an* $\varepsilon_0>0$ *such that* $Q^{(1)}+\varepsilon^{-1}Q^{(2)}$ *is irreducible (resp., weakly irreducible) for* $0<\varepsilon\le\varepsilon_0$.

Proof. First of all, suppose that $Q^{(2)}$ is weakly irreducible. Let $\boldsymbol{\nu}^\varepsilon=(\nu_0^\varepsilon,\ldots,\nu_m^\varepsilon)$ denote the equilibrium distribution of $Q^{(1)}+\varepsilon^{-1}Q^{(2)}$, i.e.,

$$\boldsymbol{\nu}^\varepsilon\big(Q^{(1)}+\varepsilon^{-1}Q^{(2)}\big)=\mathbf{0}\quad\text{and}\quad\sum_{i=0}^m \nu_i^\varepsilon=1. \tag{B.17}$$

Equivalently, we can write the above equalities in terms of matrices as

$$\boldsymbol{\nu}^\varepsilon\big(\varepsilon Q^{(1)}+Q^{(2)},\mathbf{1}\big)=(\mathbf{0},1),$$

where $\mathbf{1}=(1,\ldots,1)'$. This set of equations has a unique solution if and only if the matrix $(\varepsilon Q^{(1)}+Q^{(2)},\mathbf{1})$ is of full rank ($=m+1$). Equivalently, the determinant

$$\det\big(\varepsilon Q^{(1)}+Q^{(2)},\mathbf{1}\big)\big(\varepsilon Q^{(1)}+Q^{(2)},\mathbf{1}\big)'>0. \tag{B.18}$$

But $Q^{(2)}$ is weakly irreducible and the above determinant is not zero at $\varepsilon = 0$ in (B.18). Therefore, the continuity of the determinant with respect to ε implies that there exists an $\varepsilon_0 > 0$ such that the determinant is larger than 0 for all $0 < \varepsilon \le \varepsilon_0$. Thus, $Q^{(1)} + \varepsilon^{-1}Q^{(2)}$ is weakly irreducible for $0 < \varepsilon \le \varepsilon_0$.

Next, suppose $Q^{(2)}$ is irreducible. Then there exists a unique $\boldsymbol{\nu}^0 = (\nu_0^0, \ldots, \nu_m^0) > 0$ that satisfies equation (B.17).

If the lemma is false, then there is a sequence of ε (still denoted by ε) such that $\boldsymbol{\nu}^\varepsilon \to \overline{\boldsymbol{\nu}}$, in which some components are 0. This implies that, in view of (B.17), $\overline{\boldsymbol{\nu}} = \boldsymbol{\nu}^0$. This is a contradiction, which completes the proof. □

Next, we prove Lemmas 4.3.1 and 4.3.2 of Chapter 4. We begin with a lemma to be used in these proofs.

Lemma B.8. *Let Assumptions* (A2) *and* (A3) *of Chapter* 4 *hold. Let* $\widetilde{\tau}$ *be a Markov time with respect to the Markov chain* $\widetilde{\boldsymbol{k}}(\cdot)$, *that is,* $\{\widetilde{\tau} \le t\} \in \sigma(\widetilde{\boldsymbol{k}}(s) : 0 \le s \le t)$. *Then for any linear function* $L(\cdot)$ *defined on* $\Re_+^m$, *there exists a positive constant* $\widehat{C}$ *such that, for any* $T > 0$,

$$E \exp\left(\frac{1}{\sqrt{T}} \sup_{0 \le t \le T} \left| \int_{\widetilde{\tau}}^{\widetilde{\tau}+t} [L(\widetilde{\boldsymbol{k}}(s)) - L(\widetilde{\boldsymbol{p}})]\, ds \right| \right) \le \widehat{C},$$

where $\widetilde{p} = (\widetilde{p}_1, \ldots, \widetilde{p}_m)$ *given by* (4.14).

Proof. It follows from Corollary B.2 that for any $A > 0$ and a Markov time $\widetilde{\tau}$ with respect to $\widetilde{\boldsymbol{k}}(\cdot)$, there exists a C_A such that, for any $T > 0$,

$$E \exp\left(\frac{A}{\sqrt{T+1}} \sup_{0 \le t \le T} \left| \int_{\widetilde{\tau}}^{\widetilde{\tau}+t} \left[I_{\{\widetilde{\boldsymbol{k}}(s) = \widetilde{\boldsymbol{k}}^i\}} - \widetilde{\nu}_{\widetilde{\boldsymbol{k}}^i} \right] ds \right| \right) \le C_A. \qquad \text{(B.19)}$$

From the linearity of $L(\cdot)$, it suffices to show that there exists a constant C_1 such that, for any $T > 0$ and $j = 1, \ldots, m$,

$$E \exp\left(\frac{1}{\sqrt{T+1}} \sup_{0 \le t \le T} \left| \int_{\widetilde{\tau}}^{\widetilde{\tau}+t} [\widetilde{k}_j(s) - \widetilde{p}_j]\, ds \right| \right) \le C_1. \qquad \text{(B.20)}$$

To do this, we note that

$$\begin{aligned}
&E \exp\left(\frac{1}{\sqrt{T+1}} \sup_{0 \le t \le T} \left| \int_{\widetilde{\tau}}^{\widetilde{\tau}+t} (\widetilde{k}_j(s) - \widetilde{p}_j) ds \right| \right) \\
&= E \exp\left(\frac{1}{\sqrt{T+1}} \sup_{0 \le t \le T} \left| \int_{\widetilde{\tau}}^{\widetilde{\tau}+t} \sum_{i=1}^{p} \widetilde{k}_j^i \left[I_{\{\widetilde{\boldsymbol{k}}(s) = \widetilde{\boldsymbol{k}}^i\}} - \widetilde{\nu}_{\widetilde{\boldsymbol{k}}^i} \right] ds \right| \right) \\
&\le E \exp\left(\frac{1}{\sqrt{T+1}} \sum_{i=1}^{p} \widetilde{k}_j^i \sup_{0 \le t \le T} \left| \int_{\widetilde{\tau}}^{\widetilde{\tau}+t} \left[I_{\{\widetilde{\boldsymbol{k}}(s) = \widetilde{\boldsymbol{k}}^i\}} - \widetilde{\nu}_{\widetilde{\boldsymbol{k}}^i} \right] ds \right| \right).
\end{aligned}$$

Using the Schwarz inequality we get

$$E\exp\left(\frac{1}{\sqrt{T+1}}\sum_{i=1}^{p}\widetilde{k}_j^i \sup_{0\le t\le T}\left|\int_{\widetilde{\tau}}^{\widetilde{\tau}+t}\left[I_{\{\widetilde{\boldsymbol{k}}(s)=\widetilde{\boldsymbol{k}}^i\}}-\widetilde{\nu}_{\widetilde{\boldsymbol{k}}^i}\right]ds\right|\right)$$
$$\le\prod_{i=1}^{p}\left[E\exp\left(\frac{\sum_{i=1}^{p}\widetilde{k}_j^i}{\sqrt{T+1}}\sup_{0\le t\le T}\left|\int_{\widetilde{\tau}}^{\widetilde{\tau}+t}\left[I_{\{\widetilde{\boldsymbol{k}}(s)=\widetilde{\boldsymbol{k}}^i\}}-\widetilde{\nu}_{\widetilde{\boldsymbol{k}}^i}\right]ds\right|\right)\right]^{\frac{\widetilde{k}_j^i}{\sum_{i=1}^{p}\widetilde{k}_j^i}}.$$

This together with (B.19) imply (B.20). □

Proof of Lemma 4.3.1. For simplicity in exposition, we will write $\theta, \widehat{\theta}$, $\boldsymbol{x}^\ell(t)$, and $\boldsymbol{u}^\ell(t)$ ($\ell = 0, 1$) instead of $\theta(s, \boldsymbol{x}, \overline{\boldsymbol{x}}), \widehat{\theta}(s, \boldsymbol{x}, \overline{\boldsymbol{x}})$, $\boldsymbol{x}^\ell(t|s, \boldsymbol{x})$, and $\boldsymbol{u}^\ell(\boldsymbol{x}^\ell(t|s, \boldsymbol{x}), \widetilde{\boldsymbol{k}}(t))$, respectively, in the proofs of this lemma and Lemma 4.3.2.

First we sketch the idea of the proof. Lemma 4.3.1 follows from the fact that the probability of the event $\{\theta - s > t\}$ decreases exponentially as t increases. To prove this fact, we divide the interval $(s,\ s + t)$ into two parts $(s,\ s + \delta t)$ and $(s + \delta t,\ s + t)$, in such a way that the behavior of all coordinates of $\boldsymbol{x}^0(\cdot)$ is defined mainly by the behavior on the interval $(s + \delta t,\ s + t)$. Using Lemma B.8 we show that all coordinates are positive on the interval $(s + \delta t,\ s + t)$ with a probability which exponentially tends to 1. If all coordinates are positive on the interval $(s + \delta t,\ s + t)$, then $\boldsymbol{u}^0(t) = \widetilde{\boldsymbol{k}}(t)$ on this interval, and using Lemma B.8 we can estimate the shift in process $\boldsymbol{x}^0(t)$.

Let $\bar{c}_j = \max_{1\le i\le p}\widetilde{k}_j^i$ ($j = 1, \ldots, m$) and $\bar{c}_{m+1} = \widetilde{p}_{m+1} = z$. Noting (4.15), we can choose $\delta > 0$ such that

$$(\widetilde{p}_j - \widetilde{p}_{j+1})(1-\delta) - \delta\bar{c}_{j+1} > 0 \quad \text{for all } 1 \le j \le m.$$

Let $b_j = (\widetilde{p}_j - \widetilde{p}_{j+1})(1-\delta) - \delta\bar{c}_{j+1}$ for $j = 1, \ldots, m$, and let $M_1 = 0$, and $M_j = M$ for $j = 2, \ldots, m$. By the definition of $\theta(s, \boldsymbol{x}, \overline{\boldsymbol{x}})$,

$$\begin{aligned}
P(\theta - s > t) &\le \sum_{j=1}^{m} P(x_j^0(s+t) < M_j + \overline{x}_j)\\
&\le P\left(\inf_{s+\delta t\le v\le s+t} x_1^0(v) = 0\right)\\
&\quad + \sum_{j=1}^{m-2} P\left(\bigcap_{j_1=1}^{j}\left\{\inf_{s+\delta t\le v\le s+t} x_{j_1}^0(v) > 0\right\}\right.\\
&\qquad\qquad \left.\bigcap\left\{\inf_{s+\delta t\le v\le s+t} x_{j+1}^0(v) = 0\right\}\right) \qquad \text{(B.21)}\\
&\quad + \sum_{j=1}^{m} P\left(\bigcap_{j_1=1}^{m-1}\left\{\inf_{s+\delta t\le v\le s+t} x_{j_1}^0(v) > 0\right\}\right.\\
&\qquad\qquad \left.\bigcap\left\{x_j^0(s+t) < M_j + \overline{x}_j\right\}\right).
\end{aligned}$$

First we estimate the first term on the right-hand side of (B.21). Note that $u_2^0(v) \le \widetilde{k}_2(v)$. Thus using Lemma B.8 we get

$$\begin{aligned}
&P\Big(\inf_{s+\delta t\le v\le s+t} x_1^0(v)=0\Big)\\
&\le P\Big(\inf_{\delta t\le v\le t}\int_s^{s+v}[\widetilde{k}_1(r)-\widetilde{k}_2(r)]\,dr\le 0\Big)\\
&\le P\Big(\inf_{\delta t\le v\le t}\int_s^{s+v}[\widetilde{k}_1(r)-\widetilde{k}_2(r)-(\widetilde{p}_1-\widetilde{p}_2)]\,dr\le -\delta(\widetilde{p}_1-\widetilde{p}_2)t\Big)\\
&\le P\Big(\sup_{0\le v\le t}\Big|\int_s^{s+v}[(\widetilde{k}_1(r)-\widetilde{k}_2(r))-(\widetilde{p}_1-\widetilde{p}_2)]\,dr\Big|\ge \delta(\widetilde{p}_1-\widetilde{p}_2)t\Big)\\
&\le C_1\exp\big(-\delta(\widetilde{p}_1-\widetilde{p}_2)\sqrt{t}\big),
\end{aligned}\tag{B.22}$$

for some positive constant C_1. If

$$\inf_{s+\delta t\le v\le s+t} x_{j_1}^0(v)>0,\quad \text{for all } 1\le j_1\le j,\quad 1\le j\le m-2,$$

then $u_{j+1}^0(v)=\widetilde{k}_{j+1}(v)$ and $u_{j+2}^0(v)\le\widetilde{k}_{j+2}(v)$ for $v\in(s+\delta t,\ s+t)$. So, just as in the proof of (B.22), we can show that, for $j=1,\ldots,m-2$,

$$\begin{aligned}
&P\Big(\bigcap_{j_1=1}^{j}\Big\{\inf_{s+\delta t\le v\le s+t} x_{j_1}^0(v)>0\Big\}\bigcap\Big\{\inf_{s+\delta t\le v\le s+t} x_{j+1}^0(v)=0\Big\}\Big)\\
&\qquad\le C_2\exp\big\{-\delta(\widetilde{p}_{j+1}-\widetilde{p}_{j+2})\sqrt{t}\big\},
\end{aligned}\tag{B.23}$$

for some positive constant C_2. Now we consider the members of the last sum on the right-hand side of (B.21). According to the definition of $u_j^0(t)$,

$$\begin{aligned}
&P\Big(\bigcap_{j_1=1}^{m-1}\Big\{\inf_{s+\delta t\le v\le s+t} x_{j_1}^0(v)>0\Big\}\cap\{x_j^0(s+t)<M_j+\overline{x}_j\}\Big)\\
&\le P\Big(x_j-\delta t\overline{c}_{j+1}+\int_{s+\delta t}^{s+t}[\widetilde{k}_j(r)-\widetilde{k}_{j+1}(r)]\,dr<M_j+\overline{x}_j\Big)\\
&\le P\Big(\int_{s+\delta t}^{s+t}[\widetilde{k}_j(r)-\widetilde{k}_{j+1}(r)-(\widetilde{p}_j-\widetilde{p}_{j+1})]\,dr<(M_j+\overline{x}_j-x_j)^+-b_jt\Big).
\end{aligned}\tag{B.24}$$

Applying Lemma B.8 we have, from (B.24),

$$P\left(\bigcap_{j_1=1}^{m-1}\left\{\inf_{s+\delta t\le v\le s+t} x_{j_1}^0(v)>0\right\}\bigcap\{x_j^0(s+t)<M_j+\overline{x}_j\}\right)$$
$$\le\begin{cases}1, & \text{for } t\le \dfrac{(M_j+\overline{x}_j-x_j)^+}{b_j},\\ C_3\exp\left(-\dfrac{b_jt-(M_j+\overline{x}_j-x_j)^+}{\sqrt{t}}\right), & \text{for } t\ge \dfrac{(M_j+\overline{x}_j-x_j)^+}{b_j},\end{cases} \tag{B.25}$$

for some constant $C_3>0$. Note that

$$E(\theta-s)^{2r}=\int_0^\infty t^{2r-1}P(\theta-s>t)\,dt. \tag{B.26}$$

By substituting (B.21), (B.22), (B.23), and (B.25) into (B.26), we complete the proof of Lemma 4.3.1. □

Proof of Lemma 4.3.2. Taking the sum of all the equations in (4.1), we have

$$\sum_{j=1}^m x_j^1(t)=\sum_{j=1}^m x_j+\int_s^t[u_1^0(v)-z]\,dv,\quad \text{for } s\le t\le\theta.$$

Consequently,

$$\sum_{j=1}^m x_j^1(\theta)\le\sum_{j=1}^m x_j+(\bar{c}_1-z)(\theta-s), \tag{B.27}$$

where $\bar{c}_1$ is given in the proof of Lemma 4.3.1. Since $u_1^1(t)=0$ for $t\in[\theta,\,\widehat{\theta}]$, we have as before that $\sum_{j=1}^m x_j^1(\widehat{\theta})=\sum_{j=1}^m x_j^1(\theta)-z(\widehat{\theta}-\theta)$. Since $\widehat{\theta}>\theta$ and $x_j^1(\widehat{\theta})\ge\overline{x}_j$, we have

$$\widehat{\theta}-\theta\le\frac{1}{z}\left(\sum_{j=1}^m x_j^1(\theta)-\sum_{j=1}^m\overline{x}_j\right)\quad\text{and}\quad\sum_{j=1}^m x_j^1(\widehat{\theta})\le\sum_{j=1}^m x_j^1(\theta). \tag{B.28}$$

From the definitions of $\widehat{\theta}$ and θ, we have

$$\begin{aligned}\overline{x}_m&=x_m^1(\widehat{\theta})=x_m^1(\theta)+\textstyle\int_\theta^{\widehat{\theta}}[u_m^1(s)-z]\,ds\\&\ge\overline{x}_m+M-z(\widehat{\theta}-\theta),\end{aligned}$$

i.e., $\widehat{\theta}-\theta\ge M/z$. This relation together with (B.27) and (B.28) proves statement (ii) of Lemma 4.3.2.

To prove statement (i), we introduce the following notations:

$$\begin{aligned}
&\widehat{\theta}(j) = \inf\{t \geq \theta \ : \ x_j^1(t) = \overline{x}_j\}, \quad j = 1, \ldots, m, \\
&\widetilde{\mathcal{B}} = \{\omega \ : \ \boldsymbol{x}^1(\widehat{\theta})(\omega) = \overline{\boldsymbol{x}}\}, \\
&\mathcal{B}(j) = \left\{\omega : \inf_{0 \leq t < \infty} \int_\theta^{\theta+t} \left[\widetilde{k}_j(v) - \widetilde{k}_{j+1}(v)\right] dv > -\frac{M}{2}\right\}, \\
&\qquad\qquad\qquad\qquad\qquad\qquad j = 2, \ldots, m-1, \\
&\mathcal{B}(m) = \left\{\omega : \inf_{0 \leq t < \infty} \int_\theta^{\theta+t} \left[\widetilde{k}_m(v) - z\right] dv > -\frac{M}{2}\right\}, \\
&\overline{\mathcal{B}} = \bigcap_{j=2}^m \mathcal{B}(j), \quad \mathcal{B}(0) = \{\omega \ : \ \boldsymbol{k}(\widehat{\theta})(\omega) = \overline{\boldsymbol{k}}\}, \quad \mathcal{B} = \widetilde{\mathcal{B}} \cap \mathcal{B}(0).
\end{aligned}$$

Note that

$$\widetilde{\mathcal{B}} = \left\{\omega : \widehat{\theta}(m)(\omega) \geq \max_{1 \leq j \leq m-1} \widehat{\theta}(j)(\omega)\right\}.$$

From the definition of $\boldsymbol{u}^1(\boldsymbol{x}, \boldsymbol{k})$ and $\boldsymbol{x}^1(t)$, it follows that if $\omega \in \overline{\mathcal{B}}$, then, for $j = 1, \ldots, m-1$,

$$u_{j+1}^1(t) = \begin{cases} \widetilde{k}_{j+1}(t), & \text{for } \theta < t \leq \widehat{\theta}(j), \\ 0, & \text{for } t > \widehat{\theta}(j), \end{cases}$$

and, for $j = 2, \ldots, m$,

$$\widehat{\theta}(j) - \widehat{\theta}(j-1) \geq \frac{x_j^1(\widehat{\theta}(j-1)) - \overline{x}_j}{\bar{c}_j} \geq \frac{M}{2\bar{c}_j}, \tag{B.29}$$

where $\bar{c}_j$ is given in the proof of Lemma 4.3.1. Therefore, $\overline{\mathcal{B}} \subseteq \widetilde{\mathcal{B}}$ and

$$P(\mathcal{B}^c) \leq \sum_{j=2}^m P(\mathcal{B}^c(j)) + P(\overline{\mathcal{B}} \cap \mathcal{B}^c(0)). \tag{B.30}$$

Note that for any q_1 with $0 < q_1 < 1$, there is a positive constant $\widehat{M}$ such that for any two Markov times τ_1 and τ_2 with respect to $\boldsymbol{k}(\cdot)$,

$$\tau_2 - \tau_1 \geq \widehat{M}, \quad a.s.$$

and

$$\max_{1 \leq j \leq p} P(\boldsymbol{k}(\tau_2) \neq \boldsymbol{k}^j | \boldsymbol{k}(\tau_1) = \boldsymbol{k}^j) < q_1. \tag{B.31}$$

Taking the conditional probability with respect to $\widehat{\theta}(m-1)$, choosing

$$M > 2\widehat{M} \max_{1\leq j\leq m} \bar{c}_j,$$

and using (B.29) with $j = m$ and (B.31), we have

$$P(\overline{\mathcal{B}} \cap \mathcal{B}^c(0)) < q_1 < 1. \tag{B.32}$$

Applying Lemma B.8 we have

$$\begin{aligned}
P((\mathcal{B}(j))^c) &\leq \sum_{n=1}^{\infty} P\Big(\int_{\theta}^{\theta+n} [\widetilde{k}_j(v) - \widetilde{k}_{j+1}(v)]\, dv < -\frac{M}{2} + \bar{c}_{j+1}\Big) \\
&\leq \sum_{n=1}^{\infty} P\Big(\Big|\int_{\theta}^{\theta+n} [(\widetilde{k}_j(v) - \widetilde{k}_{j+1}(v)) - (\widetilde{p}_j - \widetilde{p}_{j+1})]\, ds\Big| \\
&\qquad > \frac{M}{2} + n(\widetilde{p}_j - \widetilde{p}_{j+1}) - \bar{c}_{j+1}\Big) \\
&\leq C_1 \sum_{n=1}^{\infty} \exp\Big(-\frac{M/2 + n(\widetilde{p}_j - \widetilde{p}_{j+1}) - \bar{c}_{j+1}}{\sqrt{n}}\Big) \\
&\leq C_2 e^{-C_3\sqrt{M}},
\end{aligned} \tag{B.33}$$

for some positive constants C_1, C_2, and C_3. It follows from (B.30), (B.32), and (B.33) that we can choose M and q such that $P(\mathcal{B}^c) \leq q < 1$. This proves Lemma 4.3.2. □

To prove Lemmas 4.5.1 and 4.5.2, we need the following lemma.

Lemma B.9. *Let $\xi(t)$ be an ergodic Markov chain in continuous time with the finite state space $\{1, 2, \ldots, p\}$ and stationary distribution $\widehat{\nu}_j$, $j = 1, \ldots, p$. Let $\varsigma(t)$ be a process which takes values on the interval $[0,\ H]$ and satisfies*

$$\frac{d}{dt}\varsigma(t) = f(\xi(t)) + f^-(\xi(t)) I_{\{\varsigma(t)=0\}} - f^+(\xi(t)) I_{\{\varsigma(t)=H\}},$$

where $f(\cdot)$ is a function defined on $\{1, 2, \ldots, p\}$, $a^+ = \max\{a, 0\}$, and $a^- = (-a)^+$. Then:

(i) *There exist numbers π^i, $i = 1, \ldots, p$, such that*

$$P(\varsigma(t) = 0, \xi(t) = i) \to \pi^i, \quad \text{as } t \to \infty.$$

(ii) *If $\sum_{i=1}^{p} \widehat{\nu}_i f(i) > 0$, then $\pi^i \to 0$ as $H \to \infty$.*

(iii) *If $\alpha > 0$, then for any r there exists a constant $\widetilde{C}_r$ such that, for any $T \geq 0$ and any $b \geq 0$,*

$$P\left(\sup_{0\leq v\leq T}\left|\int_0^v (I_{\{\zeta(s)=0,\xi(s)=i\}} - \pi^i)\,ds\right| > \alpha(b+T)\right) < \frac{\widetilde{C}_r}{(\alpha^2(b+T))^r} \tag{B.34}$$

and

$$P\left(\sup_{0\leq v\leq T}\left|\int_0^v \left[f(\xi(s)) - \sum_{i=1}^k \widehat{\nu}_i f(i)\right] ds\right| > \alpha(b+T)\right) < \frac{\widetilde{C}_r}{(\alpha^2(b+T))^r}. \tag{B.35}$$

Proof. We begin with the proof of relation (B.35).

Let Q_ξ be an infinitesimal generator of Markov chain $\xi(t)$, and let

$$\widehat{\boldsymbol{\nu}} = \begin{pmatrix} \widehat{\nu}_1 \\ \widehat{\nu}_2 \\ \vdots \\ \widehat{\nu}_p \end{pmatrix}, \quad \boldsymbol{\lambda}(t) = \begin{pmatrix} I_{\{\xi(t)=1\}} \\ I_{\{\xi(t)=2\}} \\ \vdots \\ I_{\{\xi(t)=p\}} \end{pmatrix}, \quad \overline{P} = \begin{pmatrix} \widehat{\nu}_1 & \widehat{\nu}_2 & \cdots & \widehat{\nu}_p \\ \widehat{\nu}_1 & \widehat{\nu}_2 & \cdots & \widehat{\nu}_p \\ \vdots & \vdots & \vdots\ \cdots & \vdots \\ \widehat{\nu}_1 & \widehat{\nu}_2 & \cdots & \widehat{\nu}_p \end{pmatrix}.$$

It is well known (see Elliott [47]) that the process

$$\boldsymbol{w}(t) = \boldsymbol{\lambda}(t) - \boldsymbol{\lambda}(0) - \int_0^t Q'_\xi \boldsymbol{\lambda}(s)\,ds, \quad t \geq 0,$$

is an $\{\mathcal{F}_t\}$-martingale, where $\mathcal{F}_t = \sigma\{k(s) : s \leq t\}$ and

$$\boldsymbol{\lambda}(t) = \exp(Q'_\xi t)\boldsymbol{\lambda}(0) + \int_0^t \exp(Q'_\xi(t-s))\,d\boldsymbol{w}(s).$$

Denote $\boldsymbol{f} = (f(1), \ldots, f(p))'$ and $F(t) = \int_0^t [\exp(Q_\xi s) - \overline{P}]\boldsymbol{f}\,ds$. Using

$$\overline{P}'\boldsymbol{\lambda}(t) = \widehat{\boldsymbol{\nu}}, \quad \overline{P}'\boldsymbol{w}(t) = \boldsymbol{0},$$

we have

$$\begin{aligned} \int_0^t \left[F(\xi(s)) - \sum_{i=1}^p \widehat{\nu}^i f(i)\right] ds &= \int_0^t \boldsymbol{f}'(\lambda(s) - \widehat{\boldsymbol{\nu}})\,ds \\ &= F'(t)\boldsymbol{\lambda}(0) + \int_0^t F'(t-s)\,d\boldsymbol{w}(s). \end{aligned} \tag{B.36}$$

It follows from the exponential convergence for the Markov chain that

$$|F(t)| \le C_1, \tag{B.37}$$

for some constant $C_1 > 0$. Using the Burkholder-Davis-Gundy inequality (see Karatzas and Shreve [77, p. 166]), and (B.37), we get that, for some $C_2(r) > 0$,

$$E\left(\sup_{0\le v\le t}\left|\int_0^v F'(t-v)\,dw(t)\right|\right)^{2r} \le C_2(r)t^r. \tag{B.38}$$

Relation (B.35) follows now from (B.36), (B.37), (B.38), and the Chebyshev inequality.

For proving (i), let $\mathcal{I}^+ = \{i\ :\ i \in \{1,\ldots,p\},\ k_2^i - k_1^i \ge 0\}$. It is evident that if $i \notin \mathcal{I}^+$, then $\pi^i = 0$. For any fixed $i \in \mathcal{I}^+$, we define a sequence of Markov times

$$\begin{aligned}\alpha_0 &= \inf\{t:\ \zeta(t)=0,\ \xi(t)=i\},\\ \beta_1 &= \inf\{t:\ t>\alpha_0,\ \xi(t)\ne i\},\end{aligned}$$

and, for $n \ge 1$,

$$\begin{aligned}\alpha_n &= \inf\{t:\ t>\beta_n,\ \zeta(t)=0,\ \xi(t)=i\},\\ \beta_{n+1} &= \inf\{t:\ t>\alpha_n,\ \xi(t)\ne i\}.\end{aligned}$$

Next we define random variables

$$X_n = \beta_n - \alpha_{n-1},\quad Y_n = \alpha_n - \beta_n \quad \text{for } n \ge 1.$$

Then α_0, $\{X_n\}_{n=1}^\infty$, and $\{Y_n\}_{n=1}^\infty$ are independent, both $\{X_n\}_{n=1}^\infty$ and $\{Y_n\}_{n=1}^\infty$ are identically distributed. In addition, $\{X_n\}_{n=1}^\infty$ is exponentially distributed with parameter ν_j. We define the random variable

$$\gamma(t) = \begin{cases} 1, & \text{if } \alpha_n \le t < \beta_{n+1} \quad \text{for some } n \ge 0,\\ 0, & \text{otherwise.}\end{cases} \tag{B.39}$$

Note that in the case $k_2^i - k_1^i > 0$, we have

$$\begin{aligned} E\gamma(t) &= P(\zeta(t)=0,\xi(t)=i)\\ &\ge P\left(\xi(s)=i \text{ for all } t-\frac{H}{k_2^i-k_1^i}\le s\le t\right)\\ &= P\left(\xi\left(t-\frac{H}{k_2^i-k_1^i}\right)=i\right)\exp\left\{\frac{\nu_i H}{k_2^i-k_1^i}\right\}.\end{aligned} \tag{B.40}$$

By the renewal theorem we get

$$\begin{aligned} \pi^i &= \lim_{t\to\infty} P(\zeta(t)=0,\xi(t)=i) \\ &= \lim_{t\to\infty} E\gamma(t) = \frac{EX_n}{EX_n+EY_n}, \quad n\ge 1. \end{aligned} \tag{B.41}$$

It follows from (B.40) that $\pi^i>0$ for $i\in\mathcal{I}^+$. This completes the proof of statement (i).

To prove (ii) we shall show below that there exist $\delta>0$ and $T_0>0$ such that, for all $t\ge T_0$ and $0\le s\le t$,

$$P(\alpha_1-\beta_1>s)\ge\delta, \quad \text{if } H>t\widehat{k}, \tag{B.42}$$

where $\widehat{k}=\max_i(k_1^i-k_2^i)$. It follows from (B.42) that for any $H\ge T_0\widehat{k}$, we have

$$\begin{aligned} EY_1 = E(\alpha_1-\beta_1) &= \int_0^\infty P(\alpha_1-\beta_1>t)\,dt \\ &\ge \int_0^{H/\widehat{k}} P(\alpha_1-\beta_1>s)\,ds \ge \delta H/\widehat{k}. \end{aligned}$$

This implies that $EY_1\to\infty$ as $H\to\infty$. Hence,

$$\pi^i=\frac{EY_1}{EX_1+EY_1}\to 0, \quad \text{as } H\to\infty.$$

To compete the proof of statement (ii), it suffices to prove (B.42) with $s=t$. To this end, let $\widehat{\tau}=\inf\{s\ :\ s\ge 0,\xi(\beta_1+s)\notin\mathcal{I}^+\}$. From the irreducibility of $\xi(t)$, it follows that, for any a with $0<a<t$,

$$P(\widehat{\tau}<a/2,\ \xi(\beta_1+s)=\xi(\beta_1+\widehat{\tau}) \text{ for all } \widehat{\tau}\le s<\widehat{\tau}+a)\ge\delta(a)>0.$$

Note that $x(\beta_1+a)\ge a/\widehat{k}$, and if $H>t/\widehat{k}$, then $x(\beta_1+s)<H$, for all $a\le s\le t$. Therefore,

$$\begin{aligned} &P(\alpha_1-\beta_1>t) \\ &\quad\ge \delta(a)P\left(\inf_{a\le u\le t} x(\beta_1+u)>0\right) \\ &\quad= \delta(a)P\Bigg(\inf_{a\le u\le t}\bigg\{x(\beta_1+a) \\ &\qquad\qquad + \int_{\beta_1+a}^{\beta_1+u}\left[f(\xi(s))+f^-(\xi(s))I_{\{x(s)=0\}}\right]ds\bigg\}\ge 0\Bigg) \qquad\qquad \text{(B.43)} \\ &\quad\ge \delta(a)P\left(\inf_{a\le u\le t}\left(\int_{\beta_1+a}^{\beta_1+u} f(\xi(s))\,ds\right)\ge -\frac{a}{\widehat{k}}\right) \\ &\quad\ge \delta(a)P\left(\inf_{a\le u\le\infty}\left(\int_{\beta_1+a}^{\beta_1+u} f(\xi(s))\,ds\right)\ge -\frac{a}{\widehat{k}}\right). \end{aligned}$$

Just as in the proof of (B.55), one can show that the probability in the right-hand side of (B.43) tends to 1 as $a \to \infty$. So, we can take a such that this probability is greater than some fixed positive value. This completes the proof of (ii).

The proof of (B.34) is analogous to the proof of (B.35). It also uses the finiteness of all moments of Y_1 and the rate of convergence in the renewal theorem. This completes the proof of statement (iii) of Lemma B.9. □

Proof of Lemma 4.5.1. We will only consider the case $s = 0$. We write $\theta, \widehat{\theta}$, $\boldsymbol{x}^\ell(t)$, and $\boldsymbol{u}^\ell(t)$ ($\ell = 0, 1$) instead of $\theta(s, \boldsymbol{x}, \overline{\boldsymbol{x}})$, $\widehat{\theta}(s, \boldsymbol{x}, \overline{\boldsymbol{x}})$, $\boldsymbol{x}^\ell(t|s, \boldsymbol{x})$, and $\boldsymbol{u}^\ell(\boldsymbol{x}^\ell(t|s, \boldsymbol{x}), \widetilde{\boldsymbol{k}}(t))$, respectively.

First we sketch the idea of the proof. The lemma follows from the fact that $P(\theta > t)$ decreases faster than any power function, as t increases. To prove this, we divide the interval $(0,\ t)$ into two parts $(0,\ \delta t)$ and $(\delta t,\ t)$ in such a way that the behavior of both coordinates of $\boldsymbol{x}^0(\cdot)$ is defined mainly by the behavior on the interval $(\delta t,\ t)$ for some $0 < \delta < 1$. Using Lemma B.9 we then estimate a shift of the process $\boldsymbol{x}^0(\cdot)$.

Let

$$\bar{k} = \max_{1 \le i \le p, j=1,2} \widetilde{k}^i_j.$$

Using Lemma B.9 with $f(i) = \widetilde{k}^i_1 - \widetilde{k}^i_2$ and $x(t) = x^0_1(t)$, we have

$$\pi_0 = \lim_{t \to \infty} P(x^0_1(t) = 0) \to 0 \quad \text{as } H \to \infty.$$

It follows from here and (4.70) that if H is suitably large then we can choose $0 < \delta < 1$ such that

$$(\widetilde{p}_1 - \widetilde{p}_2)(1 - \delta) - \delta \bar{k} = b_1 > 0,$$

$$(\widetilde{p}_2 - \pi_0 \bar{k})(1 - \delta) - z = b_2 > 0.$$

According to the definition of θ,

$$\begin{aligned} P(\theta > t) \le P&\left(\sup_{\delta t \le s \le t} x^0_1(s) < \overline{x}_1, \inf_{\delta t \le s \le t} x^0_2(s) \ge M + \overline{x}_2 \right) \\ &+ P\left(\inf_{\delta t \le s \le t} x^0_2(s) < M + \overline{x}_2 \right). \end{aligned} \tag{B.44}$$

The first term on the right-hand side of (B.44) is estimated next. Note that $u^0_2(s) \le \widetilde{k}_2(s)$ for all s, and $u^0_1(s) = \widetilde{k}_1(s)$ if $x^0_1(s) < H$. Thus, we have

$$
\begin{aligned}
&P\left(\sup_{\delta t\le s\le t} x_1^0(s) < \overline{x}_1, \inf_{\delta t\le s\le t} x_2^0(s) \ge M + \overline{x}_2\right)\\
&\le P\left(\sup_{\delta t\le s\le t} x_1^0(s) < \overline{x}_1\right)\\
&\le P\left(\sup_{\delta t\le s\le t} \left\{x_1 - \delta t\bar{k} + \int_{\delta t}^{s} [\widetilde{k}_1(v) - \widetilde{k}_2(v)]\, dv\right\} < \overline{x}_1\right)\\
&\le P\left(\sup_{\delta t\le s\le t} \left\{\int_{\delta t}^{s} [\widetilde{k}_1(v) - \widetilde{k}_2(v)]\, dv\right\} < (\overline{x}_1 - x_1)^+ + \delta t\bar{k}\right)\\
&\le P\Bigg(\sup_{\delta t\le s\le t} \left\{\int_{\delta t}^{s} \left[\widetilde{k}_1(v) - \widetilde{k}_2(v) - (\widetilde{p}_1 - \widetilde{p}_2)\right] dv\right\}\\
&\qquad\qquad < (\overline{x}_1 - x_1)^+ - b_1 t\Bigg)\\
&\le P\Bigg(\sup_{\delta t\le s\le t} \left|\int_{\delta t}^{s} \left[\widetilde{k}_1(v) - \widetilde{k}_2(v) - (\widetilde{p}_1 - \widetilde{p}_2)\right] dv\right|\\
&\qquad\qquad > -(\overline{x}_1 - x_1)^+ + b_1 t\Bigg).
\end{aligned}
\tag{B.45}
$$

Using (B.35) with

$$f(i) = \widetilde{k}_1^i - \widetilde{k}_2^i,\ \alpha = b_1/2, \text{ and } b = 0,$$

we have from (B.45) that, for some $C_1(r) > 0$,

$$
\begin{aligned}
&P\left(\sup_{\delta t\le s\le t} x_1^0(s) < \overline{x}_1, \inf_{\delta t\le s\le t} x_2^0(s) \ge M + \overline{x}_2\right)\\
&\qquad \le \begin{cases} 1, & \text{for } 0 \le t \le \dfrac{2(\overline{x}_1 - x_1)^+}{b_1},\\[2ex] \dfrac{C_1(r)}{t^r}, & \text{for } t > \dfrac{2(\overline{x}_1 - x_1)^+}{b_1}. \end{cases}
\end{aligned}
\tag{B.46}
$$

Now we estimate the second term on the right-hand side of (B.44).

According to the definition of $u_2^0(\cdot)$,

$$
\begin{aligned}
&P\left(\inf_{\delta t\le s\le t} x_2^0(s) < M+\overline{x}_2\right)\\
&\quad\le P\left(\inf_{\delta t\le s\le t}\left\{x_2 - tz + \int_{\delta t}^{s} u_2^0(v)\,dv\right\} < M+\overline{x}_2\right)\\
&\quad\le P\biggl(\inf_{\delta t\le s\le t}\int_{\delta t}^{s}\left[\widetilde{k}_2(v) - \bar{k}\sum_{i=1}^{p} I_{\{x_1^0(v)=0,\widetilde{\boldsymbol{k}}(v)=\widetilde{\boldsymbol{k}}^i\}}\right]dv\\
&\qquad\qquad\qquad < M+\overline{x}_2 - x_2 + tz\biggr)\\
&\quad\le P\biggl(\sup_{\delta t\le s\le t}\left|\int_{\delta t}^{s}\left[\widetilde{k}_2(v) - \widetilde{p}_2 - \bar{k}\sum_{i=1}^{p}\left(I_{\{x_1^0(v)=0,\widetilde{\boldsymbol{k}}(v)=\widetilde{\boldsymbol{k}}^i\}} - \pi^i\right)\right]\right| dv\\
&\qquad\qquad\qquad > b_2 t - (M+\overline{x}_2 - x_2)^+\biggr)\\
&\quad\le P\left(\sup_{0\le s\le t}\left|\int_0^s [\widetilde{k}_2(v) - \widetilde{p}_2)]\,dv\right| > \frac{b_2 t}{2} - (M+\overline{x}_2 - x_2)^+\right)\\
&\qquad+\sum_{i=1}^{p} P\left(\sup_{0\le s\le t}\left|\int_0^s\left[I_{\{x_1^0(v)=0,\widetilde{\boldsymbol{k}}(v)=\widetilde{\boldsymbol{k}}^i\}} - \pi^i\right]ds\right| > \frac{b_2 t}{2\bar{k}}\right),
\end{aligned}
\tag{B.47}
$$

where

$$
\pi^i = \lim_{t\to\infty} P\big(x_1^0(t) = 0, \widetilde{\boldsymbol{k}}(t) = \widetilde{\boldsymbol{k}}^i\big), \quad i = 1,\ldots,p.
$$

By Lemma B.9 we have from (B.47) that, for some $C_2(r) > 0$,

$$
\begin{aligned}
&P\left(\inf_{\delta t\le s\le t} x_2^0(s) \le M+\overline{x}_2\right)\\
&\qquad\le \begin{cases} 1, & \text{for } 0\le t\le \dfrac{4(M+\overline{x}_2 - x_2)^+}{b_2},\\[2ex] \dfrac{C_2(r)}{t^r}, & \text{for } t > \dfrac{4(M+\overline{x}_2 - x_2)^+}{b_2}.\end{cases}
\end{aligned}
\tag{B.48}
$$

Note that

$$
E\theta^r = \int_0^\infty t^{r-1} P(\theta > t)\,dt.
$$

Substituting (B.44), (B.46), and (B.48) in this relation, we complete the proof of Lemma 4.5.1. □

Proof of Lemma 4.5.2. Recall the notational convention specified in Step 4. Taking the sum of both equations in (4.59), we have

$$
\sum_{j=1}^{2} x_j^1(t) = \sum_{j=1}^{2} x_j + \int_0^t [u_1^0(v) - z]\,dv, \quad 0\le t\le\theta.
$$

Consequently,

$$\sum_{j=1}^{2} x_j^1(\theta) \leq \sum_{j=1}^{2} x_j + (\bar{k} - z)\theta, \tag{B.49}$$

where $\bar{k}$ is given in the proof of Lemma 4.5.1. Since $u_1^1(t) = 0$ for $t > \theta$, we have as before that

$$\sum_{j=1}^{2} x_j^1(\widehat{\theta}) = \sum_{j=1}^{2} x_j^1(\theta) - z(\widehat{\theta} - \theta).$$

Since $\widehat{\theta} > \theta$ and $x_j^1(\widehat{\theta}) \geq \overline{x}_j, j = 1, 2$, we have

$$\widehat{\theta} - \theta \leq \frac{1}{z} \sum_{j=1}^{2} [x_j^1(\theta) - \overline{x}_j], \quad \sum_{j=1}^{2} x_j^1(\widehat{\theta}) \leq \sum_{j=1}^{2} x_j^1(\theta). \tag{B.50}$$

From the definitions of θ and $\widehat{\theta}$, and (4.59), we have

$$\begin{aligned} \overline{x}_2 = x_2^1(\widehat{\theta}) &= x_2^1(\theta) + \textstyle\int_\theta^{\widehat{\theta}} [u_2^1(v) - z]\, dv \\ &\geq \overline{x}_2 + M - z(\widehat{\theta} - \theta), \end{aligned}$$

i.e., $\widehat{\theta} - \theta \geq M/z$. This relation together with (B.49) and (B.50) proves statement (ii) of Lemma 4.5.2.

To prove statement (i) we introduce the following notation:

$$\begin{aligned} &\widehat{\theta}(\ell) = \inf\{t \geq \theta \,:\, x_\ell^1(t) = \overline{x}_\ell\}, \quad \ell = 1, 2, \\ &\widetilde{\mathcal{B}} = \{\omega \,:\, \boldsymbol{x}^1(\widehat{\theta}) = \overline{\boldsymbol{x}}\}, \\ &\overline{\mathcal{B}} = \left\{\omega \,:\, \inf_{0 \leq t < \infty} \int_\theta^{\theta + t} [\widetilde{k}_2(v) - z]\, dv > -M/2\right\}, \\ &\mathcal{B}(0) = \{\omega \,:\, \boldsymbol{k}(\widehat{\theta}) = \overline{\boldsymbol{k}}\}, \quad \mathcal{B} = \widetilde{\mathcal{B}} \cap \mathcal{B}(0). \end{aligned}$$

Note that $\widetilde{\mathcal{B}} = \{\omega \,:\, \widehat{\theta}(2) \geq \widehat{\theta}(1)\}$. From the definition of $\boldsymbol{u}^1(\boldsymbol{x}, \widetilde{\boldsymbol{k}})$ and $\boldsymbol{x}^1(t)$, it follows that if $\omega \in \overline{\mathcal{B}}$, then

$$u_2^1(t) = \begin{cases} \widetilde{k}_2(t), & \text{for } \theta < t \leq \widehat{\theta}(1), \\ 0, & \text{for } t > \widehat{\theta}(1), \end{cases}$$

and

$$\widehat{\theta}(2) - \widehat{\theta}(1) = \frac{x_2^1(\widehat{\theta}(1)) - \overline{x}_2}{z} > \frac{M}{2z}. \tag{B.51}$$

Therefore, $\overline{\mathcal{B}} \subseteq \widetilde{\mathcal{B}}$ and

$$P(\mathcal{B}^c) \leq P(\overline{\mathcal{B}}^c) + P(\overline{\mathcal{B}} \cap \mathcal{B}^c(0)). \tag{B.52}$$

Note that there exists an $\widehat{M} > 0$ such that if η_1 and η_2 are Markov times with respect to $\widetilde{\boldsymbol{k}}(\cdot)$ and $\eta_2 - \eta_1 > \widehat{M}$, a.s., then

$$\max_{\boldsymbol{k} \in \widetilde{\mathcal{M}}} P(\widetilde{\boldsymbol{k}}(\eta_2) \neq \boldsymbol{k} | \widetilde{\boldsymbol{k}}(\eta_1) = \widetilde{\boldsymbol{k}}^i) < 1. \tag{B.53}$$

Taking the conditional probability with respect to $\widehat{\theta}(1)$, using (B.51) with $M > 2\bar{k}\widehat{M}$, and using (B.53), we have

$$P(\overline{\mathcal{B}} \cap \mathcal{B}^c(0)) < 1. \tag{B.54}$$

Let $M/2 > z$. Applying Lemma B.9 we get that, for some $C_1(r) > 0$,

$$\begin{aligned} P(\overline{\mathcal{B}}^c) &\leq \sum_{n=1}^{\infty} P\left(\int_{\theta}^{\theta+n} [\widetilde{k}_2(v) - z]\, dv < -\frac{M}{2} + z \right) \\ &\leq \sum_{n=1}^{\infty} P\left(\left| \int_{\theta}^{\theta+n} [\widetilde{k}_2(v) - \widetilde{p}_2]\, dv \right| > \frac{M}{2} + n(\widetilde{p}_2 - z) - z \right) \\ &\leq C_1(r) \sum_{n=1}^{\infty} \left(\frac{M - 2z}{2(\widetilde{p}_2 - z)} + n \right)^{-r} \to 0 \quad \text{as } M \to \infty. \end{aligned} \tag{B.55}$$

It follows from (B.52), (B.54), and (B.55), that we can choose M and q such that $P(\mathcal{B}^c) \leq q < 1$. This completes the proof of Lemma 4.5.2. □

Proof of Lemma 5.3.1. For simplicity in exposition, we will write $\theta, \widehat{\theta}$, $\boldsymbol{x}^\ell(t)$, and $\boldsymbol{u}^\ell(t) (\ell = 0, 1)$ instead of $\theta(s, \boldsymbol{x}, \widehat{\boldsymbol{x}}), \widehat{\theta}(s, \boldsymbol{x}, \widehat{\boldsymbol{x}})$, $\boldsymbol{x}^\ell(t|s, \boldsymbol{x})$, and $\boldsymbol{u}^\ell(\boldsymbol{x}^\ell(t|s, \boldsymbol{x}), \boldsymbol{k}(t))$, respectively, in the proofs of this lemma and Lemma 5.3.2.

Let $\bar{k}_j = \max_{1 \leq i \leq p} k_j^i$, $j = 1, \ldots, m_c$, $\bar{k}_{n+1} = z_n$. Furthermore, let

$$\begin{aligned} \alpha_j &= \sum_{\ell=-n_0+1}^{j-1} p_{\ell j} p_{c(\ell,j)}, \quad j = 1, \ldots, n, \\ \beta_j &= \sum_{\ell=j+1}^{n} p_{j\ell} p_{c(j,\ell)}, \quad j = 1, \ldots, n-1, \quad \beta_n = z_n. \end{aligned}$$

Based on Assumption (A2), we can choose $\delta > 0$ such that

$$(\alpha_j - \beta_j)(1 - \delta) - \delta \sum_{\ell=j+1}^{n} p_{j\ell} \bar{k}_{c(j,\ell)} > 0,$$

for all $1 \leq j \leq n-1$, we write the left-side as b_j, and

$$(\alpha_n - \beta_n)(1 - \delta) - \delta z_n > 0,$$

the left-side is written as b_n. Let $M_1 = 0$ and $M_\ell = M$ for $2 \le \ell \le n$. By the definition of $\theta(s, \boldsymbol{x}, \widehat{\boldsymbol{x}})$,

$$\begin{aligned}
P(\theta - s > t) &\le \sum_{j=1}^{n} P(x_j^0(s+t) < M_j + \widehat{x}_j) \\
&\le P\Big(\inf_{s+\delta t \le v \le s+t} x_1^0(v) = 0 \Big) \\
&\quad + \sum_{j=1}^{n-2} P\bigg(\bigcap_{i=1}^{j} \Big\{ \inf_{s+\delta t \le v \le s+t} x_i^0(v) > 0 \Big\} \\
&\qquad\qquad \bigcap \Big\{ \inf_{s+\delta t \le v \le s+t} x_{j+1}^0(v) = 0 \Big\} \bigg) \\
&\quad + \sum_{j=1}^{n} P\bigg(\bigcap_{i=1}^{n-1} \Big\{ \inf_{s+\delta t \le v \le s+t} x_i^0(v) > 0 \Big\} \\
&\qquad\qquad \bigcap \big\{ x_j^0(s+t) < M_j + \widehat{x}_j \big\} \bigg).
\end{aligned} \tag{B.56}$$

First we estimate the first term on the right-hand side of (B.56). Note that

$$u_{1,j}^0(v) \le p_{1j} k_{c(1,j)}(v), \quad j = 2, ..., n.$$

Then, using Lemma B.8, we get

$$\begin{aligned}
&P\Big(\inf_{s+\delta t \le v \le s+t} x_1^0(v) = 0 \Big) \\
&\le P\bigg(\inf_{\delta t \le v \le t} \int_s^{s+v} \Big[\sum_{j=-n_0+1}^{0} p_{j1} k_{c(j,1)}(r) - \sum_{j=2}^{n} p_{1j} k_{c(1,j)}(r) \Big] dr \le 0 \bigg) \\
&\le P\bigg(\inf_{\delta t \le v \le t} \int_s^{s+v} \Big[\sum_{j=-n_0+1}^{0} p_{j1} k_{c(j,1)}(r) - \sum_{j=2}^{n} p_{1j} k_{c(1,j)}(r) \\
&\qquad\qquad - (\alpha_1 - \beta_1) \Big] dr \le -\delta(\alpha_1 - \beta_1) t \bigg) \\
&\le P\bigg(\sup_{\delta t \le v \le t} \bigg| \int_s^{s+v} \Big[\sum_{j=-n_0+1}^{0} p_{j1} k_{c(j,1)}(r) - \sum_{j=2}^{n} p_{1j} k_{c(1,j)}(r) \\
&\qquad\qquad - (\alpha_1 - \beta_1) \Big] dr \bigg| \ge \delta(\alpha_1 - \beta_1) t \bigg) \\
&\le C_1 \exp\big\{ -\delta(\alpha_1 - \beta_1) t^{1/2} \big\},
\end{aligned} \tag{B.57}$$

for some positive constant C_1. If for some j $(1 \le j \le n-2)$ such that, for all $1 \le i \le j$,

$$\inf_{s+\delta t \le v \le s+t} x_i^0(v) > 0,$$

then

$$\sum_{j=-n_0+1}^{i} u_{j,i+1}^0(v) = \sum_{j=-n_0+1}^{i} p_{j,i+1} k_{c(j,i+1)}(v),$$

and

$$\sum_{j=i+2}^{n} u_{i+1,j}^0(v) \le \sum_{j=i+2}^{n} p_{i+1,j} k_{c(i+1,j)}(v),$$

for $v \in (s+\delta t,\ s+t)$. So, just as in the proof of (B.57), we can show that, for $i = 1, ..., n-2$,

$$P\left(\bigcap_{j=1}^{i}\left\{\inf_{s+\varepsilon t \le v \le s+t} x_j^0(v) > 0\right\} \cap \left\{\inf_{s+\varepsilon t \le v \le s+t} x_{i+1}^0(v) = 0\right\}\right) \le C_2 \exp\left\{-\delta(\alpha_{i+1} - \beta_{i+1}) t^{1/2}\right\}, \tag{B.58}$$

for some positive constant C_2.

Now we consider the members of the last sum on the right-hand side of (B.56). According to the definition of $u_{i,j}^0(\cdot)$, for $i = 1, ..., n$,

$$\begin{aligned} P&\left(\bigcap_{\ell=1}^{n-1}\left\{\inf_{s+\delta t \le v \le s+t} x_\ell^0(v) > 0\right\} \bigcap \left\{x_i^0(s+t) < M_i + \widehat{x}_i\right\}\right) \\ &\le P\left(\int_{s+\delta t}^{s+t}\left(\sum_{\ell=-n_0+1}^{i-1} p_{\ell i} k_{c(\ell,i)}(r) - \sum_{\ell=i+1}^{n} p_{i\ell} k_{c(i,\ell)}(r)\right) dr \right. \\ &\qquad \left. + x_i - \delta t \sum_{\ell=i+1}^{n} p_{i\ell} \bar{k}_{c(i,\ell)} < M_i + \widehat{x}_i\right) \qquad \text{(B.59)} \\ &\le P\left(\int_{s+\delta t}^{s+t}\left[\sum_{\ell=-n_0+1}^{i-1} p_{\ell i} k_{c(\ell,i)}(r) - \sum_{\ell=i+1}^{n} p_{i\ell} k_{c(i,\ell)}(r) \right.\right. \\ &\qquad \left.\left. -(\alpha_i - \beta_i)\right] dr < M_i + \widehat{x}_i - x_i - b_i t\right), \end{aligned}$$

with the notation convenience

$$\sum_{\ell=n+1}^{n} p_{n\ell} k_{c(n,\ell)}(r) = \sum_{\ell=n+1}^{n} p_{n\ell} \bar{k}_{c(n,\ell)} = z_n. \tag{B.60}$$

Applying Lemma B.8, we have from (B.59), for $j = 1, ..., n-1$,

$$P\left(\bigcap_{i=1}^{n-1}\left\{\inf_{s+\varepsilon t\le v\le s+t} x_i^0(v) > 0\right\}\cap\left\{x_j^0(s+t) < M_j + \widehat{x}_j\right\}\right)$$

$$\le \begin{cases} 1, & \text{for } 0 \le t \le \dfrac{(M_j+\widehat{x}_j - x_j)^+}{b_j}, \\ C_3 \exp\left\{-\dfrac{b_j t - (M_j + \widehat{x}_j - x_j)^+}{t^{1/2}}\right\}, & \text{for } t \ge \dfrac{(M_j+\widehat{x}_j - x_j)^+}{b_j}, \end{cases}$$

for some constant C_3. Using

$$E[\theta(s,\boldsymbol{x},\widehat{\boldsymbol{x}}) - s]^{2r} = \int_0^\infty t^{2r-1} P(\theta(s,\boldsymbol{x},\widehat{\boldsymbol{x}}) - s > t)\, dt.$$

Substituting from (B.56), (B.57), (B.58), and the above inequality in this relation, we complete the proof of Lemma 5.3.1. □

Proof of Lemma 5.3.2. Taking the sum of all the equations in (5.5), we have

$$\sum_{j=1}^n x_j^1(t) = \sum_{j=1}^n x_j + \int_s^t \left[\sum_{i=-n_0+1}^0 \sum_{j=1}^n u_{i,j}^1(v) - z_n\right] dv,$$

for $s \le t \le \theta$. Consequently,

$$\sum_{j=1}^n x_j^1(\theta) \le \sum_{j=1}^n x_j + \left(\sum_{i=-n_0+1}^0 \sum_{j=1}^n p_{ij}\bar{k}_{c(i,j)} - z_n\right)(\theta - s). \tag{B.61}$$

Since $u_{i,j}^1(t) = 0$ for $t \in [\theta, \widehat{\theta}]$ and $i = -n+1, \ldots, 0$, we have as previously that $\sum_{j=1}^n x_j^1(\widehat{\theta}) = \sum_{j=1}^n x_j^1(\theta) - z_n(\widehat{\theta} - \theta)$. Since $\widehat{\theta} > \theta$ and $x_j^1(\widehat{\theta}) \ge \widehat{x}_j$, we have

$$\widehat{\theta} - \theta \le \frac{1}{z_n}\left(\sum_{j=1}^n x_j^1(\theta) - \sum_{j=1}^n \widehat{x}_j\right), \quad \sum_{j=1}^n x_j^1(\widehat{\theta}) \le \sum_{j=1}^n x_j^1(\theta). \tag{B.62}$$

From the definitions of $\widehat{\theta}$ and θ, we have that

$$\begin{aligned} \hat{x}_n = x_n^1(\widehat{\theta}) &= x_n^1(\theta) + \int_\theta^{\widehat{\theta}} \left[\sum_{j=-n_0+1}^{n-1} u_{j,n}^1(v) - z_n\right] dv \\ &\ge \widehat{x}_n + M - z_n(\widehat{\theta} - \theta), \end{aligned}$$

which implies $\widehat{\theta} - \theta \ge M/z_n$. This relation together with (B.61) and (B.62) proves statement (ii) of Lemma 5.3.2.

To prove statement (i), we introduce the following notations:

$$\widehat{\theta}(j) = \inf\{t \geq \theta \;:\; x_j^1(t) = \widehat{x}_j\}, \quad j = 1, \ldots, n,$$
$$\widetilde{\mathcal{B}} = \{\omega \;:\; \boldsymbol{x}^1(\widehat{\theta})(\omega) = \widehat{\boldsymbol{x}}\},$$
$$\mathcal{B}(j) = \left\{\omega \;:\; \inf_{0\leq t<\infty} \int_\theta^{\theta+t} \left[\sum_{\ell=-n_0+1}^{j-1} p_{\ell j} k_{c(\ell,j)}(v) - \sum_{\ell=j+1}^{n} p_{j\ell} k_{c(j,\ell)}(v)\right] dv > -\frac{M}{2}\right\}, \quad j = 2, \ldots, n-1,$$
$$\mathcal{B}(n) = \left\{\omega \;:\; \inf_{0\leq t<\infty} \int_\theta^{\theta+t} \left[\sum_{\ell=-n_0+1}^{n-1} p_{\ell n} k_{c(\ell,n)}(v) - z_n\right] dv > -\frac{M}{2}\right\},$$
$$\overline{\mathcal{B}} = \bigcap_{j=2}^{n} \mathcal{B}(j), \quad \mathcal{B}(0) = \{\omega \;:\; \boldsymbol{k}(\widehat{\theta})(\omega) = \widehat{\boldsymbol{k}}\}, \quad \mathcal{B} = \widetilde{\mathcal{B}} \cap \mathcal{B}(0).$$

Note that
$$\widetilde{\mathcal{B}} = \left\{\omega \;:\; \widehat{\theta}(n)(\omega) \geq \max_{1\leq j\leq n-1} \widehat{\theta}(j)(\omega)\right\}.$$

From the definition of $\boldsymbol{u}^1(\boldsymbol{x}, \boldsymbol{k})$ and $\boldsymbol{x}^1(t)$, it follows that if $\omega \in \overline{\mathcal{B}}$, then, for $j = 1, \ldots, n-1$,

$$\sum_{\ell=j+1}^{n} u_{j,\ell}^1(t) = \begin{cases} \sum_{\ell=j+1}^{n} p_{j\ell} k_{c(j,\ell)}(v), & \text{for } \theta < t \leq \widehat{\theta}(j), \\ 0, & \text{for } t > \widehat{\theta}(j), \end{cases}$$

and, for $j = 2, \ldots, n$,

$$\widehat{\theta}(j) - \widehat{\theta}(j-1) \geq \frac{x_j^1(\widehat{\theta}(j-1)) - \widehat{x}_j}{\sum_{\ell=j+1}^{n} p_{j\ell} \overline{k}_{c(j,\ell)}} \geq \frac{M}{2\sum_{\ell=j+1}^{n} p_{j\ell} \overline{k}_{c(j,\ell)}} \tag{B.63}$$

with the convention given by (B.60). Therefore, $\overline{\mathcal{B}} \subseteq \widetilde{\mathcal{B}}$ and

$$P(\mathcal{B}^c) \leq \sum_{j=2}^{n} P(\mathcal{B}^c(j)) + P(\overline{\mathcal{B}} \cap \mathcal{B}^c(0)). \tag{B.64}$$

Note that for any q_1 with $0 < q_1 < 1$, there is a positive constant $\widehat{M}$ such that for any two Markov times τ_1 and τ_2 with respect to $\boldsymbol{k}(t)$ satisfying

$$\tau_2 - \tau_1 \geq \widehat{M}, \quad a.s.,$$

$$\max_{1\le i\le p} P(\boldsymbol{k}(\tau_2) \neq \boldsymbol{k}^i | \boldsymbol{k}(\tau_1) = \boldsymbol{k}^i) < q_1.$$

Taking the conditional probability with respect to $\widehat{\theta}(n-1)$ and choosing

$$M > 2\widehat{M}\left(z_n \vee \max_{1\le j\le n-1}\left\{\sum_{\ell=j+1}^{n} p_{j\ell}\bar{k}_{c(j,\ell)}\right\}\right),$$

and using (B.63) with $j=n$ and (5.19), we have

$$P(\overline{\mathcal{B}} \cap \mathcal{B}^c(0)) < q_1 < 1. \tag{B.65}$$

Applying Lemma B.8, we have

$$\begin{aligned} P((\mathcal{B}(i))^c) &\le \sum_{\ell=1}^{\infty} P\Bigg(\int_{\theta}^{\theta+\ell}\left[\sum_{j=-n_0+1}^{i-1} p_{ji}k_{c(j,i)}(v) - \sum_{j=i+1}^{n} p_{ij}k_{c(i,j)}(v)\right] dv \\ &\qquad\qquad < -\frac{M}{2} + \sum_{j=i+1}^{n} p_{ij}\bar{k}_{c(i,j)}\Bigg) \\ &\le \sum_{\ell=1}^{\infty} P\Bigg(\Bigg|\int_{\theta}^{\theta+\ell}\Bigg[\sum_{j=-n_0+1}^{i-1} p_{ji}k_{c(j,i)}(v) - \sum_{j=i+1}^{n} p_{ij}k_{c(i,j)}(v) \\ &\qquad\qquad -(\alpha_i-\beta_i)\Bigg] dv\Bigg| > \frac{M}{2} + \ell(\alpha_i-\beta_i) - \sum_{j=i+1}^{n} p_{ij}\bar{k}_{c(i,j)}\Bigg) \\ &\le C_1\sum_{\ell=1}^{\infty}\exp\left(-\frac{M+\ell(\alpha_i-\beta_i)-\sum_{j=i+1}^{n}p_{ij}\bar{k}_{c(i,j)}}{\ell^{1/2}}\right) \\ &\le C_2 e^{-C_3\sqrt{M}}, \end{aligned} \tag{B.66}$$

for some positive constants C_1, C_2, and C_3. It follows from (B.64), (B.65), and (4.15) that we can choose M and q such that $P(\mathcal{B}^c) \le q < 1$. This proves Lemma 5.3.2. □

Appendix C
Convex Sets and Functions

In this appendix we define convex functions, superdifferentials, subdifferentials, and their properties.

Definition C.1. A real-valued function $f(\boldsymbol{x})$ is *convex* if, for every $\boldsymbol{x}_1$ and $\boldsymbol{x}_2$ and $\delta \in [0,1]$,

$$f(\delta \boldsymbol{x}_1 + (1-\delta)\boldsymbol{x}_2) \leq \delta f(\boldsymbol{x}_1) + (1-\delta) f(\boldsymbol{x}_2).$$

If the inequality above is strict whenever $\boldsymbol{x}_1 \neq \boldsymbol{x}_2$ and $0 < \delta < 1$, then $f(\boldsymbol{x})$ is *strictly convex*. □

Definition C.2. A function $f(\boldsymbol{x}, \boldsymbol{y})$ is *jointly convex* if

$$\begin{aligned} f(\delta \boldsymbol{x}_1 + (1-\delta)\boldsymbol{x}_2, \delta \boldsymbol{y}_1 + (1-\delta)\boldsymbol{y}_2) & \\ \leq \delta f(\boldsymbol{x}_1, \boldsymbol{y}_1) + (1-\delta) f(\boldsymbol{x}_2, \boldsymbol{y}_2). & \end{aligned} \tag{C.1}$$

Moreover, $f(\boldsymbol{x}, \boldsymbol{y})$ is *strictly jointly convex in* $\boldsymbol{x}$ if the inequality (C.1) holding as an equality for some $0 < \delta < 1$ implies $\boldsymbol{x}_1 = \boldsymbol{x}_2$. □

Lemma C.1. *Let $f(\boldsymbol{x})$ be a convex function. Then:*

(i) *$f(\boldsymbol{x})$ is locally Lipschitz and therefore continuous.*

(ii) *$f(\boldsymbol{x})$ is differentiable a.e.*

(iii) *Let $\{f_\rho(x) : \rho > 0\}$ be a family of convex functions defined on a convex set $\mathcal{D} \subset \Re^n$. If $\{f_\rho(x) : \rho > 0\}$ is locally bounded, then it is locally uniformly Lipschitz continuous.*

Proof. Proofs for (i) and (ii) can be found in in Clarke [34, Theorem 2.5.1]. To prove part (iii), let $\mathcal{D}_1$ and $\mathcal{D}_2$, such that $\mathcal{D}_1 \subset \mathcal{D}_2 \subset \Re^n$, be balls of radii r and $r+1$ in $\Re^n$, respectively. For $x, x' \in \mathcal{D}_1 \cap \mathcal{D}$ and $0 \leq \lambda \leq 1$, we have, by the definition of convex functions,

$$\begin{aligned} f_\rho(x) - f_\rho(x') &= f_\rho\left(\lambda x' + \frac{(1-\lambda)(x-\lambda x')}{1-\lambda}\right) - f_\rho(x') \\ &\leq \lambda f_\rho(x') + (1-\lambda) f_\rho\left(\frac{x-\lambda x'}{1-\lambda}\right) - f_\rho(x') \qquad \text{(C.2)} \\ &= (1-\lambda)\left[f_\rho\left(\frac{x-\lambda x'}{1-\lambda}\right) - f_\rho(x')\right]. \end{aligned}$$

Without loss of generality, we may take $|x - x'| \leq 1$. Let $\lambda = 1 - |x - x'|$. We can then write (C.2) as

$$\begin{aligned} f_\rho(x) - f_\rho(x') &\leq |x - x'| \left[f_\rho(x' + (x - x')/|x - x'|) - f_\rho(x')\right] \\ &\leq |x - x'| \cdot 2 \sup_{y \in \mathcal{D}_2} |f_\rho(y)|. \end{aligned}$$ □

Definition C.3. Given a function $f(\boldsymbol{x})$ (convex or not), the *superdifferential* $D^+ f(\boldsymbol{x})$ and the *subdifferential* $D^- f(\boldsymbol{x})$ of $f(\boldsymbol{x})$ are defined, respectively, as follows:

$$\begin{aligned} D^+ f(\boldsymbol{x}) &= \left\{ \boldsymbol{r} \in \Re^n : \limsup_{\boldsymbol{h} \to 0} \frac{f(\boldsymbol{x}+\boldsymbol{h}) - f(\boldsymbol{x}) - \langle \boldsymbol{h}, \boldsymbol{r} \rangle}{|\boldsymbol{h}|} \leq 0 \right\}, \\ D^- f(\boldsymbol{x}) &= \left\{ \boldsymbol{r} \in \Re^n : \liminf_{\boldsymbol{h} \to 0} \frac{f(\boldsymbol{x}+\boldsymbol{h}) - f(\boldsymbol{x}) - \langle \boldsymbol{h}, \boldsymbol{r} \rangle}{|\boldsymbol{h}|} \geq 0 \right\}. \end{aligned}$$ □

Definition C.4. The real-valued function $f(\boldsymbol{x})$ is said to have a *directional derivative* $\partial_{\boldsymbol{p}} f(\boldsymbol{x})$ along direction $\boldsymbol{p} \in \Re^n$, defined by

$$\partial_{\boldsymbol{p}} f(\boldsymbol{x}) = \lim_{\delta \to 0} \frac{f(\boldsymbol{x} + \delta \boldsymbol{p}) - f(\boldsymbol{x})}{\delta},$$

whenever the limit exists. □

Definition C.5. The *convex hull* of a set $\mathcal{D} \subset \Re^n$ is

$$\mathrm{co}(\mathcal{D}) = \left\{ \sum_{i=1}^{\ell} \alpha_i \boldsymbol{x}_i : \sum_{i=1}^{\ell} \alpha_i = 1, \alpha_i \geq 0,\ \boldsymbol{x}_i \in F,\ 0 \leq i \leq \ell \right\}.$$ □

Definition C.6. The *convex closure* $\overline{\mathrm{co}}(\mathcal{D})$ of a set $\mathcal{D} \subset \Re^n$ is the closure of $\mathrm{co}(\mathcal{D})$. □

Lemma C.2. *Let $f(\boldsymbol{x})$ be a convex function. Then:*

(i) *$f(\boldsymbol{x})$ is differentiable if and only if $D^+f(\boldsymbol{x})$ and $D^-f(\boldsymbol{x})$ are both singletons. In this case, $D^+f(\boldsymbol{x}) = D^-f(\boldsymbol{x}) = \{f_{\boldsymbol{x}}(\boldsymbol{x})\}$.*

(ii) *If $f(\boldsymbol{x})$ is convex, then $D^+f(\boldsymbol{x})$ is empty unless $f(\boldsymbol{x})$ is differentiable and $D^-f(\boldsymbol{x})$ coincides with the subdifferential in the sense of convex analysis, i.e.,*

$$D^-f(\boldsymbol{x}) = \overline{\text{co}}(\Gamma(\boldsymbol{x})),$$

where

$$\Gamma(\boldsymbol{x}) = \left\{\boldsymbol{r} = \lim_{n\to\infty} f_{\boldsymbol{x}}(\boldsymbol{x}_n) : \boldsymbol{x}_n \to \boldsymbol{x} \text{ and } f(\cdot) \text{ is differentiable at } \boldsymbol{x}_n\right\}.$$

(iii) *A convex function $f(\boldsymbol{x})$ is differentiable at $\boldsymbol{x}$ if and only if $D^-f(\boldsymbol{x})$ is a singleton.*

(iv) *If a convex function $f(\boldsymbol{x})$ is differentiable on $\Re^n$, then it is continuously differentiable on $\Re^n$.*

(v) *For every $\boldsymbol{x} \in \Re^n$ and $\boldsymbol{p} \in \Re^n$,*

$$\partial_{\boldsymbol{p}} f(\boldsymbol{x}) = \max_{\boldsymbol{r}\in D^-f(\boldsymbol{x})} \langle \boldsymbol{r}, \boldsymbol{p}\rangle.$$

Proof. See Clarke [34, Theorem 2.5.1] for the proof. □

Lemma C.3. *Let $\mathcal{D}$ be a nonempty convex set in $\Re^n$. Then:*

(i) ri$(\mathcal{D})$ *is convex in $\Re^n$ having the same affine hull, and hence the same dimension as $\mathcal{D}$;*

(ii) *$z \in \text{ri}(\mathcal{D})$ if and only if for each $x \in \mathcal{D}$, there exists a $\delta < 1$ such that $(1-\delta)x + \delta z \in \mathcal{D}$.*

Proof. See Rockafellar [109, Theorems 6.2 and 6.4]. □

Lemma C.4. *Given a relative open convex set $\mathcal{D}$, let $\{f_i | i \in I\}$ be an arbitrary family of convex functions that are finite and pointwise bounded on $\mathcal{D}$. Let $\widehat{\mathcal{D}}$ be a closed bounded subset of $\mathcal{D}$. Then $\{f_i | i \in I\}$ is uniformly bounded on $\widehat{\mathcal{D}}$ and equi-Lipschitz relative to $\widehat{\mathcal{D}}$. The pointwise boundedness condition above can be replaced by these two conditions:*

(i) *There exists a subset $\mathcal{D}'$ of $\mathcal{D}$ such that $\mathcal{D} \subset \text{conv}(\mathcal{D}'|\mathcal{D})$ and $\sup\{f_i(x)| i \in I\}$ is finite for $x \in \mathcal{D}'$.*

(ii) *There exists at least one $x \in \mathcal{D}$ such that $\inf\{f_i(x)|i \in I\}$ is finite.*

Proof. See Rockafellar [109, Theorem 10.6]. □

Appendix D

Viscosity Solutions of HJB Equations

When the state space in an optimal control problem is continuous, the classical dynamic programming argument leads to an equation known as the Hamilton-Jacobi-Bellman (HJB) equation that the value function for the problem must satisfy. The equation involves at least the first derivative of the value function, even though such a derivative might not exist at certain points. In these cases, a useful concept that is often used is that of a viscosity solution of the HJB equation. (The concept was introduced by Crandall and Lions [37].) In what follows, we briefly describe the concept and some of the related results. For more information and discussion on viscosity solutions, the reader is referred to the book by Fleming and Soner [56].

Let Θ be a finite index set and let $W(\cdot,\cdot): \Re^n \times \Theta \to \Re^1$ be a given function. Let F denote a real-valued function on

$$\Omega_F := \Re^n \times \Theta \times \Re \times \Re^p \times \Re^{n \times p},$$

where p is the number of elements in Θ. We define the concept of a viscosity solution of the following equation:

$$F\left(\boldsymbol{x}, \boldsymbol{\theta}, \lambda, W(\boldsymbol{x}, \cdot), \frac{\partial W(\boldsymbol{x}, \boldsymbol{\theta})}{\partial \boldsymbol{x}}\right) = 0. \tag{D.1}$$

Definition D.1. $(\lambda, W(\boldsymbol{x}, \boldsymbol{\theta}))$ is a *viscosity solution* of equation (D.1) if the following hold:

(a) $W(\boldsymbol{x}, \boldsymbol{\theta})$ is continuous in $\boldsymbol{x}$ and $|W(\boldsymbol{x}, \boldsymbol{\theta})| \leq C(1 + |\boldsymbol{x}|^{\beta})$, where β is a positive constant;

(b) for any $\boldsymbol{\theta}_0 \in \Theta$,

$$F\left(\boldsymbol{x}_0, \boldsymbol{\theta}_0, \lambda, W(\boldsymbol{x}_0, \cdot), \frac{d\phi(\boldsymbol{x}_0)}{d\boldsymbol{x}}\right) \leq 0,$$

whenever $\phi(\boldsymbol{x}) \in C^1$ is such that $W(\boldsymbol{x}, \boldsymbol{\theta}_0) - \phi(\boldsymbol{x})$ has a local maximum at $\boldsymbol{x} = \boldsymbol{x}_0$; and

(c) for any $\boldsymbol{\theta}_0 \in \Theta$,

$$F\left(\boldsymbol{x}_0, \boldsymbol{\theta}_0, \lambda, W(\boldsymbol{x}_0, \cdot), \frac{d\psi(\boldsymbol{x}_0)}{d\boldsymbol{x}}\right) \geq 0,$$

whenever $\psi(\boldsymbol{x}) \in C^1$ is such that $W(\boldsymbol{x}, \boldsymbol{\theta}_0) - \psi(\boldsymbol{x})$ has a local minimum at $\boldsymbol{x} = \boldsymbol{x}_0$.

If (a) and (b) (resp., (a) and (c)) hold, we say that $W(\boldsymbol{x}, \boldsymbol{\theta})$ is a *viscosity subsolution* (resp., *viscosity supersolution*). □

As defined in Appendix C, let $D^+W(\boldsymbol{x}, \boldsymbol{\theta})$ and $D^-W(\boldsymbol{x}, \boldsymbol{\theta})$ denote the superdifferential and subdifferential of $W(\boldsymbol{x}, \boldsymbol{\theta})$ for each fixed $\boldsymbol{\theta}$. We can now provide another equivalent definition of a viscosity solution of (D.1); see Crandall, Evans, and Lions [36].

Definition D.2. We say that $(\lambda, W(\boldsymbol{x}, \boldsymbol{\theta}))$ is a *viscosity solution* of equation (D.1) if the following hold:

(a) $W(\boldsymbol{x}, \boldsymbol{\theta})$ is continuous in $\boldsymbol{x}$ and $|W(\boldsymbol{x}, \boldsymbol{\theta})| \leq C(1 + |\boldsymbol{x}|^\beta)$, where β is a positive constant;

(b) for all $\boldsymbol{r} \in D^+W(\boldsymbol{x}, \boldsymbol{\theta})$,

$$F(\boldsymbol{x}, \boldsymbol{\theta}, \lambda, W(\boldsymbol{x}, \cdot), \boldsymbol{r}) \leq 0; \quad \text{and} \tag{D.2}$$

(c) for all $\boldsymbol{r} \in D^-W(\boldsymbol{x}, \boldsymbol{\theta})$,

$$F(\boldsymbol{x}, \boldsymbol{\theta}, \lambda, W(\boldsymbol{x}, \cdot), \boldsymbol{r}) \geq 0. \tag{D.3}$$

Moreover, $(\lambda, W(\boldsymbol{x}, \boldsymbol{\theta}))$ is said to be a *viscosity subsolution* (resp., *supersolution*) of (D.1) if (a) and (b) (resp., (a) and (c)) hold. □

Since we deal mostly with convex value functions in this book (except in Chapters 1, 2, and 11), it would be appropriate to provide an equivalent definition of a viscosity solution in the convex case; see Crandall, Evans, and Lions [36].

Definition D.3. We say that $(\lambda, W(\boldsymbol{x}, \boldsymbol{\theta}))$, where $W(\boldsymbol{x}, \boldsymbol{\theta})$ is convex with respect to $\boldsymbol{x}$ for each fixed $\boldsymbol{\theta} \in \Theta$, is a *viscosity solution* to equation (D.1) if the following hold:

(a) $W(\boldsymbol{x}, \boldsymbol{\theta})$ is continuous in $\boldsymbol{x}$ and $|W(\boldsymbol{x}, \boldsymbol{\theta})| \leq C(1+|\boldsymbol{x}|^{\beta})$, where β is a positive constant;

(b) for all $\boldsymbol{r} \in D^{-}W(\boldsymbol{x}, \boldsymbol{\theta})$,

$$F(\boldsymbol{x}, \boldsymbol{\theta}, \lambda, W(\boldsymbol{x}, \cdot), \boldsymbol{r}) \geq 0; \quad \text{and} \tag{D.4}$$

(c) for all $\boldsymbol{x}$ at which $W(\boldsymbol{x}, \boldsymbol{\theta})$ is differentiable,

$$F(\boldsymbol{x}, \boldsymbol{\theta}, \lambda, W(\boldsymbol{x}, \cdot), \boldsymbol{r}) = 0.$$

Moreover, $(\lambda, W(\boldsymbol{x}, \boldsymbol{\theta}))$ is said to be a *viscosity subsolution* (resp., *supersolution*) of (D.1) if (a) and (b) (resp., (a) and (c)) hold. □

Theorem D.1. (Uniqueness Theorem). *Assume, for some $\beta > 0$,*

$$|g(\boldsymbol{x}, \boldsymbol{u})| \leq C(1+|\boldsymbol{x}|^{\beta}), \quad \textit{and}$$
$$|g(\boldsymbol{x}_1, \boldsymbol{u}) - g(\boldsymbol{x}_2, \boldsymbol{u})| \leq C(1+|\boldsymbol{x}_1|^{\beta}+|\boldsymbol{x}_2|^{\beta})|\boldsymbol{x}_1-\boldsymbol{x}_2|.$$

Then the HJB *equation*

$$\lambda = \min_{0 \leq u_1+\cdots+u_n \leq k} \left\{ \left\langle (\boldsymbol{u}-\boldsymbol{z}), \frac{\partial W(\boldsymbol{x}, k)}{\partial \boldsymbol{x}} \right\rangle + g(\boldsymbol{x}, \boldsymbol{u}) \right\} + QW(\boldsymbol{x}, \cdot)(k) \tag{D.5}$$

has a unique solution λ, where Q is the generator of a finite-state Markov chain.

Proof. We shall base our proof on Ishii [73] and Soner [129]. For convenience in notation, we only prove the theorem in the one-dimensional case. The control constraint in this case is given by $0 \leq u \leq k$.

Let $(\lambda_1, W_1(x, k))$ and $(\lambda_2, W_2(x, k))$ be two viscosity solutions to the HJB equation (D.5). We need to show that $\lambda_1 = \lambda_2$.

Let $\eta(x) = \exp((1+|x|^2)^{1/2})$. For any $0 < \delta < 1$ and $0 < \alpha < 1$, we consider a function

$$\begin{aligned}\Phi(x_1, x_2, k) = & \; [W_1(x_1, k) - W_2(x_2, k)] \\ & -\frac{1}{\delta}|x_1 - x_2|^2 - \alpha(\eta(x_1) + \eta(x_2)).\end{aligned}$$

Then $\Phi(x_1, x_2, k)$ has a global maximum at a point (x_1^0, x_2^0, k_0), since Φ is continuous and $\lim_{|x_1|+|x_2| \to +\infty} \Phi(x_1, x_2, k) = -\infty$ for each $k \in \mathcal{M}$. This means, in particular, that

$$\Phi(x_1^0, x_1^0, k_0) + \Phi(x_2^0, x_2^0, k_0) \leq 2\Phi(x_1^0, x_2^0, k_0).$$

Therefore,

$$\begin{aligned}
&[W_1(x_1^0,k_0)-W_2(x_1^0,k_0)]-\alpha[\eta(x_1^0)+\eta(x_1^0)]\\
&\qquad+[W_1(x_2^0,k_0)-W_2(x_2^0,k_0)]-\alpha[\eta(x_2^0)+\eta(x_2^0)]\\
&\quad\le 2[W_1(x_1^0,k_0)-W_2(x_2^0,k_0)]\\
&\quad\quad -\frac{2}{\delta}|x_1^0-x_2^0|^2-2\alpha[\eta(x_1^0)+\eta(x_2^0)],
\end{aligned}$$

this implies that

$$\frac{2}{\delta}|x_1^0-x_2^0|^2\le[W_1(x_1^0,k_0)-W_1(x_2^0,k_0)]+[W_2(x_1^0,k_0)-W_2(x_2^0,k_0)].$$

In view of the polynomial growth of $W_1(x,k)$ and $W_2(x,k)$ (see Definition D.1(a)),

$$|x_1^0-x_2^0|\le\sqrt{\delta C_1(1+|x_1^0|^\beta+|x_2^0|^\beta)}, \tag{D.6}$$

for some $C_1>0$. The choice of (x_1^0,x_2^0,k_0) also implies $\Phi(0,0,k_0)\le\Phi(x_1^0,x_2^0,k_0)$. This yields

$$\begin{aligned}
\alpha[\eta(x_1^0)+\eta(x_2^0)]\le\ &(\lambda_1-\lambda_2)+W_1(x_1^0,k_0)-W_2(x_2^0,k_0)\\
&-\frac{1}{\delta}|x_1^0-x_2^0|^2-\Phi(0,0,k_0)\\
\le\ &C(1+|x_1^0|^\beta+|x_2^0|^\beta).
\end{aligned}$$

Thus, there exists a constant C_α (independent of δ) such that

$$|x_1^0|+|x_2^0|\le C_\alpha. \tag{D.7}$$

Since $x\mapsto\Phi(x,x_2^0,k_0)$ takes its maximum at $x=x_1^0$, we have, according to Definition D.1 of viscosity solutions,

$$\begin{aligned}
\lambda_1\le\min_{0\le u\le k_0}&\left\{(u-z)\cdot\left(\frac{2}{\delta}(x_1^0-x_2^0)+\alpha\frac{d\eta(x_1^0)}{dx}\right)\right.\\
&\left.+g(x_1^0,u)+QW_1(x_1^0,\cdot)(k_0)\right\}.
\end{aligned} \tag{D.8}$$

Similarly, since $x\mapsto-\Phi(x_1^0,x,k_0)$ takes its minimum at $x=x_2^0$, we have

$$\begin{aligned}
\lambda_2\ge\min_{0\le u\le k_0}&\left\{(u-z)\cdot\left(\frac{2}{\delta}(x_1^0-x_2^0)-\alpha\frac{d\eta(x_2^0)}{dx}\right)\right.\\
&\left.+g(x_2^0,u)+QW_2(x_2^0,\cdot)(k_0)\right\}.
\end{aligned} \tag{D.9}$$

Combining the two inequalities (D.8) and (D.9), we obtain

$$\begin{aligned}
\lambda_1 - \lambda_2 \le & \min_{0\le u\le k_0} \Bigg\{ (u-z)\cdot\left(\frac{2}{\delta}(x_1^0 - x_2^0) + \alpha\frac{d\eta(x_1^0)}{dx}\right) + g(x_1^0, u) \\
& \qquad\qquad + QW_1(x_1^0,\cdot)(k_0)\Bigg\} \\
& - \min_{0\le u\le k_0} \Bigg\{ (u-z)\cdot\left(\frac{2}{\delta}(x_1^0 - x_2^0) - \alpha\frac{d\eta(x_2^0)}{dx}\right) + g(x_2^0, u) \\
& \qquad\qquad + QW_2(x_2^0,\cdot)(k_0)\Bigg\} \\
\le & \sup_{0\le u\le k_0} \left\{ \alpha(u-z)\cdot\left[\frac{d\eta(x_1^0)}{dx} + \frac{d\eta(x_2^0)}{dx}\right] + g(x_1^0,u) - g(x_2^0,u)\right\} \\
& + QW_1(x_1^0,\cdot)(k_0) - QW_2(x_2^0,\cdot)(k_0).
\end{aligned} \tag{D.10}$$

In view of (D.6) and (D.7), there exists a subsequence of $\delta \to 0$ (still denoted by δ) and x_0 such that $x_1^0 \to x_0$ and $x_2^0 \to x_0$. Using this fact, and letting $\delta \to 0$ in (D.10), we get

$$\begin{aligned}
\lambda_1 - \lambda_2 \le & \sup_{0\le u\le k_0} \left\{ 2\alpha(u-z)\cdot\frac{d\eta(x_0)}{dx}\right\} \\
& + QW_1(x_0,\cdot)(k_0) - QW_2(x_0,\cdot)(k_0).
\end{aligned} \tag{D.11}$$

Note that, for all x and $k \in \mathcal{M}$,

$$\begin{aligned}
&[W_1(x,k) - W_2(x,k)] - 2\alpha\eta(x) \\
&\quad = \Phi(x,x,k) \le \Phi(x_1^0, x_2^0, k_0) \\
&\quad \le [W_1(x_1^0,k_0) - W_2(x_2^0,k_0)] - \alpha[\eta(x_1^0) + \eta(x_2^0)].
\end{aligned}$$

Letting $\delta \to 0$ and recalling the fact that $x_1^0 \to x_0$ and $x_2^0 \to x_0$, we obtain

$$\begin{aligned}
&[W_1(x,k) - W_2(x,k)] - 2\alpha\eta(x) \\
&\quad \le [W_1(x_0,k_0) - W_2(x_0,k_0)] - 2\alpha\eta(x_0).
\end{aligned} \tag{D.12}$$

In particular, taking $x = x_0$, we have

$$W_1(x_0,k) - W_2(x_0,k) \le W_1(x_0,k_0) - W_2(x_0,k_0).$$

Thus,

$$\begin{aligned}
&QW_1(x_0,\cdot)(k_0) - QW_2(x_0,\cdot)(k_0) \\
&= \sum_{k\ne k_0} q_{k_0k}[W_1(x_0,k) - W_1(x_0,k_0) - W_2(x_0,k) + W_2(x_0,k_0)] \le 0.
\end{aligned} \tag{D.13}$$

Combine (D.11), (D.12), and (D.13) to obtain

$$\begin{aligned}\lambda_1 - \lambda_2 &\leq \sup_{0 \leq u \leq k_0} \left\{ 2\alpha(u - z) \cdot \frac{d\eta(x_0)}{dx} \right\} \\ &\leq \sup_{0 \leq u \leq k_0} \left\{ 2a(u - z) \cdot \eta(x_0) \right\}.\end{aligned}$$

The second inequality above is due to the fact that $\eta_x(x) \leq a\eta(x)$. Letting $\alpha \to 0$, we obtain

$$\lambda_1 - \lambda_2 \leq 0.$$

Similarly, we can obtain the opposite inequality and conclude $\lambda_1 = \lambda_2$. □

Theorem D.2. *Assume the conditions of Theorem* D.1. *Let $V^\rho(\boldsymbol{x}, k)$ denote the viscosity solution to the equation*

$$\begin{aligned}\rho V^\rho(\boldsymbol{x}, k) = &\min_{0 \leq u_1 + \cdots + u_n \leq k} \left\{ \left\langle (\boldsymbol{u} - \boldsymbol{z}), \frac{\partial V^\rho(\boldsymbol{x}, k)}{\partial \boldsymbol{x}} \right\rangle + g(\boldsymbol{x}, \boldsymbol{u}) \right\} \\ &+ QV^\rho(\boldsymbol{x}, \cdot)(k).\end{aligned} \tag{D.14}$$

Assume that when $\rho \to 0$, $\rho V^\rho(\boldsymbol{x}, k) \to \lambda$ and $(V^\rho(\boldsymbol{x}, k) - V^\rho(0, 0)) \to W(\boldsymbol{x}, k)$ locally uniformly in x. Then (λ, W) is a viscosity solution to the equation

$$\lambda = \min_{0 \leq u_1 + \cdots + u_n \leq k} \left\{ \left\langle (\boldsymbol{u} - \boldsymbol{z}), \frac{\partial W(\boldsymbol{x}, k)}{\partial \boldsymbol{x}} \right\rangle + g(\boldsymbol{x}, \boldsymbol{u}) \right\} + QW(\boldsymbol{x}, \cdot)(k). \tag{D.15}$$

Moreover, λ is uniquely determined by (D.15).

Proof. We only prove the one-dimensional case, i.e., $x \in \Re^1$ with control constraint $0 \leq u \leq k$. Fix $k_0 \in \mathcal{M}$ and let $\phi(x) \in C^1$ be such that $W(x, k_0) - \phi(x)$ has a strict local maximum at x_0, i.e., for some neighborhood $\mathcal{N}(x_0)$ of x_0,

$$W(x_0, k_0) - \phi(x_0) = \max_{x \in \mathcal{N}(x_0)} \{W(x, k_0) - \phi(x)\}.$$

Then, for each $\rho > 0$, there exists $x_\rho \in \mathcal{N}(x_0)$ such that

$$V^\rho(x_\rho, k_0) - \phi(x_\rho) = \max_{x \in \mathcal{N}(x_0)} \{V^\rho(x, k_0) - \phi(x)\}.$$

Moreover, $x_\rho \to x_0$.

Since $V^\rho(\cdot, \cdot)$ is a viscosity subsolution to (D.14), we have

$$\rho V^\rho(x_\rho, k_0) \leq \min_{0 \leq u \leq k_0} \left\{ (u - z)\phi_x(x_\rho) + g(x_\rho, u) + QV^\rho(x_\rho, \cdot)(k_0) \right\}.$$

Sending $\rho \to 0$, we can show that $(\lambda, W(x, k))$ is a viscosity subsolution to (D.15).

We now show that $(\lambda, W(x,k))$ is a viscosity supersolution to (D.15). Similarly as above, we fix $k_0 \in \mathcal{M}$ and let $\psi(x) \in C^1$ be such that $W(x,k_0) - \psi(x)$ has a strict local minimum at x_0, i.e., for some neighborhood $\mathcal{N}(x_0)$ of x_0,

$$W(x_0,k_0) - \psi(x_0) = \min_{x \in \mathcal{N}(x_0)} \{W(x,k_0) - \psi(x)\}.$$

Then, for each $\rho > 0$, there exists $x_\rho \in \mathcal{N}(x_0)$ such that

$$V^\rho(x_\rho,k_0) - \psi(x_\rho) = \min_{x \in \mathcal{N}(x_0)} \{V^\rho(x,k_0) - \psi(x)\}.$$

Moreover, $x_\rho \to x_0$.

Since $V^\rho(\cdot,\cdot)$ is a viscosity supersolution to (D.14), we have

$$\rho V^\rho(x_\rho,k_0) \geq \min_{0 \leq u \leq k_0} \{(u-z)\psi_x(x_\rho) + g(x_\rho,u) + QV^\rho(x_\rho,\cdot)(k_0)\}.$$

Sending $\rho \to 0$, we have that $(\lambda, W(x,k))$ is a viscosity supersolution to (D.15).

Hence, we have shown that $(\lambda, W(x,k))$ is a viscosity solution to (D.15). □

Theorem D.3. *Let $\mathcal{O}^n \subseteq \Re^n$ be an open set and let $\eta \in C(\mathcal{O}^n)$ be differentiable at $\boldsymbol{x}_0 \in \mathcal{O}^n$. Then there exist $\phi_+ \in C^1(\mathcal{O}^n)$ and $\phi_- \in C^1(\mathcal{O}^n)$, such that $\partial\phi_+(\boldsymbol{x}_0)/\partial\boldsymbol{x} = \partial\phi_-(\boldsymbol{x}_0)/\partial\boldsymbol{x} = \partial\eta(\boldsymbol{x}_0)/\partial\boldsymbol{x}$, and $\eta - \phi_+$ (resp., $\eta - \phi_-$) has a strict local maximum (resp., minimum) value of zero at $\boldsymbol{x}_0$.*

Proof. See Crandall, Evans, and Lions [36, Lemma 1.1]. □

Appendix E
Discounted Cost Problems

In this appendix we present the elementary properties of value functions and optimal feedback controls. We present these results by considering a simple but typical model. Results for other such models can be obtained similarly.

Let $k(t) \in \mathcal{M} = \{0, 1, \ldots, m\}$, $t \geq 0$, denote a Markov chain with generator Q. Let us consider the one-dimensional optimal control problem given as follows. The system equation and the control constraints are

$$\frac{d}{dt}x(t) = u(t) - z, \quad x(0) = x \text{ and } 0 \leq u(t) \leq k(t).$$

Definition E.1. A control process (production rate) $u(\cdot) = \{u(t) \in \Re_+ : t \geq 0\}$ is called *admissible* with respect to the initial capacity k, if: (i) $u(\cdot)$ is adapted to the filtration $\{\mathcal{F}_t\}$ with $\mathcal{F}_t = \sigma\{k(s) : 0 \leq s \leq t\}$, the σ-field generated by $k(\cdot)$; and (ii) $0 \leq u(t)(\omega) \leq k(t)(\omega)$ for all $t \geq 0$ and $\omega \in \Omega$. □

Let $\mathcal{A}(k)$ denote the set of admissible control processes with the initial condition $k(0) = k$.

Definition E.2. A function $u(x, k)$ is called an *admissible feedback control*, or simply *feedback control*, if: (i) for any given initial x, the equation

$$\frac{d}{dt}x(t) = u(x(t), k(t)) - z, \quad x(0) = x,$$

has a unique solution; and (ii) $\{u(t) = u(x(t), k(t)),\ t \geq 0\} \in \mathcal{A}(k)$. □

Remark E.1. The class of admissible controls contains feedback controls and other controls termed *nonfeedback* controls. Nonfeedback controls can be *partial open-loop controls* or *open-loop controls* depending on whether they respond to some (but not all) of the states in a systematic way or not. In this book all nonfeedback controls will be referred to as open-loop controls. □

Remark E.2. Rather than define a set of admissible feedback control functions, we shall express the admissibility condition (ii) as simply $u(x,k) \in \mathcal{A}(k)$, with a slight abuse of notation. □

The problem is to choose an admissible control $u(\cdot)$ so as to minimize the objective function

$$J^\rho(x,k,u(\cdot)) = E\int_0^\infty e^{-\rho t} g(x(t),u(t))\,dt,$$

where x and k are the initial values of $x(t)$ and $k(t)$, respectively, and $g(x,u)$ is a function of x and u.

Let $V^\rho(x,k)$ denote the value function of the problem, i.e.,

$$V^\rho(x,k) = \inf_{u(\cdot)\in\mathcal{A}(k)} J^\rho(x,k,u(\cdot)).$$

Then we have the following lemma.

Lemma E.1. (i) *If $g(x,u)$ is jointly convex, then $V^\rho(x,k)$ is convex in x for each $k \in \mathcal{M}$.*

(ii) *If $g(x,u)$ is locally Lipschitz, i.e.,*

$$|g(x,u) - g(\widetilde{x},u)| \le C(1 + |x|^{\beta_g} + |\widetilde{x}|^{\beta_g})|x - \widetilde{x}|$$

for some constant C and β_g, then $V^\rho(x,k)$ is also locally Lipschitz, i.e.,

$$|V^\rho(x,k) - V^\rho(\widetilde{x},k)| \le C(1 + |x|^{\beta_g} + |\widetilde{x}|^{\beta_g})|x - \widetilde{x}|.$$

Proof. To show (i), it suffices to show that $J^\rho(\cdot,k,\cdot)$ is jointly convex. For any initial values x and $\widetilde{x}$ and any admissible controls $u(\cdot)$ and $\widetilde{u}(\cdot)$, let $x^1(t)$ and $x^2(t)$, $t \ge 0$, denote the trajectories corresponding to $(x,u(\cdot))$ and $(\widetilde{x},\widetilde{u}(\cdot))$. Then, for any $\delta \in [0,1]$,

$$\begin{aligned}
&\delta J^\rho(x,k,u(\cdot)) + (1-\delta)J^\rho(\widetilde{x},k,\widetilde{u}(\cdot)) \\
&\quad = E\int_0^\infty e^{-\rho t}[\delta g(x(t),u(t)) + (1-\delta)g(\widetilde{x}(t),\widetilde{u}(t))]\,dt \\
&\quad \ge E\int_0^\infty e^{-\rho t} g(\widehat{x}(t),\widehat{u}(t))\,dt,
\end{aligned}$$

where $\widehat{u}(t) := \delta u(t)+(1-\delta)\widetilde{u}(t)$ and $\widehat{x}(t)$, $t \geq 0$, denotes the state trajectory with initial value $\delta x + (1-\delta)\widetilde{x}$ and control $\widehat{u}(\cdot)$. Thus,

$$\begin{aligned}\delta J^\rho(x,k,u(\cdot)) + (1-\delta)J^\rho(\widetilde{x},k,\widetilde{u}(\cdot))\\ \geq J^\rho(\delta x + (1-\delta)\widetilde{x}, k, \delta u(\cdot) + (1-\delta)\widetilde{u}(\cdot)).\end{aligned}$$

This means that $J^\rho(\cdot,k,\cdot)$ is jointly convex. Therefore, $V^\rho(x,k)$ is convex.

We now show (ii). Let $u(\cdot)$ denote an admissible control and let $x(\cdot)$ and $\widetilde{x}(\cdot)$ denote the state trajectories under $u(\cdot)$ with initial values x and $\widetilde{x}$, respectively. Then

$$|x(t) - \widetilde{x}(t)| = |x - \widetilde{x}|, \quad |x(t)| \leq C_1(t + |x|),$$

and

$$|\widetilde{x}(t)| \leq C_1(t + |\widetilde{x}|),$$

for some $C_1 > 0$. In view of the local Lipschitz assumption, we can show that there exists a constant C_2 independent of $u(\cdot)$, x, and $\widetilde{x}$ such that

$$|J^\rho(x,k,u(\cdot)) - J^\rho(\widetilde{x},k,u(\cdot))| \leq C_2(1 + |x|^{\beta_g} + |\widetilde{x}|^{\beta_g})|x - \widetilde{x}|.$$

It follows that

$$\begin{aligned}|V^\rho(x,k) - V^\rho(\widetilde{x},k)| &\leq \sup_{u(\cdot)\in\mathcal{A}(k)} |J^\rho(x,k,u(\cdot)) - J^\rho(\widetilde{x},k,u(\cdot))|\\ &\leq C_2(1 + |x|^{\beta_g} + |\widetilde{x}|^{\beta_g})|x - \widetilde{x}|.\end{aligned}$$ □

Lemma E.2. *Assume that $g(x,u)$ is jointly convex. Then $V^\rho(x,k)$ is the unique viscosity solution to the Hamilton-Jacobi-Bellman* (HJB) *equation*

$$\rho V^\rho(x,k) = \min_{0\leq u\leq k}\left[(u-z)\frac{\partial V^\rho(x,k)}{\partial x} + g(x,u)\right] + QV^\rho(x,\cdot)(k). \quad \text{(E.1)}$$

Proof. First of all, Theorem D.1 implies the uniqueness of $V^\rho(x,k)$. We need to show only that $V^\rho(x,k)$ is a viscosity solution to (E.1). In view of Definition D.1, we shall show that $V^\rho(x,k)$ is both a viscosity subsolution and a viscosity supersolution.

For any fixed k_0 and $x_0 \in \Re^1$, let $\phi(\cdot) \in C^1$ be such that $V^\rho(x,k_0) - \phi(x)$ attains its maximum at $x = x_0$ in a neighborhood $\mathcal{N}(x_0)$. Let τ denote the first jump time of $k(\cdot)$. We consider the control $u(t) = u$ for $0 \leq t < \tau$, where u, $0 \leq u \leq k_0$, is a constant. Moreover, let $\theta \in (0,\tau]$ be such that $x(t)$ starts at x_0 and stays in $\mathcal{N}(x_0)$ for $0 \leq t \leq \theta$. Define

$$\psi(x,k) = \begin{cases} \phi(x) + V^\rho(x_0,k_0) - \phi(x_0), & \text{if } k = k_0,\\ V^\rho(x,k), & \text{if } k \neq k_0.\end{cases}$$

Then by Dynkin's formula and the fact that $k(\theta) = k_0$, we have, for $0 \leq t \leq \theta$,

$$\begin{aligned} &Ee^{-\rho\theta}\psi(x(\theta), k(\theta)) - V^\rho(x_0, k_0) \\ &\quad = E\int_0^\theta \left[-\rho\psi(x(t), k_0) + \frac{d\phi(x(t))}{dx} \cdot (u(t) - z) + Q\psi(x(t), \cdot)(k_0) \right] dt. \end{aligned} \tag{E.2}$$

Note also that $x(t) \in \mathcal{N}(x_0)$ for $0 \leq t \leq \theta$. Thus, by our definition of $\phi(\cdot)$,

$$\phi(x(t)) \geq V^\rho(x(t), k_0) - (V^\rho(x_0, k_0) - \phi(x_0)), \quad \text{for } 0 \leq t \leq \theta. \tag{E.3}$$

Then, replacing $\psi(x(t), k_0)$ in (E.2) by the right-hand side of (E.3) and noting that $V^\rho(x_0, k_0) - \phi(x_0)$ is a constant, we have

$$\begin{aligned} &E[e^{-\rho\theta}V^\rho(x(\theta), k_0)] - V^\rho(x_0, k_0) \\ &\quad \leq E\int_0^\theta \left[-\rho V^\rho(x(t), k_0) + \frac{d\phi(x(t))}{dx} \cdot (u(t) - z) + QV^\rho(x(t), \cdot)(k_0) \right] dt. \end{aligned} \tag{E.4}$$

Furthermore, by the optimality principle,

$$V^\rho(x_0, k_0) \leq E\left[\int_0^\theta e^{-\rho t} g(x(t), u(t))\, dt + e^{-\rho\theta} V^\rho(x(\theta), k(\theta)) \right]. \tag{E.5}$$

Combining (E.4) and (E.5), we obtain

$$\begin{aligned} 0 \leq E\int_0^\theta e^{-\rho t} &\left[g(x(t), u(t)) - \rho V^\rho(x(t), k_0) + \frac{d\phi(x(t))}{dx} \cdot (u(t) - z) \right. \\ &\left. + QV^\rho(x(t), \cdot)(k_0) \right] dt. \end{aligned}$$

By letting $\theta \to 0$, we can conclude

$$\min_{0 \leq u \leq k_0} \left[(u - z)\frac{d\phi(x_0)}{dx} + g(x_0, u) \right] + QV^\rho(x_0, \cdot)(k_0) - \rho V^\rho(x_0, k_0) \geq 0.$$

Thus, $V^\rho(x, k)$ is a viscosity subsolution.

We next show that $V^\rho(x, k)$ is a viscosity supersolution. Suppose it is not. Then there exist k_0, x_0, and $\delta_0 > 0$ such that, for all $0 \leq u \leq k_0$,

$$(u - z)\frac{d\phi(x)}{dx} + g(x, u) + QV^\rho(x, \cdot)(k_0) - \rho V^\rho(x, k_0) \geq \delta_0 \tag{E.6}$$

in a neighborhood $\mathcal{N}(x_0)$, where $\phi(\cdot) \in C^1$ is such that $V^\rho(x, k_0) - \phi(x)$ attains its minimum at x_0 in the neighborhood $\mathcal{N}(x_0)$. Then, for all $x \in \mathcal{N}(x_0)$,

$$V^\rho(x, k_0) \geq \phi(x) + (V^\rho(x_0, k_0) - \phi(x_0)). \tag{E.7}$$

For any $0 \le u \le k_0$, let θ_0 denote a small number such that $x(t)$ starts at $x = x_0$ and $x(t)$ stays in $\mathcal{N}(x_0)$ for $0 \le t \le \theta_0$. Note that θ_0 depends on the control $u(\cdot)$. However, since $u(t) - z$ is always bounded, there exists a constant $\theta_1 > 0$ such that $\theta_0 \ge \theta_1 > 0$. Let τ denote the first jump time of the process $k(\cdot)$. Then, for $0 \le \theta \le \min\{\theta_0, \tau\}$,

$$\begin{aligned} &J^\rho(x_0, k_0, u(\cdot)) \\ &\quad \ge E\left\{\int_0^\theta e^{-\rho t} g(x(t), u(t))\,dt + e^{-\rho\theta} V^\rho(x(\theta), k(\theta))\right\} \\ &\quad \ge E\left\{\int_0^\theta e^{-\rho t}\left[\delta_0 - (u(t) - z)\cdot \frac{d\phi(x(t))}{dx} + \rho V^\rho(x(t), k_0)\right.\right. \\ &\qquad\qquad \left.\left. - QV^\rho(x(t), \cdot)(k_0)\right] dt + e^{-\rho\theta} V^\rho(x(\theta), k(\theta))\right\}. \end{aligned} \tag{E.8}$$

Now we can use the differentiability of $\phi(x)$ together with (E.7) to show

$$\begin{aligned} V^\rho(x_0, k_0) \le\ & E\left\{\int_0^\theta e^{-\rho t}\left[\rho V^\rho(x(t), k(t)) - (u(t) - z)\cdot \frac{d\phi(x(t))}{dx}\right.\right. \\ & \qquad \left.\left. - QV^\rho(x(t), \cdot)(k_0)\right] dt + e^{-\rho\theta} V^\rho(x(\theta), k(\theta))\right\}. \end{aligned}$$

Thus,

$$J^\rho(x_0, k_0, u(\cdot)) \ge V^\rho(x_0, k_0) + \delta_0 E \int_0^\theta e^{-\rho t}\,dt \ge V^\rho(x_0, k_0) + \eta,$$

where $\eta = \delta_0 E \int_0^{\theta_1 \wedge \tau} e^{-\rho t}\,dt > 0$. This means that

$$V^\rho(x_0, k_0) \ge V^\rho(x_0, k_0) + \eta,$$

which is a contradiction. This shows that $V^\rho(x, k)$ is a viscosity supersolution.

Thus, $V^\rho(x, k)$ is a viscosity solution to equation (E.1). □

Lemma E.3 (Verification Theorem). *Let $W^\rho(x, k) \in C^1(\Re)$ such that $|W^\rho(x, k)| \le C(1 + |x|^{\beta_g})$ and*

$$\rho W^\rho(x, k) = \min_{0 \le u \le k}\left[(u - z)\frac{W^\rho(x, k)}{\partial x} + g(x, u) + QW^\rho(x, \cdot)(k)\right].$$

Then, we have the following:

(i) *$W^\rho(x, k) \le J^\rho(x, k, u(\cdot))$ for any $0 \le u(t) \le k(t)$.*

(ii) *Suppose that there are $u^*(t)$ and $x^*(t)$ which satisfy*

$$\frac{d}{dt} x^*(t) = u^*(t) - z, \quad \text{with } x^*(0) = x,$$

$$r^*(t) = \partial W^\rho(x^*(t), k(t))/\partial x,$$

and

$$\begin{aligned}&\min_{0\le u\le k}\ [(u-z)r^*(t) + g(x^*(t),u) + QW^\rho(x^*(t),\cdot)(k(t))]\\&\quad = (u^*(t)-z)r^*(t) + g(x^*(t),u^*(t)) + QW^\rho(x^*(t),\cdot)(k(t)),\end{aligned}$$

a.e. in t with probability one. Then,

$$W^\rho(x,k) = V^\rho(x,k) = J^\rho(x,k,u^*(\cdot)).$$

Proof. We sketch the proof here, as further details are available in Fleming and Rishel [53].

For $T < \infty$, we have the usual dynamic programming relation

$$V^\rho(x,k) \le E\left[\int_0^T e^{-\rho t} g(x(t),u(t))\,dt + e^{-\rho T} V^\rho(x(T),k(T))\right]. \tag{E.9}$$

Note that $|x(t)| = C(x+t)$, for some $C > 0$. We obtain (i) as $T \to \infty$. Using the polynomial growth condition assumed in the lemma, inequality (E.9) becomes an equality. □

Lemma E.4. *For any $\boldsymbol{x} \in \mathcal{X}_H$, the value function $V^{\rho_k}(\boldsymbol{x},\boldsymbol{k})$ of the discounted cost problem satisfies*

$$\rho_\ell V^{\rho_\ell}(\boldsymbol{x},\boldsymbol{k}) = \inf_{\boldsymbol{u}\in\mathcal{U}(\boldsymbol{x},\boldsymbol{k})} \{\partial_{A\boldsymbol{u}+Bz} V^{\rho_\ell}(\boldsymbol{x},\boldsymbol{k}) + g(\boldsymbol{x},\boldsymbol{u})\} + QV^{\rho_\ell}(\boldsymbol{x},\boldsymbol{k}).$$

Proof. See Presman, Sethi, and Suo [99]. □

Appendix F
Miscellany

In this appendix we present miscellaneous results needed in this book.

Theorem F.1 (Arzelà-Ascoli Theorem). *Let $\mathcal{N}(R) = \{\boldsymbol{x} : |\boldsymbol{x}| \leq R\}$. Let $f_n(\boldsymbol{x})$ denote a sequence of continuous functions defined on $\mathcal{N}(R)$. If*

$$\sup_n \sup_{\boldsymbol{x} \in \mathcal{N}(R)} |f_n(\boldsymbol{x})| < \infty,$$

and

$$\sup_{n \geq 1, |\boldsymbol{x} - \widetilde{\boldsymbol{x}}| \leq \delta} |f_n(\boldsymbol{x}) - f_n(\widetilde{\boldsymbol{x}})| \to 0 \quad \text{as } \delta \to 0,$$

then there exists a continuous function $f(\boldsymbol{x})$ defined on $\mathcal{N}(R)$ and a subsequence $\{n_\ell\}$ such that

$$\sup_{\boldsymbol{x} \in \mathcal{N}(R)} |f_{n_\ell}(\boldsymbol{x}) - f(\boldsymbol{x})| \to 0 \text{ as } \ell \to \infty.$$

Proof. See Yosida [150] for a proof. □

Lemma F.1. *Let $f(t, \boldsymbol{x})$ denote a Borel measurable function on $[0, \infty) \times \Re^n$. Assume*

$$\langle f(t, \widetilde{\boldsymbol{x}}) - f(t, \widehat{\boldsymbol{x}}), (\widetilde{\boldsymbol{x}} - \widehat{\boldsymbol{x}}) \rangle \leq 0,$$

for all $t \in [0, \infty)$ and $\widetilde{\boldsymbol{x}}, \widehat{\boldsymbol{x}} \in \Re^n$. Then the ordinary differential equation

$$\frac{d}{dt}\boldsymbol{x}(t) = f(t, \boldsymbol{x}(t)), \quad \boldsymbol{x}(0) = \boldsymbol{x},$$

has at most one solution.

Proof. See Hartman [69, Theorem 6.2] for its proof. □

Lemma F.2 (Rademacher's Theorem). *Let $f(\boldsymbol{x})$ be a Lipschitz function on $X \subset \Re^n$, i.e.,*

$$|f(\widetilde{\boldsymbol{x}}) - f(\widehat{\boldsymbol{x}})| \leq C|\widetilde{\boldsymbol{x}} - \widehat{\boldsymbol{x}}|, \quad \widetilde{\boldsymbol{x}}, \widehat{\boldsymbol{x}} \in X,$$

for some $C > 0$. Then $f(\boldsymbol{x})$ is differentiable at almost every interior point of X.

Proof. See Federer [50, p. 216] for a proof. □

Lemma F.3. *If $f(t)$ is a nonnegative Borel measurable function defined on $[0, \infty)$, then*

$$\limsup_{\rho \to 0} \rho \int_0^\infty e^{-\rho t} f(t)\, dt \leq \limsup_{T \to \infty} \frac{1}{T} \int_0^T f(t)\, dt. \tag{F.1}$$

Proof. We write

$$\limsup_{T \to \infty} \frac{1}{T} \int_0^T f(t)\, dt = M.$$

Without loss of generality, we may assume $M < \infty$. For each $\delta > 0$, there exists T_0 such that, for each $T > T_0$,

$$\int_0^T f(t)\, dt \leq (M + \delta)T.$$

Then,

$$\begin{aligned}
&\rho \int_0^\infty e^{-\rho t} f(t)\, dt \\
&= \int_0^\infty f(t) \int_t^\infty \rho^2 e^{-\rho s}\, ds\, dt \\
&= \int_0^\infty \rho^2 e^{-\rho s} \int_0^s f(t)\, dt\, ds \\
&= \int_{T_0}^\infty \rho^2 e^{-\rho s} \int_0^s f(t)\, dt\, ds + \int_0^{T_0} \rho^2 e^{-\rho s} \int_0^s f(t)\, dt\, ds \qquad \text{(F.2)} \\
&\leq \int_{T_0}^\infty \rho^2 e^{-\rho s} (M + \delta) s\, ds + \int_0^{T_0} \rho^2 e^{-\rho s} \int_0^s f(t)\, dt\, ds \\
&\leq \int_0^\infty \rho^2 e^{-\rho s} (M + \delta) s\, ds + \int_0^{T_0} \rho^2 e^{-\rho s} \int_0^s f(t)\, dt\, ds \\
&= (M + \delta) + \int_0^{T_0} \rho^2 e^{-\rho s} \int_0^s f(t)\, dt\, ds.
\end{aligned}$$

Obviously the second term on the right-hand side of (F.2) goes to 0 as $\rho \to 0$. Therefore, the left-hand side of (F.2) does not exceed $(M + \delta)$. Since δ can be arbitrarily small, we have inequality (F.1). □

References

[1] Adiri, I. and Ben-Israel, A. (1966). An extension and solution of Arrow-Karlin type production models by the Pontryagin maximum principle, *Cahiers du Centre d'Etudes de Recherche Opérationnelle*, **8**, 147-158.

[2] Akella, R., Choong, Y.F., and Gershwin, S.B. (1984). Performance on hierarchical production scheduling policy, *IEEE Transactions on Components, Hybrids, and Manufacturing Technology*, **7**, 225-240.

[3] Akella, R. and Kumar, P.R. (1986). Optimal control of production rate in a failure-prone manufacturing system, *IEEE Transactions on Automatic Control*, **AC-31**, 116-126.

[4] Alexander, C. (1967). *Synthesis of Forms*, Harvard University Press, Cambridge, MA.

[5] Alvarez, R., Dallery, Y., and David, R. (1994). Experimental study of the continuous flow model of production lines with unreliable machines and finite buffers, *Journal of Manufacturing Systems*, **13**, 221-234.

[6] Arrow, K.J., Karlin, S., and Scarf, H. (1958). *Studies in the Mathematical Theory of Inventory and Production*, Stanford University Press, Stanford, CA.

[7] Auger, P. (1989). *Dynamics and Thermodynamics in Hierarchically Organized Systems*, Pergamon Press, Oxford, England.

[8] Bai, S.X. (1989). Scheduling manufacturing systems with work-in-process inventory control, Ph.D. Thesis, Operations Research Center, Massachusetts Institute of Technology, Cambridge, MA.

[9] Bai, S.X. and Gershwin, S.B. (1990). Scheduling manufacturing systems with work-in-process inventory, *Proceedings of the IEEE Conference on Decision and Control*, Dec. 5-7, Honolulu, HI, pp. 557-564.

[10] Bai, S.X. and Gershwin, S.B. (1994). Scheduling manufacturing systems with work-in-process inventory control: Multiple-part-type systems, *International Journal of Production Research*, **32**, 365-385.

[11] Bai, S.X. and Gershwin, S.B. (1995). Scheduling manufacturing systems with work-in-process inventory control: Single-part-type systems, *IIE Transactions*, **27**, 599-617.

[12] Basak, G.B., Bisi, A., and Ghosh, M.K. (1997). Controlled random degenerate diffusions under long-run average cost, *Stochastic and Stochastic Reports*, **61**, 121-140.

[13] Basar, T. and Bernhard, P. (1991). *H^∞-Optimal Control and Related Minimax Design Problems*, Birkhäuser, Boston, MA.

[14] Bensoussan, A. (1988). *Perturbation Methods in Optimal Control*, Wiley, New York, NY.

[15] Bensoussan, A., Crouhy, M., and Proth, J-M. (1983). *Mathematical Theory of Production Planning*, North-Holland, New York, NY.

[16] Bensoussan, A. and Nagai, H. (1991). An ergodic control problem arising from the principal eigenfunction of an elliptic operator, *Journal of the Mathematical Society of Japan*, **43**, 49-65.

[17] Bensoussan, A., Sethi, S.P., Vickson, R., and Derzko, N. (1984). Stochastic production planning with production constraints, *SIAM Journal on Control and Optimization*, **22**, 6, 920-935.

[18] Beyer, D. (1994). An inventory model with Wiener demand process and positive lead time, *Optimization*, **29**, 181-193.

[19] Bielecki, T.R. and Filar, J.A. (1991). Singularly perturbed Markov control problem: Limiting average cost, *Annals of Operations Research*, **28**, 153-168.

[20] Bielecki, T.R. and Kumar, P.R. (1988). Optimality of zero-inventory policies for unreliable manufacturing systems, *Operations Research*, **36**, 532-541.

[21] Boukas, E.K. (1991). Techniques for flow control and preventive maintenance in manufacturing systems, *Control and Dynamic Systems*, **48**, 327-366.

[22] Boukas, E.K. and Haurie, A. (1990). Manufacturing flow control and preventive maintenance: A stochastic control approach, *IEEE Transactions on Automatic Control*, **AC-35**, 1024-1031.

[23] Boukas, E.K., Haurie, A., and Van Delft, Ch. (1991). A turnpike improvement algorithm for piecewise deterministic control, *Optimal Control Applications & Methods*, **12**, 1-18.

[24] Boukas, E.K., Zhang, Q., and Yin, G. (1995). Robust production and maintenance planning in stochastic manufacturing systems, *IEEE Transactions on Automatic Control*, **AC-40**, 1098-1102.

[25] Boukas, E.K., Zhang, Q., and Zhu, Q. (1993). Optimal production and maintenance planning of flexible manufacturing systems, *Proceedings of the 32nd IEEE Conference on Decision and Control*, Dec. 15-17, San Antonio, TX.

[26] Boukas, E.K., Zhu, Q., and Zhang, Q. (1994). A piecewise deterministic Markov process model for flexible manufacturing systems with preventive maintenance, *Journal of Optimization Theory and Applications*, **81**, 259-275.

[27] Bramson, M. and Dai, J.G. (2001). Heavy traffic limits for some queueing networks. *The Annals of Applied Probability*, **11**, 49-90.

[28] Caramanis, M. and Sharifnia, A. (1991). Optimal manufacturing flow controller design, *International Journal of Flexible Manufacturing Systems*, **3**, 321-336.

[29] Caromicoli, C.A., Willsky, A.S., and Gershwin, S.B. (1988). Multiple time scale analysis of manufacturing systems, in *Analysis and Optimization of Systems, Proceedings of the Eighth International Conference on Analysis and Optimization of Systems*, Antibes, France.

[30] Chartrand, G. and Lesniak, L. (1986). *Graphs and Digraphs*, 2nd ed., Wadsworth & Brooks/Cole Advanced Books & Software, Monterey, CA.

[31] Chen, H. and Mandelbaum, A. (1994). Hierarchical modelling of stochastic networks: Part I: Fluid models, in *Stochastic Modeling and Analysis of Manufacturing Systems*, Yao, D.D. (Ed.), Springer Series in Operations Research, Springer-Verlag, New York, NY.

[32] Chen, H. and Mandelbaum, A. (1994). Hierarchical modelling of stochastic networks: Part II: Strong approximations, in *Stochastic Modeling and Analysis of Manufacturing Systems*, Yao, D.D. (Ed.), Springer Series in Operations Research, Springer-Verlag, New York, NY.

[33] Chow, Y.S. and Teicher, H. (1988). *Probability Theory*, Springer-Verlag, New York, NY.

[34] Clarke, F. (1983). *Optimization and Non-Smooth Analysis*, Wiley-Interscience, New York, NY.

[35] Chung, K.L. (1960). *Markov Chains with Stationary Transition Probabilities*, Springer-Verlag, Berlin, Germany.

[36] Crandall, M.G., Evans, L.C., and Lions, P.L. (1984). Some properties of viscosity solutions of Hamilton-Jacobi equations, *Transactions of the American Mathematical Society*, **282**, 487-501.

[37] Crandall, M.G. and Lions, P.L. (1983). Viscosity solutions of Hamilton-Jacobi equations, *Transactions of the American Mathematical Society*, **277**, 1-42.

[38] Crandall, M.G. and Lions, P.-L. (1987). Remarks on the existence and uniqueness of unbounded viscosity solutions of Hamilton-Jacobi equations, *Illinois Journal of Mathematics*, **31**, 665-688.

[39] Dallery, Y. and Gershwin, S.B. (1992). Manufacturing flow line systems: A review of models and analytical results, *Queueing Systems*, **12**, 3-94.

[40] Darakananda, B. (1989). Simulation of manufacturing process under a hierarchical control structure, Master's Thesis, EECS, Massachusetts Institute of Technology, Cambridge, MA.

[41] David, R., Xie, X.L., and Dallery, Y. (1990). Properties of continuous models of transfer lines with unreliable machines and finite buffers, *IMA Journal of Mathematics Applied in Business and Industry*, **6**, 281-308.

[42] Davis, M.H.A. (1993). *Markov Models and Optimization*, Chapman & Hall, New York, NY.

[43] Delebecque, F. and Quadrat, J. (1981). Optimal control for Markov chains admitting strong and weak interactions, *Automatica*, **17**, 281-296.

[44] Eleftheriu, M.N. (1989). On the analysis of hedging point policies of multi-stage production manufacturing systems, Ph.D. Thesis, Department of Electrical, Computer and Systems Engineering, Rensselaer Polytechnic Institute, Troy, NY.

[45] Eleftheriu, M.N. and Desrochers, A.A. (1991). An approximation schema for the estimation of buffer sizes for manufacturing facilities, *IEEE Transactions on Robotics and Automation*, **7**, 4, 551-561.

[46] Elliott, R.J. (1982). *Stochastic Calculus and Applications*, Springer-Verlag, New York, NY.

[47] Elliott, R.J. (1985). Smoothing for a finite state Markov process, *Springer Lecture Notes in Control and Information Sciences*, **69**, 199-206.

[48] Ethier, S.N. and Kurtz, T.G. (1986). *Markov Processes: Characterization and Convergence*, Wiley, New York, NY.

[49] Evans, L.C. and Souganidis, P.E. (1984). Differential games and representation formulas for solutions of Hamilton-Jacobi-Isaacs equations, *Indiana University Mathematical Journal*, **33**, 773-797.

[50] Federer, H. (1969). *Geometric Measure Theory*, Springer-Verlag, New York, NY.

[51] Feng, Y. and Yan, H. (2000). Optimal production control in a discrete manufacturing system with unreliable machines and random demands, *IEEE Transactions on Automatic Control*, **AC-35**, 2280-2296.

[52] Fleming, W.H. and McEneaney, W.M. (1995). Risk sensitive control on an infinite horizon, *SIAM Journal on Control and Optimization*, **33**, 1881-1921.

[53] Fleming, W.H. and Rishel, R.W. (1975). *Deterministic and Stochastic Optimal Control*, Springer-Verlag, New York, NY.

[54] Fleming, W.H., Sheu, S.J., and Soner, H.M. (1987). A remark on the large deviations of an ergodic Markov process, *Stochastics*, **22**, 187-199.

[55] Fleming, W.H., Sethi, S.P., and Soner, H.M. (1987). An optimal stochastic production planning problem with randomly fluctuating demand, *SIAM Journal on Control and Optimization*, **25**, 6, 1494-1502.

[56] Fleming, W.H. and Soner, H.M. (1992). *Controlled Markov Processes and Viscosity Solutions*, Springer-Verlag, New York, NY.

[57] Fleming, W.H. and Souganidis, P.E. (1989). On the existence of value functions of two-player, zero-sum stochastic different games, *Indiana University Mathematics Journal*, **38**, 293-314.

[58] Fleming, W.H. and Zhang, Q. (1998). Risk-sensitive production planning of a stochastic manufacturing system, *SIAM Journal on Control and Optimization*, **36**, 1147-1170.

[59] Fong, N.T. and Zhou, X.Y. (1996). Hierarchical production policies in stochastic two-machine flowshops with finite buffers, *Journal of Optimization Theory and Applications*, **89**, 681-712.

[60] Gershwin, S.B. (1989). Hierarchical flow control: A framework for scheduling and planning discrete events in manufacturing systems, *Proceedings of the IEEE, Special Issue on Dynamics of Discrete Event Systems*, **77**, 1, 195-209.

[61] Gershwin, S.B. (1994). *Manufacturing Systems Engineering*, Prentice-Hall, Englewood Cliffs, NJ.

[62] Gershwin, S.B., Akella, R., and Choong, Y.F. (1985). Short-term production scheduling of an automated manufacturing facility, *IBM Journal of Research and Development*, **29**, 14, 392-400.

[63] Ghosh, M.K., Aropostathis, A., and Marcus, S.I. (1993). Optimal control of switching diffusions with application to flexible manufacturing systems, *SIAM Journal on Control and Optimization*, **31**, 5, 1183-1204.

[64] Glover, K. and Doyle, J.C. (1988) State space formulae for all stabilizing controllers that satisfy an H^∞ norm bound and relations to risk sensitivity, *Systems Control Letters*, **11**, 167-172.

[65] Glynn, P.W. (1990). Diffusion approximations, in *Stochastic Models*, Vol. 2, in the series *Handbooks in Operations Research and Management Science*, Heyman, D.P. and Sobel, M.J. (Eds.), North-Holland, Amsterdam, pp. 145-198.

[66] Harrison, J.M. (2000). Brownian models of open processing networks: Canonical representation of workload, *The Annals of Applied Probability*, **10**, 75-103.

[67] Harrison, J.M. and Van Mieghem, A. (1997). Dynamic control of Brownian networks: State space collapse and equivalent workload formulations, *The Annals of Applied Probability*, **7**, 747-771.

[68] Hartl, R.F. and Sethi, S.P. (1984). Optimal control problems with differential inclusions: Sufficiency conditions and an application to a production-inventory model, *Optimal Control Applications & Methods*, **5**, 4, 289-307.

[69] Hartman, P. (1982). *Ordinary Differential Equations*, 2nd ed., Birkhäuser-Verlag, Boston, MA.

[70] Haurie, A. and Van Delft, Ch. (1991). Turnpike properties for a class of piecewise deterministic control systems arising in manufacturing flow control, *Annals of Operations Research*, **29**, 351-373.

[71] Hu, J.Q. and Xiang, D. (1994). Structural properties of optimal production controllers in failure prone manufacturing systems, *IEEE Transactions on Automatic Control*, **AC-39**, 3, 640-642.

[72] Hu, J.Q. and Xiang, D. (1995). Monotonicity of optimal flow control for failure prone production systems, *Journal of Optimization Theory and Applications*, **86**, 57-71.

[73] Ishii, H. (1984). Uniqueness of unbounded viscosity solutions of Hamilton-Jacobi equations, *Indiana University Mathematics Journal*, **33**, 5, 721-748.

[74] Jennings, L.S., Fisher, M.E., Teo, K.L., and Goh, C.J. (1990). *MISER3 Optimal Control Software: Theory and User Manual*, EMCOSS Pty Ltd., Perth, Australia.

[75] Jennings, L.S., Sethi, S.P., and Teo, K.L. (1997). Computation of optimal production plans for manufacturing systems, *Nonlinear Analysis, Theory, Methods and Applications*, **30**, 7, 4329-4338.

[76] Johnson, S.M. (1957). Sequential production planning over time at minimum cost, *Management Science*, **3**, 4, 435-437.

[77] Karatzas, I. and Shreve, S.E. (1988) *Brownian Motion and Stochastic Calculus*, Springer-Verlag, New York, NY.

[78] Karlin, S. and Taylor, H.M. (1975). *A First Course in Stochastic Processes*, 2nd ed., Academic Press, New York, NY.

[79] Kimemia, J.G. and Gershwin, S.B. (1983). An algorithm for the computer control production in flexible manufacturing systems, *IIE Transactions* **15**, 353-362.

[80] Krichagina, E., Lou, S., Sethi, S.P., and Taksar, M.I. (1993). Production control in a failure-prone manufacturing system: Diffusion approximation and asymptotic optimality, *The Annals of Applied Probability*, **3**, 2, 421-453.

[81] Krichagina, E., Lou, S., and Taksar, M.I. (1994). Double band polling for stochastic manufacturing systems in heavy traffic, *Mathematics of Operations Research*, **19**, 560-596.

[82] Kushner, H.J. (2001). *Heavy Traffic Analysis of Controlled Queueing and Communication Networks,* Springer-Verlag, New York, NY.

[83] Kushner, H.J. and Ramachandran, K.M. (1989). Optimal and approximately optimal control policies for queues in heavy traffic, *SIAM Journal on Control and Optimization*, **27**, 1293-1318.

[84] Lasserre, J.B. (1992). An integrated model for job-shop planning and scheduling, *Management Science*, **38**, 8, 1201-1211.

[85] Lasserre, J.B. (1992). New capacity sets for the hedging point strategy, *International Journal of Production Research*, **30**, 12, 2941-2949.

[86] Lasserre, J.B. (1993). Performance evaluation for large ergodic Markov chains, *Proceedings of the XIIth World IFAC Congress 1993*, July 18-23, Sydney, Australia.

[87] Lasserre, J.B. (1994). *An Integrated Approach in Production Planning and Scheduling*, Lecture Notes in Economics and Mathematical Systems, Vol. **411**, Springer-Verlag, New York, NY.

[88] Liberopoulos, Q. (1992). Flow control of failure-prone manufacturing systems: Controller design theory and applications, Ph.D. Dissertation, Boston University.

[89] Liberopoulos, G. and Caramanis, M. (1994). Production control of manufacturing systems with production rate dependent failure rates, *IEEE Transactions on Automatic Control*, **AC-39**, 4, 889-895.

[90] Liberopoulos, G. and Hu, J.Q. (1995). On the ordering of hedging points in a class of manufacturing flow control models, *IEEE Transactions on Automatic Control*, **AC-40**, 282-286.

[91] Lieber, Z. (1973). An extension to Modigliani and Hohn's planning horizons results, *Management Science*, **20**, 319-330.

[92] Lou, S. and Kager, P.W. (1989). A robust production control policy for VLSI wafer fabrication, *IEEE Transactions on Semiconductor Manufacturing*, **2**, 4, 159-164.

[93] Lou, S. and Van Ryzin, G. (1989). Optimal control rules for scheduling jobshops, *Annals of Operations Research*, **17**, 233-248.

[94] Lou, S., Yan, H., Sethi, S.P., Gardel, A., and Deosthali, P. (1990). Hub-centered production control of wafer fabrication, *Proceedings of the IEEE Advanced Semiconductor Manufacturing Conference*, Danvers, MA.

[95] Modigliani, F. and Hohn, F. (1955). Production planning over time and the nature of the expectation and planning horizon, *Econometrica*, **23**, 46-66.

[96] Morimoto, H. and Fujita, Y. (2000). Ergodic control in stochastic manufacturing systems with constant demand, *Journal of Mathematical Analysis and Applications*, **243**, 228-248.

[97] Perkins, J.R. and Kumar, P.R. (1995). Optimal control of pull manufacturing systems, *IEEE Transactions on Automatic Control*, **AC-40**, 2040-2051.

[98] Phillips, R.G. and Kokotovic, P.V. (1981). A singular perturbation approach to modelling and control of Markov chains, *IEEE Transactions on Automatic Control*, **AC-26**, 1087-1094.

[99] Presman, E., Sethi, S.P., and Suo, W. (1997). Optimal feedback controls in dynamic stochastic jobshops, in *Mathematics of Stochastic Manufacturing Systems*, G. Yin and Q. Zhang (Eds.), Lectures in Applied Mathematics, **33**, American Mathematical Society, Providence, RI, pp. 235-252.

[100] Presman, E., Sethi, S.P. and Suo, W. (1997). Optimal feedback production planning in a stochastic N-machine flowshop with limited buffers, *Automatica*, **33**, 1899-1903.

[101] Presman, E., Sethi, S.P., and Zhang, H. (2000). Optimal production planning in stochastic jobshops with long-run average cost, in *Optimization, Dynamics, and Economic Analysis, Essays in Honor of Gustav Feichtinger*, E.J. Dockner, R.F. Hartl, M. Luptacik, and G. Sorger (Eds.), Physica-Verlag, Heidelberg, pp. 259-274.

[102] Presman, E., Sethi, S.P., Zhang, H., and Bisi, A. (2000). Optimality in two-machine flowshop with limited buffer, *Annals of Operations Research*, **98**, 333-351.

[103] Presman, E., Sethi, S.P., Zhang, H., and Zhang, Q. (1998). Analysis of average cost optimality for an unreliable two-machine flowshop, *Proceedings of the Fourth International Conference on Optimization Techniques and Applications*, Curtain University of Technology, Perth, Australia, pp. 94-112.

[104] Presman, E., Sethi, S.P., Zhang, H., and Zhang, Q. (1998). Optimality of zero-inventory policies for an unreliable manufacturing system producing two part types, *Dynamics of Continuous, Discrete and Impulsive Systems*, **4**, 485-496.

[105] Presman, E., Sethi, S.P., Zhang, H., and Zhang, Q. (2000). Optimal production planning in a N-machine flowshop with long-run average cost, in *Mathematics and Its Applications to Industry,* S.K. Malik (Ed.), Indian National Science Academy, New Delhi, pp. 121-140.

[106] Presman, E., Sethi, S.P., Zhang, H., and Zhang, Q. (2002). On optimality of stochastic N-machine flowshop with long-run average cost, in *Stochastic Theory and Control.* B. Pasik-Duncan (Ed.), Lecture Notes in Control and Information Sciences, Vol. **280**, Springer-Verlag, Berlin, pp. 399-417.

[107] Presman, E., Sethi, S.P., and Zhang, Q. (1995). Optimal feedback production planning in a stochastic N-machine flowshop, *Automatica,* **31**, 1325-1332.

[108] Rishel, R.W. (1975). Dynamic programming and minimum principles for systems with jump Markov distributions, *SIAM Journal on Control,* **13**, 380-391.

[109] Rockafellar, R.T. (1972). *Convex Analysis,* Princeton University Press, Princeton, NJ.

[110] Samaratunga, C., Sethi, S.P., and Zhou, X.Y. (1997). Computational evaluation of hierarchical production control policies for stochastic manufacturing systems, *Operations Research,* **45**, 258-274.

[111] Sethi, S.P., Soner, H.M., Zhang, Q., and Jiang, J. (1992). Turnpike sets and their analysis in stochastic production planning problems, *Mathematics of Operations Research,* **17**, 932-950.

[112] Sethi, S.P., Suo, W., Taksar, M.I., and Yan, H. (1998). Optimal production planning in a multi-product stochastic manufacturing system with long-run average cost, *Discrete Event Dynamic Systems: Theory and Applications,* **8**, 37-54.

[113] Sethi, S.P., Suo, W., Taksar, M.I., and Zhang, Q. (1997). Optimal production planning in a stochastic manufacturing system with long-run average cost, *Journal of Optimization Theory and Applications,* **92**, 161-188.

[114] Sethi, S.P. and Thompson, G.L. (2000). *Optimal Control Theory: Applications to Management Science and Economics,* 2nd ed., Kluwer Academic, Boston, MA.

[115] Sethi, S.P. and Thompson, G.L. (1981). Simple models in stochastic production planning, in *Applied Stochastic Control in Econometrics and Management Science,* Bensoussan, A., Kleindorfer, P., and Tapiero, C. (Eds.), North-Holland, New York, NY, pp. 295-304.

[116] Sethi, S.P., Yan, H., Zhang, H., and Zhang, Q. (2001). Turnpike set analysis in stochastic manufacturing systems with long-run average cost, in *Optimal Control and Partial Differential Equations, In honour of Professor Alain Bensoussan's 60th Birthday,* J.L. Menaldi, E. Rofman, and A. Sulem (Eds.), IOS Press, Amsterdam, pp. 414-423.

[117] Sethi, S.P. and Zhang, H. (1999). Hierarchical production controls for a stochastic manufacturing system with long-run average cost: Asymptotic optimality, in *Stochastic Analysis, Control, Optimization and Applications, A Volume in Honor of W.H. Fleming,* W.M. McEneany, G. Yin, and Q. Zhang (Eds.), Systems & Control: Foundations & Applications, Birkhäuser, Boston, MA, pp. 621-637.

[118] Sethi, S.P. and Zhang, H. (1999). Average-cost optimal policies for an unreliable flexible multiproduct machine, *The International Journal of Flexible Manufacturing Systems,* **11**, 147-157.

[119] Sethi, S.P., Zhang, H., and Zhang, Q. (1997). Hierarchical production control in a stochastic manufacturing system with long-run average cost, *Journal of Mathematical Analysis and Applications*, **214**, 151-172.

[120] Sethi, S.P., Zhang, H., and Zhang, Q. (1998). Minimum average cost production planning in stochastic manufacturing systems, *Mathematical Models and Methods in Applied Sciences*, **8**, 1251-1276.

[121] Sethi, S.P., Zhang, H., and Zhang, Q. (2000). Optimal production rates in a deterministic two-product manufacturing system, *Optimal Control Applications and Methods*, **21**, 125-135.

[122] Sethi, S.P., Zhang, H., and Zhang, Q. (2000). Hierarchical production control in a stochastic N-machine flowshop with long-run average cost. *Journal of Mathematical Analysis and Applications*, **251**, 285-309.

[123] Sethi, S.P., Zhang, H., and Zhang, Q. (2000). Hierarchical production control in dynamic stochastic jobshops with long-run average cost, *Journal of Optimization Theory and Applications*, **106**, 231-264.

[124] Sethi, S.P., Zhang, H., and Zhang, Q. (2000). Hierarchical production control in a stochastic N-machine flowshop with limited buffers, *Journal of Mathematical Analysis and Applications*, **246**, 28-57.

[125] Sethi, S.P. and Zhang, Q. (1994). *Hierarchical Decision Making in Stochastic Manufacturing Systems*, Birkhäuser, Boston, MA.

[126] Sethi, S.P., Zhang, Q., and Zhou, X.Y. (1992). Hierarchical controls in stochastic manufacturing systems with machines in tandem, *Stochastics and Stochastics Reports*, **41**, 89-118.

[127] Sethi, S.P. and Zhou, X.Y. (1994). Stochastic dynamic job shops and hierarchical production planning, *IEEE Transactions on Automatic Control*, **AC-39**, 2061-2076.

[128] Sharifnia, A. (1988). Production control of a manufacturing system with multiple machine states, *IEEE Transactions on Automatic Control*, **AC-33**, 620-625.

[129] Soner, H.M. (1986). Optimal control with state space constraints II, *SIAM Journal on Control and Optimization*, **24**, 1110-1122.

[130] Sprzeuzkouski, A.Y. (1967). A problem in optimal stock management, *Journal of Optimization Theory and Applications*, **1**, 232-241.

[131] Srivatsan, N. (1993). Synthesis of optimal policies for stochastic manufacturing systems, Ph.D. Thesis, Department of Mechanical Engineering, Massachusetts Institute of Technology, Cambridge, MA.

[132] Srivatsan, N., Bai, S.X., and Gershwin, S.B. (1994). Hierarchical real-time integrated scheduling of a semiconductor fabrication facility, in *Computer-Aided Manufacturing/Computer-Integrated Manufacturing*, Part 2 of Vol. 61 in the series *Control and Dynamic Systems*, Leondes, C.T. (Ed.), Academic Press, New York, NY, pp. 197-241.

[133] Srivatsan, N. and Dallery, Y. (1998). Partial characterization of optimal hedging point policies in unreliable two-part-type manufacturing systems, *Operations Research*, **46**, 36-45.

[134] Sworder, D.D. (1969). Feedback control of a class of linear systems with jump parameters, *IEEE Transactions on Automatic Control*, **AC-14**, 9-14.

[135] Thompson, G.L. and Sethi, S.P. (1980). Turnpike horizons for production planning, *Management Science*, **26**, 229-241.

[136] Uzsoy, R., Lee, C.Y., and Martin-Vega, L.A. (1992). A review of production planning and scheduling of models in the semiconductor industry, Part I: System characteristics, performance evaluation and production planning, *IIE Transactions on Scheduling and Logistics*, **24**, 47-61.

[137] Uzsoy, R., Lee, C.Y., and Martin-Vega, L.A. (1996). A review of production planning and scheduling of models in the semiconductor industry, Part II: Shop floor control, *IIE Transactions on Scheduling and Logistics*, **26**, 44-55.

[138] Van Ryzin, G., Lou, S., and Gershwin, S.B. (1991). Scheduling jobshops with delays, *International Journal of Production Research*, **29**, 7, 1407-1422.

[139] Van Ryzin, G., Lou, S., and Gershwin, S.B. (1993). Production control for a tandem two-machine system, *IIE Transactions*, **25**, 5, 5-20.

[140] Veatch, M. (1992). Queueing control problems for production/inventory systems, Ph.D. Dissertation, O.R. Center, MIT.

[141] Veinott, A.F. (1964). Production planning with convex costs: A parametric study, *Management Science*, **10**, 3, 441-460.

[142] Wein, L.M. (1990). Scheduling networks of queues: Heavy traffic analysis of a two-station network with controllable inputs, *Operations Research*, **38**, 1065-1078.

[143] Whitt, W. (2001). *Stochastic-Process Limits: An Introduction to Stochastic-Process Limits and their Application to Queues*, Springer-Verlag, New York, NY.

[144] Whittle, P. (1990) *Risk-Sensitive Optimal Control*, Wiley, New York, NY.

[145] Xie, X.L. (1991). Hierarchical production control of a flexible manufacturing system, *Applied Stochastic Models and Data Analysis*, **7**, 343-360.

[146] Yan, H., Lou, S., Sethi, S.P., Gardel, A., and Deosthali, P. (1996). Testing the robustness of two-boundary control policies in semiconductor manufacturing, *IEEE Transactions on Semiconductor Manufacturing*, **9**, 2, 285-288.

[147] Yan, H., Yin, G., and Lou, S. (1994). Using stochastic optimization to determine threshold values for control of unreliable manufacturing systems, *Journal of Optimization Theory and Applications*, **83**, 511-539.

[148] Yin, G. and Zhang, Q. (1994). Near optimality of stochastic control in systems with unknown parameter processes, *Applied Mathematics and Optimization*, **29**, 263-284.

[149] Yin, G. and Zhang, Q. (1998). *Continuous–Time Markov Chains and Applications*, Springer-Verlag, New York, NY.

[150] Yosida, K. (1980). *Functional Analysis*, 6th ed., Springer-Verlag, New York, NY.

[151] Zhang, Q. (1995). Risk sensitive production planning of stochastic manufacturing systems: A singular perturbation approach, *SIAM Journal on Control and Optimization*, **33**, 498-527.

[152] Zhang, Q. and Yin, G. (1994). Turnpike sets in stochastic manufacturing systems with finite time horizon, *Stochastics and Stochastics Reports*, **51**, 11-40.

[153] Zhou, X.Y. and Sethi, S.P. (1994). A sufficient condition for near optimal stochastic controls and its application to manufacturing systems, *Applied Mathematics and Optimization*, **29**, 67-92.

Author Index

Subject Index

Copyright Permissions

Charles Street, Providence, RI 02904, USA.

"Optimal feedback controls in dynamic stochastic jobshops," by Presman, E., Sethi, S.P., and Suo, W., *Lectures in Applied Mathematics*, **33**, 235-252. Copyright ©1997, American Mathematical Society, 201 Charles Street, Providence, RI 02904, USA.

"Hierarchical production control in a stochastic manufacturing system with long-run average cost," by Sethi, S.P., Zhang, H., and Zhang, Q., *Journal of Mathematical Analysis and Applications*, **214**, 151-172. Copyright ©1997, Elsevier, P.O. Box 800, Oxford OX5 1GB, UK.

"Existence of optimal feedback production plans in stochastic flowshops with limited buffers," by Presman, E., Sethi, S.P., and Suo, W., *Automatica*, **33**, 1899-1903. Copyright ©1997, Elsevier, P.O. Box 800, Oxford OX5 1GB, UK.

"Optimal production planning in a multi-product stochastic manufacturing system with long-run average cost," by Sethi, S.P., Suo, W., Taksar, M.I., and Yan, H., *Discrete Event Dynamic Systems: Theory and Applications*, **8**, 37-54. Copyright ©1998, Kluwer Academic/Plenum Publishers, 233 Spring Street, New York, NY 10013, USA.

"Risk-sensitive production planning of a stochastic manufacturing systems," by Fleming, W.H. and Zhang, Q., *SIAM Journal on Control and Optimization*, **36**, 1147-1170. Copyright ©1998, Society for Industrial and Applied Mathematics, 3600 University City Science Center, Philadelphia, PA 19104, USA.

"Minimum average cost production planning in stochastic manufacturing systems," by Sethi, S.P., Zhang, H., and Zhang, Q., *Mathematical Models and Methods in Applied Sciences*, **8**, 1251-1276. Copyright ©1998, World Scientific Publishing Co., Inc., Suite 202, 1060 Main Street, River Edge, NJ 07661, USA.

"Optimality of zero-inventory policies for an unreliable manufacturing system producing two part types," by Presman, E., Sethi, S.P., Zhang, H., and Zhang, Q., *Dynamics of Continuous, Discrete and Impulsive Systems*, **4**, 485-496. Copyright ©1998, Watam Press, University of Waterloo, Waterloo, Ontario N2L 3G1, Canada.

"Hierarchical production controls for a stochastic manufacturing system with long-run average cost: Asymptotic optimality," by Sethi, S.P. and Zhang, H., in *Stochastic Analysis, Control, Optimization and Applications, A Volume in Honor of W.H. Fleming* (W.M. McEneany, G.G. Yin, and Q.

Zhang, Eds.). pp. 621-637, Copyright ©1999, Springer-Verlag, Tiergartenstr. 17, D-69121 Heidelberg, Germany.

"Average-cost optimal policies for an unreliable flexible multiproduct machine," by Sethi, S.P. and Zhang, H., *International Journal of Flexible Manufacturing Systems*, **11**, 147-157. Copyright ©1999, Kluwer Academic/Plenum Publishers, 233 Spring Street, New York, NY 10013, USA.

"Optimal production planning in stochastic jobshops with long-run average cost," by Presman, E. and Sethi, S.P., and Zhang, H., in *Optimization, Dynamics, and Economic Analysis, Essays in Honor of Gustav Feichtinger* (E.J. Dockner, R.F. Hartl, M. Luptacik, and G. Sorger, Eds.), pp. 259-274. Copyright ©2000, Springer-Verlag, Tiergartenstr. 17, D-69121 Heidelberg, Germany.

"Hierarchical production control in a stochastic N-machine flowshop with limited buffers," by Sethi, S.P., Zhang, H., and Zhang, Q., *Journal of Mathematical Analysis and Applications*, **246**, 28-57. Copyright ©2000, Elsevier, P.O. Box 800, Oxford OX5 1GB, UK.

"Average cost optimality for an unreliable two-machine flowshop with limited internal buffer," by Presman, E., Sethi, S.P., Zhang, H., and Bisi, A., *Annals of Operations Research*, **98**, 333-351. Copyright ©2000, Kluwer Academic/Plenum Publishers, 233 Spring Street, New York, NY 10013, USA.

"Optimal production planning in a stochastic N-machine flowshop with long-run average cost," by Presman, E., Sethi, S.P., Zhang, H., and Zhang, Q., in *Mathematics and Its Applications to Industry* (S.K. Malik, Ed.), 121-140. Copyright ©2000, Indian National Science Academy, Bahadur Shah Zafar Marg, New Delhi 110002, India.

"Hierarchical production control in a stochastic N-machine flowshop with long-run average cost," by Sethi, S.P., Zhang, H., and Zhang, Q., *Journal of Mathematical Analysis and Applications*, **251**, 285-309. Copyright ©2000, Elsevier, P.O. Box 800, Oxford OX5 1GB, UK.

"Hierarchical production control in dynamic stochastic jobshops with long-run average cost," by Sethi, S.P., Zhang, H., and Zhang, Q., *Journal of Optimization Theory and Applications*, **106**, 231-264. Copyright ©2000, Kluwer Academic/Plenum Publishers, 233 Spring Street, New York, NY 10013, USA.

"Optimal production rates in a deterministic two-product manufacturing system," by Sethi, S.P., Zhang, H., and Zhang, Q., *Optimal Control*

Applications and Methods, **21**, 125-135. Copyright ©2000, John Wiley & Sons, Inc., 111 River Street, Hoboken, NJ 07030, USA.

"Turnpike set analysis in stochastic manufacturing systems with long-run average cost," by Sethi, S.P., Yan, H., Zhang, H., and Zhang, Q., in *Optimal Control and Partial Differential Equations, In Honour of Professor Alain Bensoussan's 60th Birthday* (J.L. Menaldi, E. Rofman, and A. Sulem, Eds.), pp. 414-423. Copyright ©2001, IOS Press, Nieuew Hemweg 6B, 1013 BG Amsterdam, Netherlands.

"On optimality of stochastic N-machine flowshop with long-run average cost," by Presman, E., Sethi, S.P., Zhang, H., and Zhang, Q., in *Stochastic Theory and Control* (B. Pasik-Duncan, Ed.), Lecture Notes in Control and Information Sciences, **Vol. 280**, pp. 399-417. Copyright ©2002, Springer-Verlag, Tiergartenstr. 17, D-69121 Heidelberg, Germany.